Ashraf M.T. Elewa

Migration of Organisms

Climate • Geography • Ecology

Ashraf M. T. Elewa
(Editor)

Migration of Organisms

Climate • Geography • Ecology

With 67 Figures

Springer

Dr. Ashraf M. T. Elewa
Professor
Minia University
Faculty of Science
Geology Department
Egypt
E-mail: *aelewa@link.net*

Library of Congress Control Number: 2005927792

ISBN-10 3-540-26603-8 **Springer Berlin Heidelberg New York**
ISBN-13 978-3-540-26603-7 **Springer Berlin Heidelberg New York**

Springer is a part of Springer Science+Business Media
springeronline.com

Printed in The Netherlands

Cover design: Erich Kirchner
Production: Luisa Tonarelli
Typesetting: Camera-ready by the editor

Printed on acid-free paper 30/2132/LT – 5 4 3 2 1 0

Dedication

This book is dedicated to all people who

Believe in One God

Believe in Peace

Believe in Migration in the Way of God

To my father who died on Sunday, the 10th of April, 2005

Foreword

P. David Polly

School of Biological Sciences, Queen Mary, University of London, London E1 4NS, UK, d.polly@qmul.ac.uk

Evolutionary biology journals of the 1950s and 60s were filled with musings on the improbable. Disjunct distributions of related organisms on continents separated by hundreds of miles of open ocean or in seas divided by mountainous barriers were a historical challenge. Pregnant porcupines were rafted across the South Atlantic on mangrove trunks to explain the presence of Palaeogene hystricomorph rodents in Africa and South America. Limpets were lofted across the Isthmus of Panama in the talons of birds to connect latter day Caribbean descendants with their Miocene Pacific ancestors. By the 70s continental drift lumbered into the paradigms of palaeontology and vicariance inserted itself as the null hypothesis – the evolutionary hegira was redefined as abduction by subduction. Dramatic finds like the Triassic vertebrate *Lystrosaurus* on the Antarctic mountains that once lay within the genus's now disjunct African and Indian range confirmed the predictive power of vicariance. In some instances, the vicariant sword grew so sharp that scenarios of circuitous continental conveyers became even wilder than earlier tales of drifting diaspora. By the end of the century, however, ideological battles between migrationists and vicariants had given way to more dialectic mixed model of biogeographic history.

In the past decade, molecular phylogeography and geological isotope geochemistry have renewed interest in migration on much smaller temporal scales. Surveys of molecular diversity within species revealed that much geographic variation was phylogenetically structured. Pictures of species as panmictic gene pools have been redrawn so that speciation is no longer an event, but a never-ending narrative of the gradual breakdown in relationship. Rivers and hills, rather than oceans and mountains, separate subspecific clades. At the same time, geochemical evidence has revealed fine-scale wobbles in the Earth's climate. The Pleistocene, for example, is no longer the time when regal glaciers made stately advances and retreats across the continent, but a jumble of transient alternations of warm and cold overlain by more regular glacial and interglacial cycles. Warm spikes only hundreds of years long saw *Saiga* antelope sweep horde-like from Central Asia as far as England, only to be driven back just as quickly. Phylogeographers and palaeontologists alike now talk of putative refugia and post-glacial recolonization to explain the biogeographic history of populations of seemingly continuously distributed continental species.

The present collection by Dr Ashraf Elewa of Minia University contains papers about migration that are as timely as the subject is traditional. Migration, it seems, is an ever-moving subject, and Elewa presents us with a sampling of some of its current trajectories. Reyment discusses the interpretation, and misinterpretation, of distributions of fossil marine organisms in light of pre- and post-mortem movements. Elewa, in two papers, considers the distribution of tiny but biostratigraphically important ostracods in relation to the changing reaches of the shallow sea that periodically inundated what is now northern Africa. Petrakis and Legakis investigate problems of detecting and understanding insect migration in Mediterranean ecosystems. Longer-term migration and adaptation in cold water Pacific mollusc faunas are described by Amano. Thompson and Russell take a particularly elegant look at phylogeographic structuring in the mtDNA in salamanders in the Pacific Northwest. The Miocene Vallesian Crisis, a time when the subtropical forest faunas of Europe virtually disappeared during a cooling episode, is used by Casanovas-Vilar et al. to investigate the geographic restructuring of entire faunas and floras. The final paper of the book, by Hortal et al., contains a creative quantitative study of the provincializing effect of basin and range structures in Iberia on mammalian communities and an extrapolation by GIS modelling to see whether the same provincial structure existed in the Palaeogene. Readers will get a flavour of the latest quantitative analyses – GIS, faunal clustering, mtDNA phylogenetics – and a sense of the breadth of international research in migration and biogeography.

Table of Contents

1 Introduction

Ashraf M. T. Elewa

Geology Department, Faculty of Science, Minia University, Egypt, aelewa@link.net

Migration is commonly thought of as a mechanism by which organisms avoid unfavorable environments for more auspicious ones. Migration can be the seasonal movement of organisms from place to place or a more permanent dispersal from one region to another; either way, changing environmental conditions provide the impetus. But why do some organisms migrate and not others? How do the benefits of migration compare with the metabolic costs and hazards of the journey? Do differences between terrestrial and marine species in their interface with the environment have different effects on migration? What are the trade-offs between migration and evolutionary adaptation to local conditions?

Different species respond to environmental changes in different ways, partly based on their individual adaptations and range of tolerance. This phenomenon is most easily seen when comparing terrestrial and marine organisms, which have different relations to their environment. Terrestrial organisms are surrounded by low-density atmosphere and expend considerable energy moving and keeping themselves upright; marine organisms, in contrast, are surrounded by seawater, which has a density close to their own, making support and movement much less costly (Sverdrup et al. 2005). Thus, migration is likely to be more metabolically costly to a terrestrial animal than a marine one, all other things being equal. The factors affecting migration of land and sea species may therefore differ considerably, and one geologic episode may affect the two realms differently.

Terrestrial and marine organisms also differ in the environmental factors that affect them the most. Temperature and rainfall are major determinants on the land, but salinity is one of the most important in the sea. The difference in the salt content of seawater and bodily fluids affects the transport of water molecules through cell membranes in a process called osmosis; changes in salinity can have disastrous effects on the physiological balance of an organism. Temperature variation affects marine organisms also, but in different ways than terrestrial ones. Changes in temperature affect water density, and therefore buoyancy and the ability of an organism to maintain its optimal position in the water column, positions that are important in terms of ambient pressure and available nutrients. Pressure in deep waters is so great that organisms must avoid it or evolve special adaptations to survive. Most life in the marine realm requires carbon dioxide and oxygen, whose concentrations vary with depth. Oceans can be divided into an upper photic zone, where sufficient sunlight penetrates to support photosynthesis, and a lower aphotic zone, where the intensity of sunlight is low or does not penetrate at all, so inhibiting photosynthetic activity. The constant motion of water in the oceans affects the distribution of organisms and patterns of productivity in the seas, both by transporting the nutrients and gases necessary for life and by carrying floating or drifting organisms with the currents.

Migration often serves as a short-term solution to variable environments, but adaptation can provide a long-term solution. Adaptation can thus be viewed as a parallel way for organisms to survive. Some groups have evolved bioluminescence to bring their own light to ocean depths. Such adaptations will have evolved through selection on favourable chance variants. Because some individual organisms possess genetic combinations that are particularly valuable for survival in a local environment, they will find the environment less antagonistic than do others. For example, for many organisms living in, on, or near the sea floor, the properties of the bottom material, or substrate, are very important for survival. In the vast oceanic environment, both the water and the sea floor are divided into specific zones that have distinct characteristics and support different populations of organisms. Depending on the circumstances, selection may favour adaptations to one of these, or it may favour generalized adaptations that permit movement from one to another. As conditions change, the organisms will either cope or move, and some species will be geographically localized and others widespread. A challenge for migration studies, especially in the fossil record, is to distinguish widespread species from species that migrate between localized environments (Taylor in Boardman et al. 1987).

Since the classic works of Schmidt-Koenig (1975), Schmidt-Koenig and Keeton (1978), Gauthreaux (1980) and Adley (1981), which contain valuable information on migration and quantitative approaches for its study, no edited book, to my knowledge, has considered migration in its broadest sense across different groups of organisms. Moreover, the quantitative analyses in these other works are now aging, with many new approaches having been developed over the last decades. It is, however, worth noting that Dingle (1996) recently published an excellent textbook on migration, which covers all migratory organisms. The work focuses on the distinction between migration and dispersal and why the two should not be conflated. A second excellent book by Drake and Gatehouse (1995) is limited to insects and was an outgrowth of a symposium at the International Congress of Entomology in Beijing, China. A third recent book, was edited by Jones et al. (2004) with the title "Migrations and dispersal of marine organisms" and based on the proceedings of the 37th European Biology Symposium held in Reykjavík, Iceland, 5-9 August 2002. The main themes of the symposium were migrations and dispersal of marine organisms. Accordingly, no attention was paid to non-marine organisms.

To fill this gap, I have selected, in my opinion, a suitable collection of topics to explore migration in different groups of marine and non-marine organisms ranging from ostracodes and foraminifera (micro-invertebrates), then molluscs (macro-invertebrates), followed by insects, reptiles, amphibians, salamanders, birds, and finally mammals. Conceptually, these papers address a broad range of topics that have not been collected, I think, in any past single work:

1. The paleobiogeography and biogeography of organisms (both invertebrates and vertebrates);
2. The relation of ecological, paleoecological and other factors to migration;
3. The distinction between migration and species origination, extinction, and turnover;
4. Adaptability as a parallel way to migration for organisms to survive;
5. The ecological relation among species (predation, parasitism, commensalism, mutualism... etc.) and its effect on migration;
6. The application of modern quantitative techniques for constructing migration models.

This book should appeal to professionals and students at all levels. I hope this work represents an up to date summary of ideas in the field, as well as a useful summary of recent progress of the subject. In terms of authors, as in my previous book on morphometrics, I have invited a group of experts who present topics showing a broad diversity of organisms ranging from invertebrates to vertebrates, including popular groups such as insects, mammals, birds, reptiles and amphibians.

Additionally, as I am an ostracodologist, I have tried to highlight the importance of ostracodes for paleogeography. Pokorny in Haq and Boersma (1980) stated that marine ostracodes are not as well suited for interregional and intercontinental stratigraphic correlation as are other groups of planktonic microfossils, since benthic ostracodes have no planktonic larvae, therefore, the shallow, warm water species cannot easily cross geographic barriers (see also Keen et al. 1994; Elewa 2002). However, this limitation makes ostracodes excellent paleobiogeographical markers. Pokorny added that the deep-sea ostracodes of the Mediterranean province, studied by Benson and Sylvester-Bradley (1971), are of considerable paleogeographical interest. Examples are found in Paleocene to Middle Miocene and in Pliocene sediments from different areas of the Mediterranean province. Ostrocodes provided evidence that, in the Paleocene, there was a possibility in the southern Mediterranean for east-west migration through the Trans-Saharan Seaway (see Reyment and Reyment 1980; Reyment 1981; Elewa, chapter 3 of this volume). In the Late Miocene, the Mediterranean Sea was cut off from the Atlantic Ocean and transformed into a series of lagoons, as documented by endemic ostracode communities. At the beginning of the Pliocene, communication between the Mediterranean Sea and the Atlantic Ocean was re-established in the west, so that Atlantic euhaline species re-invaded the Mediterranean. Recent Mediterranean ostracodes are chiefly of Atlantic origin and differ from their Tertiary descendants. With the opening of the Suez Canal more than a hundred years ago, a migration route was established between the Indo-West Pacific and the Mediterranean. This is a special example of increasing human effect on the composition of naturally established regional faunas, as was emphasized by McKenzie (1973). Chapter 5 (this volume) by Elisabeth Brouwers presents valuable information about the migration of ostracodes along the northeast Pacific coast in response to temperature changes during glacial-interglacial cycles.

I would like to express my deep gratitude to all people who played an important role in the completion of this book. I especially acknowledge David Polly (UK) for writing the preface, as well as reviewing this introduction and two chapters for this book. The rest of the reviewers, an exceptional group of experts, are also thanked for their critical reviews of chapters in this book (Abdel Kireem and Bassiouni from Egypt; Blain from France; Rook from Italy; Tanabe from Japan; Campomanes, Hortal and Morales as well as an anonymous reviewer from Spain; Reyment and Savazzi from Sweden; Korner-Nievergelt from Switzerland; Whatley from UK; Cronin, Dodd, Finger and Peterson from USA; arranged alphabetically according to their countries). Thanks also to all of the contributors for devoting their time in preparing their chapters for this book. They have done excellent work and without their contributions this project would not exist. A special word of thanks is due to the publishers of Springer-Verlag for their continuous help during the several steps of editing this book. I also appreciate the great help of the staff at Minia University in Egypt.

References

Adley DJ (ed) (1981) Animal migration. Cambridge University Press

Benson RH, Sylvester-Bradley PC (1971) Deep-sea ostracods and the transformation of ocean to sea in the Tethys. In Oertli HJ (ed) Paléoécologie d'Ostracodes. Bull Centre Rech, Pau-SNPA 5: 63-91

Drake VA, AG Gatehouse (1995) Insect Migration: tracking resources through space and time. Cambridge University Press

Boardman, RS, Cheetham AH, Rowell, AJ (1987) Fossil Invertebrates. Blackwell Scientific Publications

Dingle H (1996) Migration: the Biology of Life on the Move. Oxford University Press

Gauthreaux, SAJr (ed) (1980) Animal migration, orientation and navigation. New York: Academic Press

Haq BU, Boersma A (1980) Introduction to marine micropaleontology. Elsevier (2nd edn), New York

Jones MB, Ingólfsson A, Ólafsson E, Helgason GV, Gunnarsson K, Svavarsson J (eds) (2004) Migrations and dispersal of marine organisms. Kluwer Academic Publishers, developments in hydrobiology, Dordrecht

McKenzie KG (1973) Cenozoic Ostracoda. In Hallam A (ed) Atlas of palaeobiogeography. Elsevier, Amsterdam, pp 477-487

Schmidt-Koenig K (1975) Migration and homing in animals. Springer-Verlag, Berlin

Schmidt-Koenig K, Keeton WT (eds) (1978) Animal migration, navigation, and homing. Springer-Verlag, New York

Sverdrup KA, Duxbury AC, Duxbury AB (2005) An Introduction to the World's Oceans. McGraw Hill Companies (8th edn)

2 Active migration and passive transport of marine organisms in the fossil record

Richard A. Reyment

Department of Paleozoology, Swedish Museum of Natural History, Stockholm, richard.reyment@nrm.se

2.1 Abstract

For the paleontologically oriented purposes of this review, migration can be conveniently considered from several aspects, some of which cannot be definitely identified nor definitely interpreted in the fossil record. In favourable circumstances, inference drawn from Schäfer's concept of "actuopaleontology" can provide enlightenment. Several *ad hoc* categories are briefly discussed here, to wit, seasonal migration, unidirectional migration driven by directed climatic, and or, geological factors, such as changes in the configuration of ocean basins, competition between species and the battle for survival and overpopulation, and enforced migration due to deep-sea oceanic currents. Additionally, and of particular paleontological consequence there is the contrasting circumstance of nekroplanktonic dispersal.

Keywords: Passive transport, directed migration, nekroplankton, eustasy, actuopaleontology, ammonites, ostracods.

2.2 Introduction

The problem posed of identifying cases of genuine migration in the fossil record is not as straightforward as it may seem on first thought. Migration may be viewed from several aspects. One of these concerns migratory patterns induced by seasonal factors and which is most clearly articulated in terms of the annual odysseys undertaken by many species of birds and a surprising number of cephalopod species (Schäfer 1962). Clearly, the possibilities of identifying this pattern are not great in paleontology. A second category concerns the unidirectional migration of organisms from one geographical location to another in answer to such factors as tectonic influences, climatic deterioration, ephemeral island-bridges, the opening up of new areas for colonization, predation pressure, overpopulation, and competition. A good example of unidirectional migration is that of the "locked"arrival of *Elephas antiquus* on Cyprus and the evolution of this species into the dwarf species *Elephas falconeri* (Reyment 1983a). A similar example is the island evolution of the Columbian mammoth into a pygmy form. There are many such cases to be found in the literature on island speciation for deer, rodents, hippopotami (cf. Boekschoten and Sondaar 1972; Reyment 1983a; Thaler 1973). Another category is that of nekroplanktonic dispersal whereby befitting hard parts of dead organisms are dispersed as flotsam and jetsam by surface oceanic currents.

During the Cretaceous Period, for example, large-scale changes of sea level took place. In my experience, it is seldom appreciated that the level of the ocean does not oscillate about a chronologically stable average. The relevant stochastic model is, in effect, one of oscillations about a fluctuating, moving average; dominated by tectono-eustasy (Suess 1888, p. 364; 1892, pp. 465, 768; 1909, pp. 100-102), whereby the sea level changes slowly, but upon which minor oscillations are superimposed over a long time-period in answer to changes, and fluctuations, in the volume of the oceans occasioned by the growth of mid-oceanic ridges and their eventual subsidence (Thiede 1977). Glacial eustasy is another important factor in, for example, the Pleistocene and Holocene (Mörner 1976), which, however, in relation to tectono-eustasy is of fleeting significance time-wise. Reyment and Mörner (1977) examined the interplay between factors governing sea level change for the Upper Cretaceous. Additionally, vertical movements of the crust may play an important part at the local level. An example of this is the history of the coastal sedimentary basins of the Cretaceous South Atlantic Ocean (Reyment and Tait 1972; Reyment and Dingle 1987).

For the explicit purposes of the present note, we shall make a set of arbitrary distinctions between migrational models, passive and active. It will be apparent that these "categories" encompass the migration of individuals in their lifetime as well as long-term migration of a lineage and post-mortem dispersal. I have not been so bold as to provide unique labels for each case considered but rather retained a more anecdotal form - moreover, there no hard and fast marches can be drawn between them.

1. Seasonal migration from an overwintering site to a summer location. The bird paradigm for active North-South migration.

2. Active dispersal by living marine organisms such as occurs when a new *lebensraum* is being exploited. This may be exemplified by the gradual invasion by marine organisms into epicontinental seas (i.e. very

shallow seas of wide geographical extent that spread over surfaces of slight relief) formed by rising Suessian sea level (cf. Reyment 1996) which may be exemplified by the incursions into the vast, shallow epicontinental trans-Saharan seas of molluscs and crustaceans.

A special case of this design is the way in which Cretaceous and Tertiary ostracods were able to cross the South Atlantic from western Africa to south-eastern South America by a chain of mid-Atlantic islands (Reyment 1980a, 1980b, 1983b; Thiede 1977).

3. Enforced migration of living organisms by oceanic currents.
Examples of this mode of migration are provided by ongoing research on deep sea-currents relocating the coelacanth *Latimeria* sp. to the Celebes (Sulawesi) in Indonesia together with *Nautilus pompilius.* Neither of these denizens of deeper waters are strong swimmers and cannot successfully hold their own against powerful westerly flowing submarine currents (television programme aired in 2004 in the *Discovery Science* Channel by Professor Hans W. Fricke (Max Planck Institute, Frankfurt/Main, and coworkers). This kind of migration would obviously be most difficult to prove for fossils.

4. Nekroplanktonic dispersal. The buoyant shells of dead organisms are transported by oceanic currents sometimes over thousands of kilometres from the habitat of the living animal.

5. There is also the category I call "spurious migration"; that is, well isolated occurrences of identical species of non-marine plants and animals, but for which land-separation due to continental drift and seafloor spreading are the cause. Western Africa and eastern South America provide an example of this category (Reyment and Tait 1972).

2.3 Seasonal Migration

The distribution of fossil cephalopod shells is often a source of puzzlement as to whether active or passive in origin, not least because of the paucity of information available for living representatives. Comparative information must be extrapolated obliquely, such as afforded by the following case-history for squids. Tinbergen and Verwey (1945) reported on the seasonal migratory behaviour of the squid *Loligo vulgaris* in the waters around Den Helder (Netherlands). Individuals are observed to arrive in great numbers in the Spring when they reproduce. Most animals do not live for longer periods but it is believed that some surviving individuals may overwinter in Portuguese waters. An excellent reference to marine seasonal migration of squids, and other marine animals, is the treatise by Schäfer (1962). On the subject of the seasonal migration of dibranchiate cephalopods to and from the German tidal sea (Wattenmeer) he noted that most species exhibit migratory behaviour. He listed as being seasonal migratory the species: *Sepia officinalis, Loligo forbesi, L. vulgaris, Alloteuthis subulata* and *A.media. L vulgaris* arrives in June/July and departs in September. *L. forbesi* arrives somewhat later and leaves in October/November. Many individuals die during the reproductive period after which, the carcase rises to the surface where, according to the paleontologically germane observations of Schäfer (1962), it remains floating for about four days, on average, borne up by the gases of putrefaction. Many carcases strand along the maximum tidal shoreline, others disintegrate during advanced putrefaction.

The mass occurrences of dead calamars and squids on tidal flats of northern Europe are partly attributable to episodes heavy rainfall and the concomitant drop in salinity.

My own personal observations in the Danish wattenmeer around Esbjerg may be added to this general observation. Following on a violent out-of-season storm out of the North, thousands of lifeless cephalopods were observed to be strewn widely over the tidal flats. Transferring this observation to the past supports the interpretation of mass mortality of ammonites in the Saharan Cenomano-Turonian shallow epicontinental environment as having been caused by salinity-shock in connexion with the effect heavy rainstorms on stenohaline organisms (Reyment 1980a, 2003; Courville and Thierry 1993; Mathey et al. 1991; Mathey et al. 1995).

Whether or not shell-bearing tetrabranchiate cephalopods migrated according to some seasonal pattern is a difficult question to answer. A possible clue is the observation by Talavara and Faustino (1931) made for several areas around the Philippines that *Nautilus pompilius* appears periodically with a maximum during the summer months. In the light of current knowledge, this periodicity could possibly be due to local vertical migration and not to an active geographical displacement. In any event, the energy required for seasonal migration over vast distances cannot conceivably be mustered by shell-bearing cephalopods as has been demonstrated by a method of energy bookkeeping by Reyment (1988). A potentially useful type of simulation has recently been ventilated by Brayard et al. (2004), which may in the future turn out to be adaptable to aspects of the situation outlined above.

The seasonal migration of birds that inhabit a freshwater swamp environment is thought to be a distributional vector for ostracods via eggs trapped in mud attached to some part of the body of the bird. As far as I am aware this hypothesis has never been proven by direct observation. Hartmann (1988a, p. 792) attributed the arrival of Alaskan *Cyprideis* in

Hawaii to this intermediary. In this connexion it should be borne in mind that the eggs of marine ostracods quickly dehydrate when exposed to the air and thus lose their viability, whereas those of freshwater species can survive for many years. Many of the non-marine species described by G. O. Sars were obtained from eggs in ancient mud-samples stored in the Bergen Museum.

2.4 Active unidirectional migration

Here I shall present tow case-histories, one of which (the ammonites) harbours perforce an element of speculation. The second (the ostracods) is more firmly anchored in reasonable assumptions and confirmatory observations.

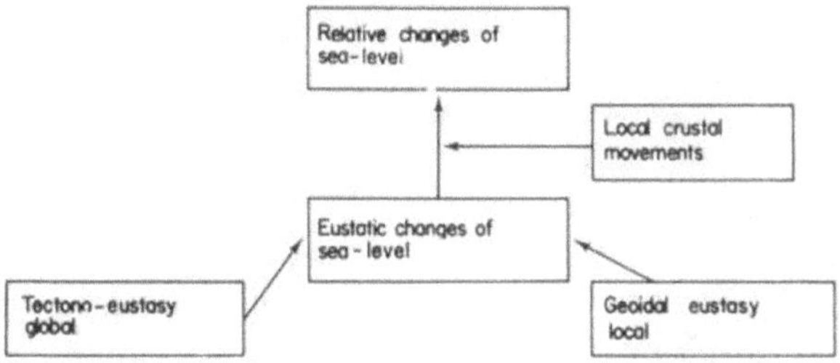

Fig. 1. Schematic representation of the factors underlying marine transgressional episodes and relative changes in sea level; these may occur singly or in various combinations. After Reyment and Mörner (1977)

In order to appreciate the forces at work it is necessary to consider how changes in sea level can and do occur. Figure 1 gives a schematic appraisal of the variables of significance for the development of epicontinental transgressional episodes. The main force driving the dispersal of marine organisms across the vast trans-Saharan inland sea was tectono-eustasy, fuelled by the growth of mid-oceanic ridges and seafloor spreading causing changes in the volume of ocean basins. The effects of geoidal eustasy (Mörner 1976) would have been negligible (glacial eustasy does not appear to have entered into the equation on an appreciable scale).

On the Atlantic side of the epicontinental transgression, local crustal movements played an important part, coupled with the growth of the mid-Atlantic ridge (Reyment and Dingle 1987). The field evidence and distribution of species points to the invasion of the sea having been mainly from the north.

2.4.1 The Ammonites of the Saharan Turonian (Cretaceous)

Reyment and Mörner (1977) dissected the elements of fluctuations in sea level in the light of factors that are directly related to the rise and fall of sea level on a mondial scale (Fig. 1). The Cretaceous sea level record for the South Atlantic region encompasses five major transgressive cycles of which three are well known to be of major global significance; the tectono-eustatic event that took place during the Late Cenomanian-Early Turonian has attracted most interest. It was on this episode that Suess (1888-1909) elaborated his classical concept of eustasy (known specifically today as tectono-eustasy). Locally, earth movements caused by faulting may be responsible for significant, though isolated, relative changes of sea level, such as occurred in the inland sedimentary basins of West Africa (Louis 1970, 1977). The Cenomano-Turonian was a particularly active phase of ridge growth and seafloor spreading connected with accelerated movement apart of South America and Africa and a paucity of geomagnetic reversals (Reyment 1974, 1976, 1980e). Gebhardt (1997) concluded from an analysis of foraminiferal associations that the water depth can hardly have exceeded 50 m in the area with which he was concerned. Viewed in a wider context (cf. Louis 1970, 1978) it is however obvious that greater depths existed off the "Saharan ramp" (Reyment 2003) and that these formed a Caspian Sea type of environment during falls of sea level related to phases of ridge-subsidence.

The seeming proliferation of "species" in Nigerian and North African vascoceratid ammonites seems to be best explained as deriving from ecophenotypic reactions to a variable and often inhospitable epicontinental environment (Courville and Thierry 1993a, 1993b; Reyment 2003). The founder forms followed the sea advancing from the Tethys and, during the acme of the epicontinental transgression, were able to enter the unfolding South Atlantic via the Nigerian coastal basin. The morphometric analyses tend to indicate that the most abundant elements of the Nigerian paleofauna consisted of two highly variable species, *Thomasites (Thomasites) gongilensis* (Woods) and *Thomasites (Bauchioceras) nigeriensis* (Woods) and that the latter evolved from the former early in the history of the Saharan mid-Cretaceous epicontinental sea (Reyment 2003). In addition to the effects of ecologically triggered polymorphism or polyphenism, sparked off by episodes of fragmentation in ephemerally isolated "Balkhash-type" basins, there are intimations of polymorphism, possibly relatable to sexual dimorphism. This eventuality deserves to be given detailed study on the basis of more comprehensive material obtained from the abundance of virtually unexplored sites in Nigeria, Niger, and Chad (Faure 1966).

2.4.2 The early Turonian ammonite genus *Benueites* in the South Atlantic Ocean

The distribution of species of *Benueites* in the late early Turonian (Upper Cretaceous) seems to accord with a model of unidirectional migration with speciation in a situation in which tectono-eustasy did not provide the main driving force and is therefore a rare example of a case-history where an interpretable migrational pattern, accompanied by evolutionary events, can be established for ammonites. The oldest occurrence of the species, *B. benuensis* Reyment and its dimorph, seems to be in Nigeria (the Wadatta limestone which directly overlies older Turonian beds with vascoceratids) near Makurdi (Reyment 1955), and, presumably, in beds of the same age in northeastern Brazil (Reyment 1971). The same species, and a new one, *B. reymenti* were recorded by Collignon (1966) from the Tarfaya Basin of southern Morocco. The same two species (and dimorphs) occur in Trinidad (Reyment 1971). *B. benueensis* (and dimorph) were figured by Reyment (1971) from Colombia together with a third new dimorphic species, *B. colombiensis* Reyment. The locations are marked schematically in Fig. 2. A plausible explanation of the distribution of the three known species of the genus would seem to lie with the oceanographical effects associated with the continued opening of the South Atlantic Ocean and a diverging pattern of northerly migration. The fact that all three species occur in an *in situ* dimorphic association speaks against the likelihood of their having been spread nekroplanktonically on a large scale.

2.4.3 Ostracoda of the North African Paleocene

During the Early Paleocene, possibly latest Maastrichtian, the last of the extensive tectono-eustatic trans-Saharan epicontinental transgressions began (Reyment 1982a). At its acme in Late Paleocene time, the sea stretched from the north of Africa to Nigeria and from the Gulf of Guinea probably through the Kontagora syneclise (Fig. 3).

An impressive paleobiological feature of this is the wide distribution of many species of ostracods. There are at least 24 West African species in common with the Paleocene of Libya (Reyment and Reyment 1980). The main thrust of the Paleocene epicontinental transgression proceeded in a manner similar to the earlier ones, that is the advance of the sea was more rapid from the north than from the south. Evidence for this lies with the indication of a multivariate progression for carapace volume for several species. An example for the abundantly occurring *Bairdia ilaroensis* Reyment and Reyment is shown schematically in Fig. 2.

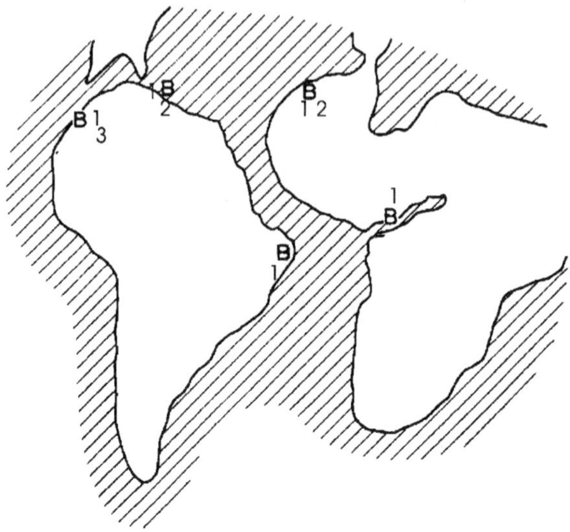

Fig. 2. Distribution of species of *Benueites* in the South Atlantic realm in the Early Turonian shown schematically. Key: 1: *Benueites benueensis* 2: *Benueites reymenti* 3: *Benueites colombiensis*

Fig. 3. Three discriminant scores for carapace volume for *Bairdia ilaroensis* showing a gradual shift from north to south. The shaded area delineates the maximum extent of the trans-Saharan Paleocene epicontinental sea

There is a statistically significant decrease in volume of the carapace over time as the transgression advanced, carrying the evolving association in its wake.

By way of comparison, ostracods were dispersed from the Tethys during the Coniacian Saharan transgression (Reyment 1980d); this incursion of the sea seems to have been of short duration.

2.4.4 South Atlantic Island-Hopping

Nigerian Paleocene and Eocene ostracods have been reported from Argentina (Bertels 1969, 1977). We shall now briefly examine a likely mechanism for this distributional pattern, granted that by Paleocene times, the southern Atlantic had attained a respectable width and under normal marine conditions it would not be possible for shallow-water benthic organisms to cross from one side to the other. A variety of "island hopping" would seem to afford a likely agent.

In the early Maastrichtian, and again in the early Paleocene, ostracods followed the last two major epicontinental transgressions across the Sahara, spreading through the Saharan system of tectonically controlled basins to the expanding South Atlantic Ocean (Reyment 1980b, 1983b).

A logical explanation for the agreement in coastal marine ostracod faunas between West Africa and eastern South America would seem to lie with the migration of shallow-water species by a chain of islands and high-grounds (Thiede 1977).

A schematic reconstruction of the situation is given in Fig. 4. The occurrence of West African shallow-water Cretaceous marine ostracod species in South America has been recorded by Reyment and Neufville (1974) and Neufville (1973).

An additional factor of importance for ostracods and some other animals relates to passive transport on algae. The passive migration of many species, including ostracods, across the Pacific Ocean, is known to occur in this manner (Hartmann 1982a, 1982b). My own observations made in Tahitian waters in 1986 confirmed the occurrence of ostracod "demes" being transported on floating algae in accordance with Hartmann's observations.

Fig. 4. Reconstruction of the averaged southern Atlantic Ocean in Maastrichtian and Paleocene time. The schematically denoted shallow region between southern Brazil and southern Africa seems to have provided a migrational route for benthic organisms, in particular ostracods. The trans-Saharan transgressions around the Cretaceous-Cenozoic boundary contain elements in common with Brazil and Argentina. Based on the paleo-oceanological results of Thiede (1977) and for faunistic aspects on Neufville (1973) and Reyment (1980b)

2.5 Imposed dispersion by submarine currents

In a recently aired television programme *(Discovery Science)* deriving from the submarine biological research of Professor Hans Fricke (*vide* p. 2), it was made reasonably apparent that individuals of the newly discovered Pacific coelacanths are transported by submarine currents that bring isolated individuals to northern Sulawesi (Indonesia) from the island complexes of the South Pacific, thence further westward, from some as yet unidentified centre of distribution in the southern Pacific Ocean. It was also illustrated that co-occurring individuals of *Nautilus pompilius* s. l. are presumably dragged from their natural habitat to Indonesian islands in the same manner (e.g. northern Sulawesi). Notwithstanding that species of *Nautilus* occur widely throughout the Pacific and Indian oceans, albeit patchily distributed (Fig. 5), it is certainly quite reasonable to accept the possibility of forced dispersal in relation to such relatively poorly endowed swimmers as the coelacanth and the pearly nautilus. It has been suggested on these grounds that the reported occurrences of living nautilus in the eastern reaches of the Indian Ocean could be the result of imposed dispersal by powerful (intermittent?) subsurface currents. Indeed, the distributional pattern being established for *Latimeria* is persuasively close to those of *Nautilus*, not least in the far western reaches of the Indian Ocean.

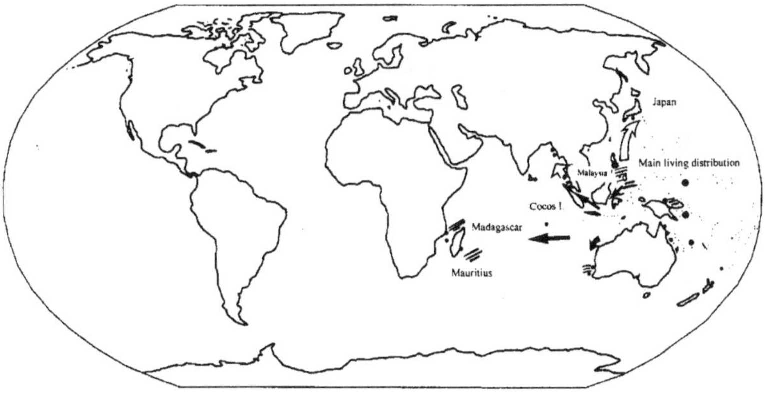

Fig. 5. Sketch map showing the distribution of living species of *Nautilus* in the Indian and Pacific Oceans and places where nekroplanktonically drifted shells occur. The black arrows indicate approximately the dispersal directions of living individuals, the open arrow marks the direction taken by floating shells borne by the Kuroshio current. The black dots mark stranding sites of shells. Based on Reyment (1973a), supplemented with information obtained from recent records

This aspect of submarine oceanology holds promise of yielding results of far-reaching significance (cf. Fricke and Hissman 1994; Fricke 1998). The tantalizing significance of this research for understanding the dispersal of fossil cephalopods with properties analogous to living nautiloids remains to be exploited.

2.6 Nekroplanktonic dispersal

The post-mortem distribution of the remains of organisms is of trivial significance in neontology but of great importance to the paleoecologist. Schäfer (1962, p. 570) discussed, and illustrated the effects of, the post-mortem distribution of gas-floated marine mammals in the Jadebucht, northern Germany. This is a potentially important factor and one that deserves more attention in the paleoecological literature than it has received. I note, in passing, that rock paintings made by past and present Australian aboriginal people correctly, and elegantly, depict the post-mortem deformation of water-transported vertebrate carcases. From the standpoint of the paleontologist it is, however, the fossil cephalopods that claim pride of precedence.

Despite the constantly accruing weight of evidence in favour of nekroplanktonic dispersal of the shells of species of the pearly nautilus, there still seems to me to be an aura of reluctance associated with accepting the practical consequences for paleobiogeographical interpretations. There is, however, a logical *non sequitur* here in that what the animals did in life has nothing to do with what happened with their remains after death, when outrageous fortune reigns supreme. The post-mortem distribution of nautiloid shells is well documented. Reyment (1958) made a survey of the existing literature, and original observations. Since then, records of stranded shells have been published by Toriyama et al. (1965) for beaches around western Thailand, Teichert (1970) for drifting shells in the Bay of Bengal and 250 km SW of Sri Lanka, Reyment (1973, pp. 35-39) for western Malaysia, Fiji, the Philippines (stranded shells and shells inhabited by hermit crabs), the Solomon Islands (study due to Mr. C. F. Fowler, reported in Reyment 1973a), Great Barrier Reef (reported by Dr. J. S. Hynd in Reyment 1973a), Mauritius, Zanzibar. The floating behaviour of various models of nekroplanktonic shell-types has been analysed and illustrated by Reyment (1980c).

Reyment (1967) discussed the post-mortem transport of rare nautiloid shells in the Paleocene proto-South Atlantic by the forerunner to the Benguella current, from southern Africa to the Niger Embayment. An analogous example of post-mortem transport has been reported by Chirat (2000) for Paleogene nautiloids. Reyment (1968) and Laufeld (1974) described preferred orientation in stranded orthocones in the Ordovician and Silurian of Sweden. These shells occur in great numbers and it is clear that they must have in life inhabited a shallow, strand-near environment.

Not only cephalopod remains are carried nekroplanktonically. Reyment (1986) recorded a floating "mummified" shell of a regular echinoid floating 100 m offshore off Estepona, Spain. This observation is analogous to the record of floating gastropod shells by Krejci-Graf (1935) and to the rafting of reef-coral and other organisms (Jokiel 1989).

The total body of evidence reporting the widespread post-mortem transport of fossil cephalopod shells is extensive, too extensive to attempt a review here. References to published information on the topic are given in Reyment (1958, 1971, 1973a, 1980c).

2.7 Spurious migration

In what must be one of the most important contributions ever to ostracodology and paleobiogeography, Krömmelbein (1970) and Krömmelbein and Wenger (1966), demonstrated that the fresh-water Neocomian ostracod faunas of northeastern Brazil and Gabon shared a common geographical origin. Of the 40 species of ostracods known from the Cocobeach Formation of Gabon and 90 species from northeastern Brazil, 33 occur in both regions. This result was fundamental to establishing the existence of a land connexion between South America and Africa in the lowermost Cretaceous and for delineating the rift-controlled freshwater inland seas in the Gabon, Congo and northeastern Brazilian contact zones. It is stimulating to observe how Krömmelbein gradually, over a spate of years, arrived by skilful deduction at the conclusion that the Brazilian and Gabon ostracod sequences had once inhabited the same sequence of rift-valley lakes and that the impression of lateral migration was not a viable concept. Transport by migrating birds as a possible mechanism was soon discredited since experience shows that birds migrate along a N-S-N axis and not in an E-W-E direction, which would be a pointless exercise from the bioclimatic standpoint (Hartmann 1988b). A referee suggested that this may not have been so in the past. Be this as it may, there is no evidence known to me that gainsays Krömmelbein (1970) and Hartmann (1988) and, moreover, the question has scant bearing on the geological facts before us. The fact that a great number of the known Wealden ostracod species of Gabon and eastern Brazil are common to both areas was used by Reyment and Tait (1972) and Reyment and Dingle (1987) in their reconstruction of the rift-valley phase of the South Atlantic Ocean. They also compiled a detailed reconstruction of the relative geographical position and comparative sedimentology of the Sergipe-Alagoas and Gabonese basins prior to the Aptian marine episode (Reyment and Tait 1972, p. 81, Fig.5).

2.8 Concluding commentary

The possibilities of unequivocally identifying migrational patterns in the fossil record vary considerably depending on such factors as the nature of the organism of interest and the degree of knowledge of the paleogeography and paleoecology of the time concerned. Brief accounts of examples involving assured migrational routes are given here for mammals, ammonites, and ostracods based on establishing a connexion in relation to water-bodies and Thiede's (1977) transient high-grounds. Also of importance is the passive transport of some species of marine ostracods over very long distances on floating "algal islands". The likelihood of enforced transport of living organisms by deep submarine currents is another feature of potential significance. Terrestrial migration is a more difficult problem and the identification of such events has not been attempted here. There are several varieties of spurious migration the best studied of which concerns mainly the nekroplantkonic dispersal of cephalopod shells. Another case of a well documented event involving crustal divergence arising from the effects of seafloor spreading result in simulating migration: a classical example of this, involving a distance of 6-7000 km, is the Krömmelbein-Wenger analysis of the Neocomian (i.e Lower Cretaceous - Wealden) freshwater ostracod sequences of Gabon and Brazil.

The vexing question remains. How can one, unequivocally, distinguish the *ad hoc* kinds of "migration", spurious or genuine, in the fossil record? To be honest, I do not know. All I can hope to suggest is that each case must be approached with an open mind. Some years ago I had a certain success in applying quantitative statistical procedures to paleobiogeography (summary in Reyment 1991) and such methods form an integral part of my arsenal of techniques. I have not taken up this aspect in this short overview because of the far-reaching range of the subject. Examples of what I believe can open new fields for the study of migration in the fossil record are given in Reyment (1983c, 1993, 1995) for applications of multivariate quantitative genetics to paleontological problems in a broad sense. Useful paleobiogeographical information can be extracted from appropriately constructed statistical analyses of the shell-chemistry of marine invertebrates (Reyment 1996, 2000). The quantitative analysis of polyphenism in fossils has likewise proven to be a valuable utensil (Reyment 2004).

References

Bertels A (1969) Estratigrafía del limite Cretacico-Tertiario en Patagonia septentrional. Rev Assoc Geol Argentina 24: 41-54

Bertels A (1977) Cretaceous Ostracoda - South Atlantic. In Swain FM (ed) Micropaleontology of Atlantic Basins and Borderlands, Elsevier, Amsterdam 271-301

Bookschoten GJ, Sondaar P (1972) On the fossil mammals of Cyprus. Proc Koninkl Nederl Akad Wetenschappen B 75: 306-330

Brayard A Héran M-.A, Costeur L, Escarguel G (2004) Triassic and Cenozoic paleobiogeography: two cases in quantitative modelling using IDL. Palaeontologia Electronica 7: 22 pp

Chirat R (2000) The so-called "cosmopolitan paleobiogeographic distribution" of Tertiary Nautilids of the genus *Aturia* Bronn 1838: the result of post-mortem transport by oceanic paleocurrents. Palaeoclimatology, Palaeogeography, Palaeoecology 157: 59-77

Collignon M (1966) Les céphalopodes crétacés du bassin côtier de Tarfaya. Notes et Memoires du service géologique du Maroc 175: 1-148

Courville P, Thierry J (1993a) Nouvelles données biostratigraphiques sur les dépôts cénomano-turoniens du Nord-Est du fossé de la Bénoue (Nigeria). Cretaceous Research 14: 385-396

Courville P, Thierry J (1993b) Sous-espèces géographiques et/ou contrôle environmental de la variablité morphologique chez *"Thomasites" gongilensis* (Woods 1911), (Ammonitina, Acanthocerataceae, Vascoceratidinae) du Turonien Inferiéur de la Haute Benoué (Nigeria). Geobios 15: 73-89

Faure H (1966) Reconnaissance géologique des formations sédimentaires post-paléozoiques du Niger Oriental. Mémoires du Bureau de Recherche Géologique et Minière 47: 629 pp

Fricke HW (1998) Im Reich der lebendigen Fossilien. In Gaede P-M (ed) Die Seele des weißen Bären, Hamburg

Fricke HW, Hissman K (1994) Home ranges and migration of the living coelacanth *Latimeria chalumnae*. Environ Biology Fishes 32: 313-339

Hartmann G (1982a) Variation in surface ornament of the valves of three ostracod species from Australia. In Bate RH, Robinson E, Sheppard LM Fossil and Recent Ostracods. British Micropalaeontological Society series 365-380. Ellis Horwood Ltd, Chichester, UK

Hartmann G (1982b) Beitrag zur Ostracodenfauna Neuseelands (mit einem Nachtrag zur Ostracodenfauna der Westküste Australiens). Mitteilungen hamb Zool Mus Inst 79: 119-150

Hartmann G (1988a) Distribution and dispersal of littoral Pacific Island Ostracoda. "Evolutionary Biology of Ostracoda". In Hanai T, Ikeya N,

Ishizaki K Proceedings of the 9th International Symposium on Ostracoda, Kodansha Ltd, Tokyo 787-795

Hartmann G (1988b) Gibt es biologische Argumente zur Entstehung der Süderdteile. Die Geowissenschaften 6(9): 270-275

Jokiel PL (1989) Rafting of reef coral and other organisms at Kwajalein Atoll. Marine Biology 101: 483-493

Kreji-Graf K (1935) Beobachtungen am Tropenstrand: 1-4. Senckenbergiana 17: 21-61

Krömmelbein K (1970) Non-marine Cretaceous ostracods and their importance for the hypothesis of "Gondwanaland". Proceedings of the Second Gondwana Symposium, South Africa (1970) 7: 617-619

Krömmelbein K, Wenger R (1966) Sur quelques analogies remarquables dans les microfaunes crétacées du Gabon et du Brésil oriental (Bahia et Sergipe). IUGS Symposium on sedimentary basins of the African coasts. Atlantic Coast, Rfl 193-196

Laufeld S (1974) Preferred orientation of orthoconic nautiloids in the Ludlovian Hemse Beds of Gotland. GFF 96: 157-162

Louis P (1970) Contribution géophysique à la connaissance géologique du Bassin du Lac Tchad. Mémoires géophysiques de la Recherche Scientifique et Téchnique Outre Mer 42: 311 pp

Louis P (1978) Gravimétrie et géologie en Afrique occidentale et centrale. Bulletin du bureau de recherches géologiques et minières 91, 53-61

Mathey B, Lang J, Pierre C, Alzouma K (1991) Effets combinés de l'eustatisme et des variations climatiques: l'exemple des premiers transgressions du Cretacé supérieur dans le Bassin des Iullemeden (Niger). CR de l'Académie des Sciences, Paris 312 ii 523-528

Mathey B, Alzouma K, Lang J, Meister C, Neraudeau D, Pascal P (1995) Unusual faunal associations during Upper Cenomanian - Lower Turonian floodings on the Niger ramp (central West Africa). Palaeogeography, Palaeoclimatology, Palaeoecology 119: 63-75

Mörner NA (1976) Eustasy and geoid changes. Journal of Geology 84: 123-151

Neufville E.MH (1973) Upper Cretaceous - Paleocene ostracods from the South Atlantic. Ph D Dissertation, University of Uppsala 205 pp

Reyment RA (1955) The Cretaceous Ammonoidea of Southern Nigeria and the southern Cameroons. Bulletin of the geological Survey of Nigeria 25: 1-112

Reyment RA (1958) Factors in the distribution of fossil cephalopods. Stockholm Contributions in Geology 1: 97-184

Reyment RA (1967) A note on Paleocene nautiloids from Nigeria. Overseas Geology and Mineral Resources 10: 47-55

Reyment RA (1968) Orthoconic nautiloids as indicators of shoreline currents. Journal of Sedimentary Petrology 38: 1387-1389

Reyment RA (1971) Vermuteter Dimorphismus bei der Ammonitengattung *Benueites*. Buleltin of the Geological Institutions, University of Uppsala NS 3: 1-18

Reyment RA (1971) Introduction to Quantitative Paleoecology. Elsevier, Amsterdam 226 pp

Reyment RA (1973a) Factors in the distribution of fossil cephalopods. Part 3. Experiments with exact models of certain shell types. Bulletin of the Geological Institutions, University of Uppsala, NS 4: 9-41
Reyment RA (1973b) L'histoire de la mer transcontinentale saharienne pendant le Cénomanien-Turonien. Bull Géol Soc France 13: 6 pp
Reyment RA (1974) Application des méthodes paléobiologiques à la théorie de la dérive des continents. Rép Géol Dynam Géogr Phys 16: 61-70
Reyment RA (1976) Geomagnetic reversal data statistically appraised. Tectonophysics 31: 73-91
Reyment RA (1980a) Biogeography of the Saharan Cretaceous and Paleocene epicontinental transgressions. Cretaceous Research 1: 299-327
Reyment RA (1980b) Paleo-oceanology and paleobiogeography of the Cretaceous South Atlantic Ocean. Oceanologica Acta 3: 127-133
Reyment RA (1980c) Floating orientations of cephalopod shell models. Paleontology 23, 931-936
Reyment RA (1980d) Beobachtungen über die Ostrakoden der Coniactransgression im Saharagebiet. Monatshefte des neuen Jahrbuchs (1980) 186-190
Reyment RA (1980e) Trends in Cretaceous and Tertiary geomagnetic reversals. Cretaceous Research 1: 27-48
Reyment RA (1982a) The ostracods of the Kalambaina Formation, northwestern Nigeria. Bull Geol Inst Univ Uppsala NS 9: 51-65
Reyment RA (1982b) Analysis of trans-specific evolution in Cretaceous ostracods. Paleobiology 8: 293-306
Reyment RA (1983a) Paleontological aspects of island biogeography: colonization and evolution of mammals on Mediterranean islands. Oikos 41: 299-306
Reyment RA (1983b) West African and North American transgressional maxima and the dispersal of marine organisms. J african Earth Sciences 1: 255-262
Reyment RA (1983c. Phenotypic evolution in microfossils. *Evolutionary Biology* 16: 293-306
Reyment RA (1986) Nekroplanktonic dispersal of echinoid shells. Palaeoclimatology, Palaeogeography, Palaeoecology 52: 357-349
Reyment RA (1988) A foraging model for shelled cephalopods pp. 687-703 In Wiedmann J, Kullmann J (eds) Cephalopoda - present and past. Schweizerbart'sche Verlagsbuchhandlung, Stuttgart
Reyment, RA (1991) Multidimensional Paleobiology. Pergamon Press, Oxford, 377 pp + 31 pp of computer programs
Reyment RA (1995) Analysis of shape variation in *Vargula hilgendorfi* (Ostr, Crust). Revista Española de Paleontología, Homenaje al Dr. Guillermo Colom 26-30
Reyment RA (1996a) Eduard Suess (1831-1914). Terra Nova 8: 105-107
Reyment RA (1996b) Case-study of the statistical analysis of chemical compositions exemplified by ostracod shells. Environmetrics 7: 39-47
Reyment RA (2000) Protocoles Mathématiques: application à la forme, l'écologie, la géochimie des Ostracodes. Geobios 924-928

Reyment RA (2003. Morphometric analysis of variability in the shell of some Nigerian Turonian (Cretaceous) ammonites. Cretaceous Research 24: 785-503

Reyment RA (2004) Instability in principal component analysis and the quantification of polymorphism in paleontological data. Mathematical Geology 36: 629-638

Reyment RA, Dingle, RV (1987) Paleogeography of Africa during the Cretaceous Period. Palaeogeography, Palaeoclimatology, Palaeoecology 59: 93-116

Reyment RA, Mörner, NA (1977) Cretaceous transgressions and regressions exemplified by the South Atlantic. Special Papers, Paleontological Society of Japan 21: 247-261

Reyment RA, Neufville, EMH (1974) Multivariate analysis of populations split by continental drift. Math Geol 6: 173-181

Reyment RA, Reyment, ER (1980) The Paleocene trans-Saharan transgression and its ostracod fauna. Proceedings of the Second Symposium on the Geology of Libya (Tripoli 1978) 1: 245-254

Reyment RA, Tait EA (1972) Biostratigraphical dating of the early history of the South Atlantic Ocean. Phil Trans Roy Soc London, B, Biological Sciences 204: 858, 55-95

Schäfer W (1962) Aktuo-Paläontologie nach Studien in der Nordsee. Senckenberg-Buch 41, Verlag Waldemar Kramer, Frankfurt am Main 666 pp

Suess E (1888-1909) Das Antlitz der Erde. Tempsky, Prague and Vienna, and Freytag, Leipzig 1: 778 pp, 2: 704 pp, 3a: 508 pp, 3b: 789 pp

Talavara F, Faustino A (1931) Industrial shells of the Philippines. Philippine Journal of Science 3: 321-350

Teichert C (1970) Drifted *Nautilus* shells in the Bay of Bengal. J Paleont 44: 1129-1130

Thaler L (1973) Nanisme et gigantisme insulaire. La Recherche 37: 741-750

Thiede J (1977) Subsidence of aseismic ridges: evidence from sediments on the Rio Grande Rise (southwest Atlantic Ocean). Bull Amer Assoc Petrol Geol 61: 929-940

Tinbergen L, Verwey J (1945) Zur Biologie von *Loligo vulgaris* Lam. Arch Néerland Zool 71: 213-226

Toriyama R, Sato T, Hamada T, Komalarjun P (1965) *Nautilus pompilius* drifts on the west coast of Thailand. Japan J Geol Geogr 36: 149-161

3 Paleoecology and paleogeography of Eocene ostracod faunas from the Nile Valley between Minia and Maghagha, Upper Egypt

Ashraf M. T. Elewa

Geology Department, Faculty of Science, Minia University, Egypt, aelewa@link.net

3.1 Abstract

Sixty-four ostracod species and morphs in 130 samples from five sections along the Nile Valley between Minia and Maghagha cities could define the paleoenvironmental conditions that prevailed during the Eocene in the study area. Cluster analysis based on the Jaccard coefficient of similarity (the paired group average method) led to the distinction of three biofacies of ostracod assemblages, each of them with its particular paleoenvironment. The Minia biofacies represents the early Eocene, and is characterized by deposition in an inner to outer shelf environment. The Maghagha biofacies is middle Eocene and represents a restricted shallow water environment, possibly with more or less reduced salinity. At the base of the sequence, these restricted conditions were accompanied by reduced, dissolved oxygen conditions. At the top, calm water conditions prevailed. The Qarara biofacies is also middle Eocene in age and was deposited in turbulent water conditions, gradually shallowing over time. On the other hand, the Samalut Formation, which was barren of microfossils, can be attributed, according to its lithological composition, to a particular type of facies representing a shallow marine environment. The ostracod species of the study area provide evidence of distinct migration of ostracods during the Paleogene along the southern shores of the Tethys.

Keywords: Eocene, Ostracoda, paleoecology, paleobiogeography, Nile Valley, Egypt.

3.2 Introduction

Eocene microfossils of the Nile Valley have been the subject of different authors (Bishay 1961, 1966; Bassiouni 1969a, 1969b, 1971; El Dawoody 1971; Bassiouni et al. 1977; Cronin and Khalifa 1979; Khalifa and Cronin 1979; Aref 1982; Boukhary, Blondeau and Ambroise 1982; Boukhary and Abdel Malik 1983; Azab 1984; Haggag and Anan 1987; Zalat 1987; Haggag 1989; Hussein 1992, 1994; El Dawy 1997; Elewa 1997, 1998a, b. None of the previous studies, despite the work of Elewa (2004) on the Eocene of Gebel Mokattam, provide a detailed study of the ostracod taxonomy and stratigraphy as well as the usefulness of these ostracods for interpreting the prevailed paleoenvironments during the deposition of the Eocene rocks in the study area.

The present work focuses on the taxonomy, stratigraphy and the use of ostracod assemblages to interpret the paleoenvironments prevalent during the deposition of five surface sections exposed along the Nile valley of Egypt, between Minia and Maghagha, as a step towards reconstructing a complete view of the paleoenvironmental conditions associated with the Eocene deposition in the Nile Valley. These sections were examined for their ostracod content and for biostratigraphically significant planktic foraminiferal species. Moreover, the identified ostracod species show a significant paleogeographical distribution in the surrounding countries of North Africa and the Middle East.

3.3 Material and methods

One hundred and thirty samples were collected by the author from five sections (Figs. 1, 2) representing the lower Eocene Minia Formation and the middle Eocene Samalut Formation, Maghagha and Qarara Formations. The studied samples were washed through a 250 µm mesh sieve screen and dried. These samples yielded 64 ostracod species and morphs. By discarding samples containing less than three ostracod species and neglecting those that are represented by less than three specimens in the studied samples, 52 samples containing 38 ostracod species and two morphs were subjected to the analysis. Cluster analysis based on the Jaccard coefficient of similarity (the paired group average method) was applied to the data matrix of 40 ostracod species and morphs in 52 samples to discriminate the existing biofacies within the studied ostracods. The ostracod specimens used in the study are deposited at the Geology Department, Faculty of Science, Minia university, in the personal collection of the author. The computer program used in this study is included in a software package called PAST, version 0.94 (2002).

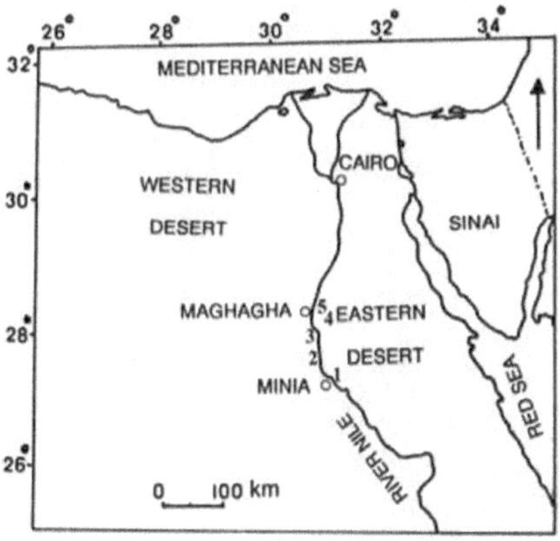

Fig. 1. Location map of the studied sections. Numbers 1 to 5 stand for the studied sections

3.4 Stratigraphy

Said (1990) stated that Eocene outcrops cover about 21% of the surface area of Egypt. He mentioned that these Eocene outcrops span several thousand metres in thickness and consist almost exclusively of carbonates, occassionally with a mixture of clastic material of variable amounts. They are principally exposed in four major areas of Egypt: the Nile Valley, the Eastern Desert, the Western Desert and Sinai.

In the Nile Valley, the Eocene rocks are represented by the Thebes Formation of early Eocene age (Said 1960); the Minia Formation of early Eocene age, according to Boukhary, Toumarkine, Khalifa and Aref (1982), Boukhary and Abdel Malik (1983), based on larger and planktonic foraminifera, and Elewa (1998b) based on ostracods. The Minia Formation was previously stated by Said (1960) to belong to be middle Eocene in age. The Samalut Formation (Bishay 1961), the Maghagha Formation (Bishay 1966), the Qarara Formation (Bishay 1961), the El Fashn Formation (Bishay 1966) and the Beni Suef Formation (Bishay 1966) are all of middle Eocene age. While, the Mokattam Group is of late middle Eocene age (see Said 1990 and Elewa 2004) and the Maadi Group is of late middle to late Eocene ages (Strougo and Boukhary 1987 and Said 1990).

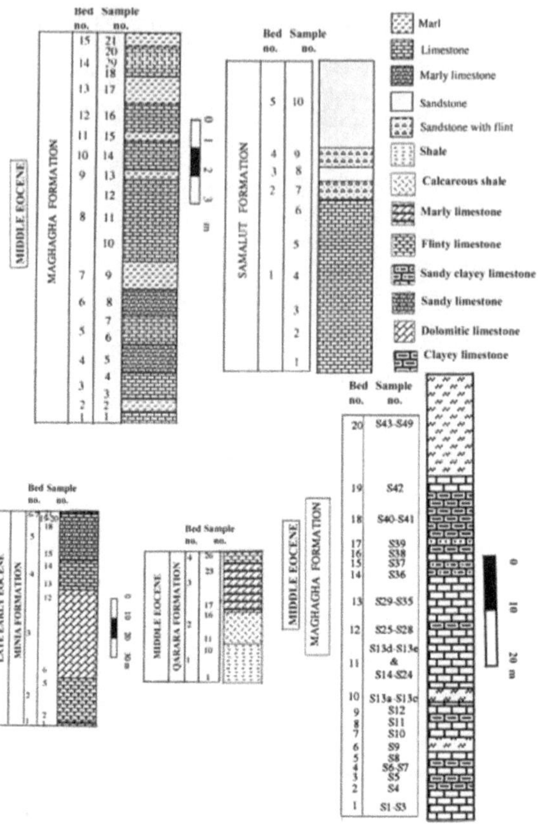

Fig. 2. Sampled sections from the Eocene deposits of the Nile Valley, between Minia and Maghagha. Numbers 1 to 5 are same as for Fig. 1

3.4.1 The studied rock units

The Minia Formation

This formation was named by Said (1960) to represent the alviolinid, snow white limesones rich in *Nummulites* and corals of a section exposed at Zawiet Sawada, opposite to Minia city on the Nile Valley.

The Samalut Formation

The Samalut Formation is characterized by a creamy fine-grained limestone with scattered flint nodules at the base and hard dolomitic laminated limestone at the top. The name of this formation was first proposed by Bishay (1961) to a section opposite to Samalut city, which is considered as its type locality.

The Maghagha Formation

This term was introduced by Bishay (1966) to designate the chalk-marl complex exposed in the Maghagha district on the Nile Valley in substitution for the informal "B" formation of Barker (1945).

The Qarara Formation

The term Qarara Formation was used by Bishay (1966) for the sequence of Gebel Qarara that forms the northern boundary of the Maghagha district ("C" formation of Barker (1945)). This sequence is composed of calcareous quartz sandstones and clays very rich in *Nummulites* and macrofossils.

The following is a description of the studied sections and their biostratigraphic zonation according to planktonic foraminiferal assemblages.

3.4.2 The studied sections

Section 1, representing the type locality of the Minia Formation

Section 1 crops-out at Sawada village opposite East Minia city on the eastern bank of the Nile Valley (Fig. 1). The basal two metres of the section is composed of massive limestone facies rich in *Nummulites* and corals (sample 1 in Fig. 2). Following upward, a 22 m thickness of greyish white, moderately hard limestones with flint intercalations (samples 2-5). Above which are about 45 m of dolomitic limestone facies (samples 6-12); 15 m of white, very hard limestone rich in nummulites (samples 13-14); 23 m of yellowish white, sandy limestone with nummulites (samples 15-18); one m of white, hard limestone (samples 19-20); 0.5 metres of yellowish white, marly limestone with nummulites (sample 21) (see Fig. 2).

Section 1 has very rare planktonic foraminifera, however, several planktonic foraminiferal species (e.g., *Morozovella aragonensis* (Nuttall 1930), *Acarinina pentacamerata pentacamerata* (Subbotina 1947), and *Subbotina inaequispira* (Subbotina 1953)) indicate this section is the late early Eocene in age (P8-9 of the updated planktic foraminiferal biostratigraphic scheme of Berggren et al. 1995).

Section 2, representing the Samalut Formation

This section is located at the northwestern side of Samalut city, on the western side of the Nile Valley (Fig. 1). It is composed of 6 m thickness of a very hard, white limestone (samples 1-6), followed by a bed of sandstone with flint bands comprising a thickness of about 2 m (sample 7-8). The section is capped by a bed of ferruginous sandstone with a thickness of 3 m (samples 9-10) (see Fig. 2).

Section 2 is barren of microfossils, although, Boukhary and Abdel Malik (1983) assigned the Samalut Formation to middle Eocene age (early to middle Lutetian) based on *Nummulites* and planktonic foraminifera. Their identified planktonic species include: *Turborotalia cerroazulensis frontosa* (Subbotina 1953), *T. cerroazulensis possagnoensis* (Toumarkine and Bolli 1970), *T. cerroazulensis pomeroli* (Toumarkine and Bolli 1970), *Globigerinatheka subconglobata subconglobata* (Shutskaya 1958), *Morozovella lehneri* (Cushman and Jarvis 1929), *Truncorotaloides topilensis* Cushman 1925, *Truncorotaloides rohri* Brönnimann and Bermúdez 1953, and *Globigerinoides higginsi* Bolli 1957. This assemblage suggests that the Samalut Formation ranges from zones P10 to P14.

Section 3, representing the Maghagha Formation

Section 3 is located opposite Bahnasa village, to the west of Beni Mazar city (Fig. 1). It consists of yellowish white, moderately hard limestone facies in alternation with yellow to yellowish brown, soft to moderately hard marl and marly limestone facies. The total thickness is about 14 m (see Fig. 2).

The following planktonic foraminiferal species were identified by the author from this section: *Acarinina broedermanni* (Cushman and Bermúdez 1949), *Globigerinatheka subconglobata subconglobata* (Shutskaya 1958), *Morozovella lehneri* (Cushman and Jarvis 1929), and *Truncorotaloides topilensis* Cushman 1925. This association indicates that the section is entirely within the *Morozovella lehneri* Zone (P11).

Section 4 (Gebel El Sheikh Fadl), representing the type locality of the Maghagha Formation

Section 4 is exposed in the Maghagha district on the eastern bank of the Nile Valley, opposite El Sheikh Fadl village. This section is composed of alternating beds of white, moderately hard, chalky limestone and grayish white to yellowish white, soft to moderately hard marl comprising a thickness of 81.30 m (see Fig. 2).

This formation includes the *Acarinina aspensis* Zone to the *Truncorotaloides rohri* Zone (P10-14), based on the following planktonic foraminiferal association: *Acarinina aspensis* (Colom 1954), *Acarinina colomi* (Bermudéz 1961), *Acarinina pentacamerata pentacamerata* (Subbotina 1947), *Acarinina pentacamerata camerata* Khalilov 1956, *Acarinina broedermanni* (Cushman and Bermúdez 1949), *Globigerinatheka subconglobata subconglobata* (Shutskaya 1958), *Morozovella lehneri* (Cushman and Jarvis 1929), *Truncorotaloides topilensis* Cushman 1925, and *Truncorotaloides rohri* Brönnimann and Bermúdez 1953.

Section 5 (Gebel Qarara), representing the type locality of the Qarara Formation

Section 5 is located at the northern boundary of the Maghagha district and is represented by 20 m of greyish yellow to yellowish brown, moderately hard shales at the base, followed upward by about 16 m of grey to yellowish green, friable, calcareous shales and 25 m of yellowish white, hard, marly limestone facies. The sequence is capped by 6 m of limestones with flints representing the El Fashn Formation (see Fig. 2). This section contains very rare planktonic foraminifera, however, the Qarara Formation has been assigned by Bishay (1966) to the middle Eocene (late Lutetian) based on *Nummulites*, and by Omara et al. (1977), by the occurrence of *Truncorotaloides rohri* Bronnimann and Bermúdez 1953.

The distribution chart of the most important identified Eocene ostracods as correlated with the planktonic foraminiferal zones (the P-Zone System) is shown in Figure 3.

3.5 Biofacies analysis of ostracod assemblages

It is necessary to treat the data statistically in order to elucidate the depositional environments of the Eocene strata of the Maghagha-Minia district based on the analysis of ostracod assemblages.

Cluster analysis based on the Jaccard coefficient of similarity (the paired group average method) led to the distinction of three biofacies of ostracod assemblages (see Fig. 4), representing three particular paleoenvironments.

Also, See table 1 for list of ostracod species and morphs used in cluster analysis.

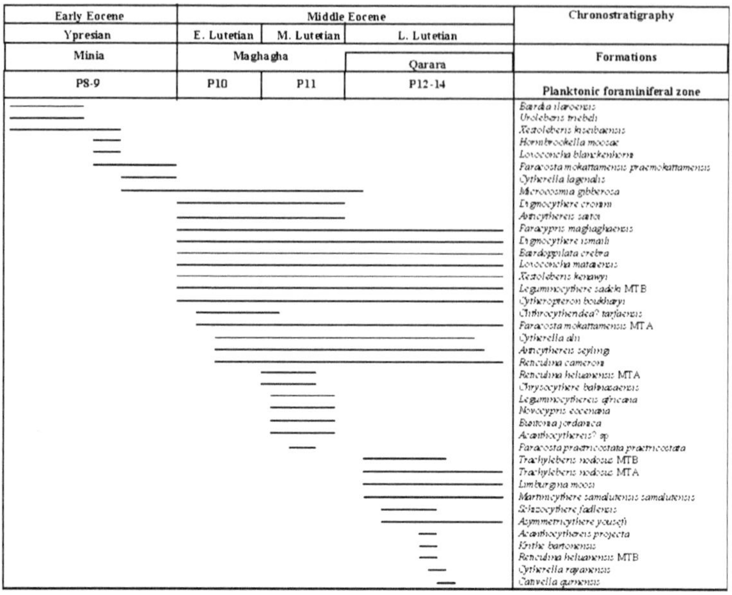

Fig. 3. Distribution chart of the most important Eocene ostracods of the study area

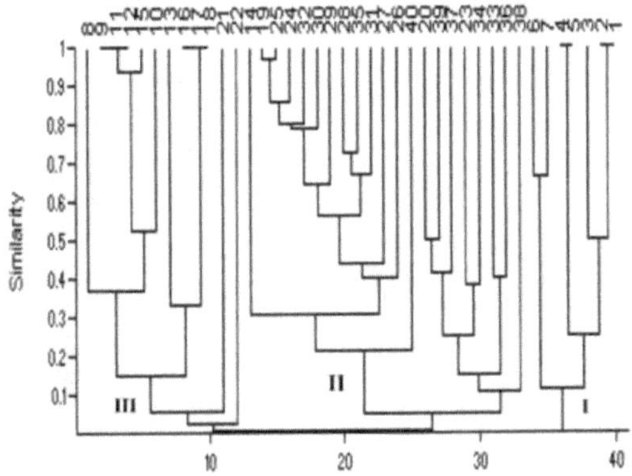

Fig. 4. Cluster analysis based on the Jaccard coefficient of similarity (the paired group average method), applied to 40 ostracod species and morphs from the study area

Table 1. List of ostracod species and morphs used in cluster analysis. Numbers are same as in Fig. 4

Species	Numbers (as for Fig. 4)
Bairdia ilaroensis Reyment and Reyment	1
Uroleberis triebeli Bassiouni and Luger	2
Xestoleberis kiseibaensis Bassiouni and Luger	3
Hornibrookella moosae Bassiouni and Luger	4
Loxoconcha blanckenhorni Bassiouni and Luger	5
Paracosta mokattamensis praemokattamensis (Bassiouni)	6
Cytherella lagenalis Marlière	7
Microcosmia gibberosa Bassiouni and Luger	8
Trachyleberis nodosus MTA Bassiouni	9
Trachyleberis nodosus MTB Bassiouni	10
Martinicythere samalutensis samalutensis Bassiouni	11
Limburgina moosi Bassiouni	12
Schizocythere fadlensis Cronin and Khalifa	13
Bairdoppilata crebra (Deltel)	14
Asymmetricythere yousefi Bassiouni	15
Acanthocythereis projecta Bassiouni	16
Krithe bartonensis (Jones)	17
Reticulina heluanensis MTB (Bassiouni)	18
Xestoleberis kenawyi Khalifa and Cronin	19
Novocypris eocenana Ducasse	20
Cytherella rayanensis Elewa	21
Cativella qurnensis Bassiouni	22
Cytherella alii Elewa	23
Digmocythere ismaili (Bassiouni)	24
Loxoconcha mataiensis Khalifa and Cronin	25
Clithrocytheridea? tarfaensis Khalifa and Cronin	26
Digmocythere cronini Elewa	27
Anticythereis seylingi Cronin and Khalifa	28
A. saitoi Khalifa and Cronin	29
Leguminocythereis sadeki MTB Bassiouni	30
Reticulina cameroni (Khalifa and Cronin)	31
Cytheropteron boukharyi Khalifa and Cronin	32
Reticulina heluanensis MTA (Bassiouni)	33
Leguminocythereis africana Bassiouni	34
Paracosta mokattamensis MTA (Bassiouni)	35
Chrysocythere bahnasaensis Elewa	36
Buntonia jordanica Bassiouni	37
Paracosta praetricostata praetricostata Bassiouni	38
Acanthocythereis sp.	39
Paracypris maghaghaensis Khalifa and Cronin	40

Biofacies I is represented by *Bairdia ilaroensis* Reyment and Reyment (sp. 1), *Uroleberis triebeli* Bassiouni and Luger 1990 (sp. 2), *Xestoleberis kiseibaensis* Bassiouni and Luger 1990 (sp. 3), *Hornibrookella moosae* Bassiouni and Luger 1990 (sp. 4), *Loxoconcha blanckenhorni* Bassiouni and Luger 1990 (sp. 5), *Paracosta mokattamensis praemokattamensis*

(Bassiouni 1969a) (sp. 6), *Cytherella lagenalis* Marlière 1958 (sp. 7). This biofacies is completely within the Minia Formation. Bassiouni and Luger (1990) considered this ostracod assemblage to represent the inner to outer shelf environment.

Biofacies II is the most noticeable cluster in the dendrogram and consists of the following species: *Bairdoppilata crebra* (Deltel 1964) (sp. 14), *Xestoleberis kenawyi* Khalifa and Cronin 1979 (sp. 19), *Novocypris eocenana* Ducasse 1967 (sp. 20), *Cytherella alii* Elewa 1997 (sp. 23), *Digmocythere ismaili* (Bassiouni 1971) (sp. 24), *Loxoconcha mataiensis* Khalifa and Cronin 1979 (sp. 25), *Clithrocytheridea*? *tarfaensis* Khalifa and Cronin 1979 (sp. 26), *Digmocythere cronini* Elewa 2003 (sp. 27), *Anticythereis seylingi* Cronin and Khalifa 1979 (sp. 28), *A. saitoi* Khalifa and Cronin 1979 (sp. 29), *Leguminocythereis sadeki* MTB Bassiouni 1969d (sp. 30), *Reticulina cameroni* (Khalifa and Cronin 1979) (sp. 31), *Cytheropteron boukharyi* Khalifa and Cronin 1979 (sp. 32), *Reticulina heluanensis* MTA (Bassiouni 1969b) (sp. 33), *Leguminocythereis africana* Bassiouni 1969d (sp. 34), *Paracosta mokattamensis* MTA (Bassiouni 1969b) (sp. 35), *Chrysocythere bahnasaensis* Elewa 1997 (sp. 36), *Buntonia jordanica* Bassiouni 1969c (sp. 37), *Paracosta praetricostata praetricostata* Bassiouni 1969b (sp. 38), *Acanthocythereis* sp. (sp. 39), *Paracypris maghaghaensis* Khalifa and Cronin 1979 (sp. 40). All these species are found in the Maghagha Formation of sections 3, 4. This formation was generally dominated by restricted shallow water, possibly with more or less reduced salinity (dominance of *Loxoconcha*, according to Whatley 1988). However, these restricted conditions were probably accompanied at the base by reduced dissolved oxygen conditions (frequent occurrence of *Cytherella* and *Bairdoppilata*, according to Whatley 1991).

Biofacies III is characterized by *Microcosmia gibberosa* Bassiouni and Luger 1990 (sp. 8), *Trachyleberis nodosus* MTA Bassiouni 1969b (sp. 9), *Trachyleberis nodosus* MTB Bassiouni 1969b (sp. 10), *Martinicythere samalutensis samalutensis* Bassiouni 1969d (sp. 11), *Limburgina moosi* Bassiouni 1969d (sp. 12), *Schizocythere fadlensis* Cronin and Khalifa 1979 (sp. 13), *Asymmetricythere yousefi* Bassiouni 1971 (sp. 15), *Acanthocythereis projecta* Bassiouni 1969b (sp. 16), *Krithe bartonensis* (Jones 1857) (sp. 17), *Reticulina heluanensis* MTB (Bassiouni 1969b) (sp. 18), *Cytherella rayanensis* Elewa 1999 (sp. 21), *Cativella qurnensis* Bassiouni 1969b (sp. 22). This biofacies is mostly distributed in the Qarara Formation of section 5. The base of this formation signifies turbulent water conditions (presence of shallow water species as *Schizocythere fadlensis* Cronin and Khalifa 1979 and *Martinicythere samalutensis samalutensis* Bassiouni 1969d together with deep water ones such as *Krithe bartonensis* (Jones 1857)). This turbulence occurred in an open deeper marine environment (occurrence of *Krithe bartonensis* (Jones 1857), according to Yassini 1979, and *Microcosmia gibberosa* Bassiouni and Luger 1990, according to Bassiouni and Luger 1990). The upper part of the Qarara Formation was dominated by shallower marine conditions (frequent occurrence of shallow marine species like: *Schizocythere fadlensis* Cronin and Khalifa 1979, *Trachyleberis nodosus* MTA Bassiouni 1969b, *Trachyleberis nodosus* MTB Bassiouni 1969b, and *Martinicythere samalutensis samalutensis* Bassiouni 1969d).

3.6 Ostracod population age-structure

Whatley (1983) stated that the differentiation between biocoenosis and thanatocoenosis is a serious problem in the reconstruction of paleoenvironments. Irizuki (1989) concluded that if the ostracods are fossilized in situ after death, their ratios of left to right valves should be 1:1.

Consequently, Elewa (1999) concluded that the biostratigraphic and paleoenvironmental reconstructions based on the Egyptian Paleogene ostracods should be revised according to whether or not these ostracod assemblages include allochthonous components.

In the present study, the ostracods have almost equal ratios of right to left valves.

Moreover, a construction of the age-structure of *Loxoconcha mataiensis* Khalifa and Cronin 1979 (the most predominant species; more than 2000 specimens) for the five samples showing the strongest turbulent conditions in the study area is given in Fig. 5 (after Elewa 1997).

The comparison with type B of Whatley (1983) shows no significant difference. Type B represents the relatively high-energy biocoenoses indicating the existence of a weak current action. This action could remove only the earlier growth stages of the ostracod carapaces.

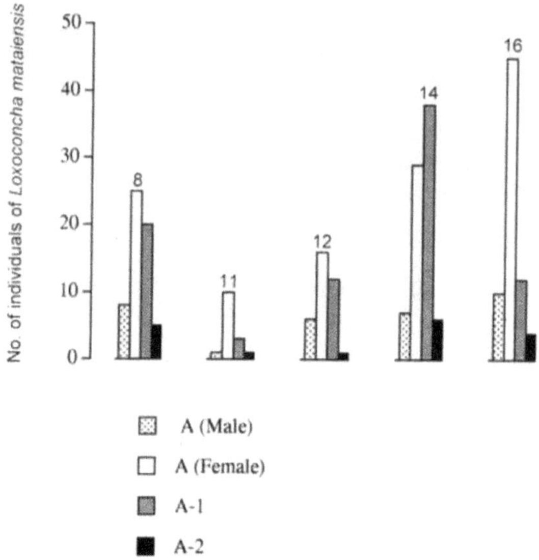

Fig. 5. Histograms showing the *Loxoconcha mataiensis* age-structure for the five samples showing the strongest turbulent conditions in the study area (after Elewa 1997). 8 to 16: sample numbers (all from Maghagha Formation; section 3 of the present study)

3.7 The depositional environment of the study area

Comparison between paleoenvironments prevelant during the middle Eocene in the four studied formations of the Nile Valley led to the recognition of three ostracod biofacies (I to III). The Samalut Formation, which is barren of microfossils, has distinct environmental conditions based on its lithology.

These data allow us to summarize the paleoenvironmental history of the Eocene of the Nile Valley between Minia and Maghagha in the following manner.

In the early Eocene, at the south of the study area, the Minia Formation (biofacies I) is shown to represent an inner to outer shelf environment. Northward, during the middle Eocene time, the Samalut Formation, according to its lithological composition, was deposited under a shallow marine environmental conditions.

To the north, the Maghagha Formation (biofacies II) was generally dominated by restricted shallow water environments, possibly with more or less reduced salinity accompanied by reduced dissolved oxygen conditions at the base, and calm water conditions at the top.

The base of the Qarara Formation (biofacies III) represents turbulent water conditions in an open deeper marine environment. The upper part of this formation was dominated by shallower marine conditions.

Overall, the scarcity or absence of the deep water planktonic foraminiferal species (e. g. *Morozovella lehneri* (Cushman and Jarvis 1929), *M. aragonensis* (Nuttall 1930) and *Orbulinoides beckmanni* (Saito 1962)), if compared with acarinids, in the studied sections supports the conclusion of Said (1990) that there was marine regression northwards, near to the Minia region, during the early and middle Eocene times.

3.8 Paleogeographic distribution of the studied ostracods

The Paleogene ostracod assemblages of the southern Tethys show clear dissimilarity with those of the northern shores. Keen et al. (1994) attributed this dissimilarity to the bottom relief of the Tethys, which was a barrier to north-south migration of ostracods, whereas, the paleogeography of the Tethys margins facilitated their east-west migration. However, Elewa (2002b) stated that the migration of ostracods between north and west Africa occurred in both directions along the southern shores of Tethys as a result of sea level oscillation.

The ostracod species of the study area supports a migration of Paleogene ostracods along the southern Tethys. Nineteen ostracod species from the study area were recorded from adjacent countries of the southern Tethys (Libya, Tunisia, Somalia, Jordan, Israel, Oman and West Africa). Figure 6 summarizes the distribution of the recognized ostracod species of the Nile Valley in the other surrounding countries. The faunal similarity between Egyptian assemblages and those from the countries of North Africa and the Middle East is stronger than with the countries of West Africa. In general, Israel has 7 species in common with Egypt; Libya (6 species); Somalia (5 species); Oman (4 species); Tunisia (3 species); Jordan (3 species). On the other hand, West Africa has only one species in common with Egypt, *Bairdia ilaroensis* Reyment and Reyment 1959. It is worth noting that only three of the 19 species in common with the adjacent countries are early Eocene, while the others are middle Eocene. All the latter species are found both in North Africa and the Middle East, indicating a tendency of the middle Eocene species to exhibit a particular provinciality in these regions simultaneously with another tendency in West Africa. Overall, the present study highlights the importance of ostracodes for paleogeography.

Species	Egypt	Libya	Tunisia	Somalia	Jordan	Israel	Oman	West Africa
Asymmetricythere yousefi	*m*E		*m*E-*l*E	*m*E				
Bairdia ilaroensis	M-*e*E	P-E		*m*E		P-E		M-P
Buntonia jordanica	*m*E				*m*E	*e*E-*m*E		
Cativella qurnensis	*m*E	*l*E						
Chrysocythere bahnasaensis	*m*E	*l*E					*m*E	
Clithrocytheridea? tarfaensis	*m*E					*m*-*l*E		
Cytheropteron boukharyi	*m*E					*m*E		
Dignocythere ismaili	*m*E					*m*E		
Hornibrookella moosae	*e*E-*m*E			*e*E-*m*E	*l*E?	*m*E		
Krithe bartonensis	*m*E						*m*E	
Leguminocythere africana	*m*E	*l*E						
Limburgina moosi	*m*E	O						
Martinicythere samalutensis samalutensis	*m*E			*m*E				
Paracosta mokattamensis MTA	*m*E		*m*E-*l*E					
Paracosta mokattamensis praemokattamensis	*e*E				*e*E			
Reticulina heluanensis MTB	*m*E		*m*E-*l*E					
Schizocythere fadlensis	*m*E						*m*E	
Trachyleberis nodosus MTA	*m*E	*l*E		*m*E			*m*E	
Xestoleberis kenawyi	*m*E					*e*E-*m*E		

Fig. 6. Distribution of common ostracods in Egypt (the study area) with similar faunas elsewhere in the region. M: Maastrichtian; P: Paleocene; E: Eocene; O: Oligocene; *e*: early; *m*: middle; *l*: late

3.9 Acknowledgements

The author would like to express his deep gratitude to Prof. Dr. M. A. Bassiouni, of the Geology Department, Faculty of Science, Ain Shams University, Egypt, for his kind help throughout the course of the present study, valuable comments as well as critical reading of the manuscript. The photographic assistance of Mrs. F. Mostafa of Minia University, Egypt, is deeply acknowledged. A special word of thanks is due to Prof. Dr. T. M. Cronin (US Geological Survey) and Prof. Dr. R. W. Whatley (Abberystwyth, UK). The critical reviewing of the manuscript by Prof. Dr. K. Finger (University of California Museum of Paleontology, USA), Prof. Dr. D. Peterson (Department of Invertebrate Zoology and Geology, California Academy of Sciences, USA) and Prof. Dr. D. Haasl (University of California Museum of Paleontology, USA) is deeply appreciated.

3.10 Taxonomic notes

Taxonomic notes of the important ostracod species are presented herein following mainly the Treatise on Invertebrate Paleontology, Part Q, Arthropoda 3, edited by Moore (1961). The abbreviations L, W, H in this chapter refer to length, width and height of representative specimens. The abbreviations M and F refer to male and female forms, respectively. Serial reference numbers (A10186 to A10220) are given only to the photographed specimens.

Subclass Ostracoda Latreille 1806
Order Platycopida Sars 1866
Suborder Platycopina Sars 1866
Family Cytherellidae Sars 1866
Genus *Cytherella* Jones 1849
Cytherella alii Elewa 1997
(Fig. 7-3)

Cytherella sp.1 Cronin and Khalifa 1979, p. 398, pl. 1, fig. 1; Khalifa and Cronin 1979, p. 173, pl. 1, fig. 1; Elewa 1994, p. 133, pl. 1, figs. 1-2.
Cytherella alii Elewa 1997, p. 369, fig. 20-1, 2.

Dimensions.-	L	W	H
	0.625	0.250	0.375

Remarks.- The present specimens compare well with the description and illustrations of the above mentioned authors. Sexual dimorphism is clear with females wider posteriorly than males.

Occurrence.- The Maghagha Formation; the middle Eocene of sections 3, 4 of the study area.

Stratigraphical range.- middle Eocene (early to late Lutetian; P10-14).

Cytherella lagenalis Marliére 1958

Cytherella lagenalis – Bassiouni and Luger 1990, p. 777, pl. 1, figs. 1-2, 4-5; Elewa 1998b, p. 9, fig. 4: 1.

Dimensions.-	L	W	H

0.750	0.325	0.450

Occurrence.- The Minia Formation; the early Eocene of section 1 of the study area.

Stratigraphical range.- late early Eocene (Ypresian; P9).

Cytherella rayanensis Elewa 1999

Cytherella sp. 2 Cronin and Khalifa 1979, p. 400, pl. 1, figs. 2-3; Morsi 1991, p. 81, pl. 1, figs. 4-5.
Cytherella rayanensis Elewa 1999, p. 49, pl. 1, figs. 1-3.

Dimensions.-	L	W	H	
	0.675	....	0.450	M
	0.800	0.375	0.500	F

Remarks.- The present species differs from *Cytherella alii* Elewa in having: 1) more numerous puncta anteriorly and posteriorly; 2) different shape in dorsal view where the maximum width in dorsal view is at the posterior end, showing a triangular shape, in *C. rayanensis* but in front of the posterior end in *C. alii*; 3) and relatively larger size of carapace. Sexual dimorphism is clear with females wider posteriorly than males.

Occurrence.- The Qarara Formation; the middle Eocene of section 5 of the study area.

Stratigraphical range.- middle Eocene (late Lutetian).

Order Podocopida Muller 1894
Suborder Podocopina Sars 1866
Superfamily Bairdiacea Sars 1866
Family Bairdiidae Sars 1866
Genus *Bairdia* Mc'Coy 1844
Bairdia ilaroensis Reyment and Reyment 1959
(Fig. 7-6)

Bairdia ilaroensis - Bassiouni and Luger 1990, p. 780, pl. 1, fig. 15; Elewa 1994, p. 137, Pl.1, fig. 9; Bassiouni and Luger 1996, p. 8, pl. 1, fig. 10-13; Honigstein and Rosenfeld in Honigstein, Rosenfeld and Benjamini 2002, p. 370, pl. 1, fig. 14.

Dimensions.-	L	W	H
	1.125	0.700	0.750

Occurrence.- The Minia Formation; the early Eocene of section 1 of the study area.

Stratigraphical range.- late early Eocene (Ypresian; P8).

Genus *Bairdoppilata* Coryell, Sample and Jennings 1935
Type species: *Bairdia martyni* Coryell, Sample and Jennings 1935
Bairoppilata crebra (Deltel 1964)
(Fig. 7-5)

Bairdia crebra Deltel 1964, p. 138, pl. 1, figs. 15-17; Blondeau 1971, pl. 1, fig. 9; Helal 1990, p. 184, pl. 8, fig. 18.
Bairdoppilata crebra Elewa 1994, p. 139, pl. 2, figs. 1-4; Elewa 1997, p. 372, fig. 20-11; Elewa 1998b, p. 9, fig. 4:8.
Bairdia crolifai Morsi in Morsi et al., 2003, p. 147, pl. 1, figs. 3-5.

Dimensions.-	L	W	H
	1.875	0.400	0.750

Remarks.- The present form was assigned to *Bairdoppilata* on the basis of the presence of a series of auxilliary teeth and sockets just above mid-height at the end margins. Morsi in Morsi et al. (2003) assigned this form to *Bairdia crolifai* n. sp., where he differentiated between *B. crebra* and his new species by the stronger convex dorsal margin and the denticulate anterior and posterior margins of the former species. Actually, Elewa (1994) and Elewa et al. (1998) studied the present form from the Eastern Desert and Fayoum, respectively, and stated that this form has anterior and posterior denticulation in the internal view (see Elewa 1994, p. 139, pl. 2, figs. 1-4). Also, Elewa (1998a) compared a representative of this form with that of Bassiouni et al. (1984) and stated that they are similar and should be assigned *Bairdoppilata crebra* based on quantitative shape analysis (Fourier analysis). Therefore, the opinion of Morsi in Morsi et al. (2003) is herein rejected and his form is considered in the present study as a junior synonym of *B. crebra.*

Occurrence.- The Maghagha Formation; the middle Eocene of sections 3, 4 of the present study. The Qarara Formation; the middle Eocene of section 5 of the study area.

Stratigraphical range.- middle Eocene (early to late Lutetian; P10-14).

Superfamily Cypridacea Baird 1845
Family Cyprididae Baird 1845
Genus *Novocypris* Ducasse 1967
Novocypris eocenana Ducasse 1967
(Fig. 7-10)

Novocypris eocenanus Ducasse 1967, p. 34, pl. 1, figs. 17-1 .
Novocypris eocenana Ducasse 1969, p. 33, pl. 2, fig. 40; Blondeau 1971, p. 30, pl. 2, figs. 12-14; Shamah 1981, pl. 21, fig. 2; Bassiouni et al. 1984, p. 182, pl. 1, fig. 2; Oertli 1985, p. 310, pl. 89, figs. 14-15; Zalat 1987, pl. 6, fig. 5; Helal 1990, p. 193, pl. 8, fig. 7; Allam et al. 1991, p. 457, pl. 2, fig. 15; Elewa 1994, p. 141, pl. 2, figs. 5-7; Elewa 1997, p. 372, fig. 20-4; Elewa 1998b, p. 9, fig. 4:5.

Dimensions.-	L	H
	0.600	0.300 M
	1.0	0.400 F
	2.0	

Remarks.- Female carapaces are wider and higher than males. The greatest width is postero-median in females in contrast to the median in males. It is worth to mention that Morsi in Morsi et al. (2003) identified a new species, he named *Paracypris naalunensis* for the forms previously named as *Novocypris eocenana* from Fayoum. In 1994, Elewa illustrated the internal view of *N. eocenana* from the Eastern Desert (Elewa 1994, p. 141, pl. 2, fig. 7), and noticed that the muscle scar pattern is typical to the genus *Novocypris* Ducasse (anterior row of four curvilinear scars with two posterior scars in front of the lower two of the anterior scars). On the other hand, Elewa et al. (1998) recognized the same external form of *N. eocenana* from Fayoum, but could not describe the internal features. Morsi assigned this last form to his new species. However, the presence of *N. eocenana* in the Eastern Desert (see Elewa 1994) and the complete resemblance of the external features of this species with that form of Fayoum (see Elewa et al. 1998) make it doubtful to use the new species of Morsi in Morsi et al. (2003) for these forms.

Occurrence.- The Maghagha Formation; the middle Eocene of section 3 of the present study. The Qarara Formation; the middle Eocene of section 5 of the study area.

Stratigraphical range.- middle Eocene (middle to late Lutetian; P11-14).

Family Candonidae Kaufmann 1900
Subfamily Paracypridinae Sars 1923
Genus *Paracypris* Sars 1866
Paracypris maghaghaensis Khalifa and Cronin 1979
(Fig. 8-2)

Paracypris sp. Cronin and Khalifa 1979, p. 400, pl. 1, fig. 23.
Paracypris maghaghaensis Khalifa and Cronin 1979, p. 174, pl. 1, figs. 3-4; Elewa 1994, p. 142, pl. 2, figs. 8-9; Elewa 1997, p. 372, fig. 20-5; Elewa 1999, pl. 1, fig. 12.

Dimensions.-	L	H
	0.525	0.275

Occurrence.- The Maghagha Formation; the middle Eocene of sections 3, 4 of the study area.

Stratigraphical range.- middle Eocene (middle to late Lutetian; P11-14).

Superfamily Cytheracea Baird 1850
Family Cytherideidae Sars 1925
Subfamily Cytherideinae Sars 1925
Genus *Clithrocytheridea* Stephenson 1936
Clithrocytheridea? *tarfaensis* Khalifa and Cronin 1979
(Fig. 9-7)

Clithrocytheridea? *tarfaensis* Khalifa and Cronin 1979, p. 180, pl. 2, figs. 13-16; Elewa 1994, p. 144, pl. 2, figs. 13-14; Elewa 1997, p. 372; Honigstein and Rosenfeld in Honigstein et al. 2002, p. 382, pl. 4, figs. 13-14.

Dimensions.-	L	H
	0.500	0.250

Remarks.- Honigstein and Rosenfeld, in Honigstein et al. (2002), have recognized this species from the middle to late Eocene deposits of the Negev area, southern Israel. Their specimens seem to have a larger carapace size than the holotype of Khalifa and Cronin and the present material.

Occurrence.- The Maghagha Formation; the middle Eocene of sections 3, 4 of the study area.

Stratigraphical range.- middle Eocene (early to late Lutetian; P10-14).

Subfamily Krithinae Mandelstam 1958

Genus *Krithe* Brady, Crosskey and Robertson 1874

Krithe bartonensis (Jones 1857)

(Fig. 7-11)

Cythere (*Cytherideis*) *bartonensis* Jones 1857, p. 50, pl. 5, fig. 23.
Krithe bartonensis - Keij 1957, p. 85, pl. 8, figs. 11-17; Ducasse 1969, p. 55, pl. 3, fig. 70; Wehrli (ed) 1969, p. 21, pl. 5, fig. 12, pl. 15, figs. 4-6; Haskins 1970, p. 13, pl. 1, figs. 5-14; Blondeau 1971, p. 82, pl. 9, fig. 6; Yassini 1979, p. 383, pl. 8, figs. 4-6; Shamah 1981, pl. 21, fig. 1; Bassiouni et al. 1984, p. 186, pl. 1, fig. 1; Morsi 1991, p. 99, pl. 3, figs. 4-5; Elewa 1994, p. 146, pl. 3, figs. 1, 2; Elewa 1998b, p. 9, fig. 4:4; Elewa 1999, pl. 1, fig. 8.
Krithe sp. Cronin and Khalifa 1979, p. 408, pl. 1, figs. 26-27.
Parakrithe? sp. Guernet et al. 1991, pl. 1, fig. 15.

Dimensions.-	L	H
	0.525	0.250

Remarks.- This species, which signifies deeper water marine conditions, is very rare in the study area indicating the dominance of the shallow water in the studied sections, except the base of the Qarara Formation. It seems that *Parakrithe*? sp. of Guernet et al. (1991; pl. 1, fig. 15) represents *Krithe bartonensis* (Jones).

Occurrence.- The Qarara Formation; the middle Eocene of section 5 of the study area.

Stratigraphical range.- middle Eocene (late Lutetian).

Family Cytheruridae Muller 1894

Subfamily Cytheropterinae Hanai 1957
Genus *Cytheropteron* Sars 1866
Cytheropteron boukharyi Khalifa and Cronin 1979
(Fig. 7-1)

Cytheropteron boukharyi Khalifa and Cronin 1979, p. 179, pl. 2, figs. 17-21; Honigstein et al. 1991, p. 102, pl. 2, fig. 3; Elewa 1994, p. 148, pl. 3, figs. 4-6; Elewa 1997, p. 372, fig. 2:7, 8.

Dimensions.-	L	H
	0.525	0.325

Occurrence.- The Maghagha Formation; the middle Eocene of sections 3, 4 of the study area.

Stratigraphical range.- middle Eocene (early to late Lutetian; P10-14).

Genus *Microcosmia* Crane 1965
Microcosmia gibberosa Bassiouni and Luger 1990
(Fig. 8-8)

Microcosmia gibberosa Bassiouni and Luger 1990, p. 801, pl. 8, figs. 1-6; Elewa 1994, p. 150, pl. 3, figs. 11-12; Elewa 1998b, p. 10, fig. 4:14.

Dimensions.-	L	H
	0.350	0.160 M
	0.325	0.150 F

Remarks.- Males are longer than females. The above measured dimensions are for middle Eocene specimens; early Eocene ones are a little smaller.

Occurrence.- The Minia Formation; the early Eocene of section 1 of the study area; The Qarara Formation; the middle Eocene of section 5 of the study area.

Stratigraphical range.- early to middle Eocene (Ypresian to late Lutetian).

Family Hemicytheridae Puri 1953
Genus *Hornibrookella* Moos 1965
Hornibrookella moosae Bassiouni and Luger 1996

Hornibrookella cf. *H.* cf. *macropora* (Bosquet) – Bassiouni 1969d, p. 214, pl. 21, figs. 8-9; Bassiouni and Luger 1990, p. 804, pl. 9, figs. 1-2, 4; Elewa 1998b, p. 12, fig. 4-11.
Hornibrookella moosae Bassiouni and Luger 1996, p. 40, pl. 12, figs. 1-5, 7-8; Honigstein and Rosenfeld in Honigstein et al. 2002, p. 376, pl. 3, fig. 9.

Dimensions.-	L	H
	0.850	0.510 F

Remarks.- Only one female carapace was recognized in the present material, which compare well with the specimens of Bassiouni and Luger (1996) from the middle Eocene of northern Somalia.

Occurrence.- The Minia Formation; the early Eocene of section 1 of the study area.

Stratigraphical range.- early Eocene (Ypresian; P9).

Genus *Limburgina* Deroo 1967
Limburgina moosi Bassiouni 1969d

Hermanites sp. 1 Salahi 1966, pl. 4, figs. 1-3.
Limburgina moosi Bassiouni 1969d, p. 211, pl. 18, figs. 1-3; Shamah 1981, pl. 22, figs. 4a-c; Bassiouni et al. 1984, p. 185, pl. 2, figs. 3a-c; Zalat 1987, pl. 7, figs. 7a-d; Helal 1990, p. 210, pl. 10, figs. 13-15; Allam, Shamah and Zalat 1991, pl. 2, fig. 13; Morsi 1991, p. 96, pl. 3, figs. 1-3; Elewa 1994, p. 153, pl. 3, figs. 13-14; Elewa 1998b, p. 10, fig. 4:6.

Dimensions.-	L	W	H
	0.700	0.275	0.350

Occurrence.- The Qarara Formation; the middle Eocene of section 5 of the study area.

Stratigraphical range.- middle Eocene (late Lutetian).

Genus *Martinicythere* Bassiouni 1969c
Martinicythere samalutensis samalutensis Bassiouni 1969d
(Fig. 9-6)

Martinicythere samalutensis samalutensis Bassiouni 1969d, p. 217, pl. 18, figs. 8-11, pl. 19, figs. 8-10; Haggag 1979, pl. 11, fig. 15; Helal 1990, p. 217, pl. 10, figs. 4-6, pl. 12, fig. 13; Boukhary et al. 1993, p. 206, pl. 4, figs. 5-6; Elewa 1994, p. 154, pl. 4, figs. 1-3.
Martinicythere samalutensis - Cronin and Khalifa 1979, p. 404, pl. 1, figs. 11-1.

Dimensions.-	L	W	H
	0.600	0.300	0.375

Occurrence.- The Qarara Formation; the middle Eocene of section 5 of the study area.

Stratigraphical range.- Middle Eocene (late Lutetian).

Family Loxoconchidae Sars 1925
Genus *Loxoconcha* Sars 1866
Loxoconcha mataiensis Khalifa and Cronin 1979
(Fig. 7-2; Fig. 9-1)

Loxoconcha mataiensis Khalifa and Cronin 1979, p. 179, pl. 1, figs. 21-22; Boukhary et al. 1982, p. 62, pl. 2, figs. 7-8; Aref 1982, p. 179, pl. 22, figs. 7-8; Elewa 1994, p. 156, pl. 4, figs. 4-6; Elewa 1997, p. 373, fig. 20: 9, 10.

Dimensions.-	L	W	H
	0.500	-------	0.250 M
	0.400	0.200	0.275 F

Remarks.- Male carapaces are more elongate and with a more acute postero-dorsal end than female carapaces.

Occurrence.- The Maghagha Formation; the middle Eocene of sections 3, 4 of the study area.

Stratigraphical range.- middle Eocene (early to late Lutetian; P10-14).

Loxoconcha blanckenhorni Bassiouni and Luger 1990
(Fig. 9-3)

Loxoconcha blanckenhorni Bassiouni and Luger 1990, p. 808, pl. 10, figs. 10-11, 13-14.

Dimensions.-	L	H
	0.370	0.19 M
	0.340	0.21 F

Remarks.- Male carapaces are more elongate and thinner with subrectangular shape in lateral view than female carapaces, which have the typical rohmboidal shape in lateral view like the holotype.

Occurrence.- The Minia Formation; the early Eocene of section 1 of the study area.

Stratigraphical range.- early Eocene (Ypresian; P9).

Family Trachyleberididae Sylvester-Bradley 1948
Subfamily Schizocytherinae Mandelstam 1960
Genus *Schizocythere* Triebel 1950
Schizocythere fadlensis Cronin and Khalifa 1979
(Fig. 8-6)

Schizocythere fadlensis Cronin and Khalifa 1979, p. 406-408, pl. 2, figs. 14-15; Boukhary et al. 1993, p. 194, pl. 5, fig. 9; Elewa 1994, p. 160, pl. 5, figs. 1-4; Elewa 1998b, p. 12, fig. 4: 3.

Schizocythere sp. Guernet et al. 1991, pl. 1, figs. 6, 8.

Dimensions.-	L	W	H
	0.425	0.250	0.275

Remarks.- The figured specimens of Guernet et al. (1991, pl. 1, figs. 6, 8) as *Schizocythere* sp. seems to represent *Schizocythere fadlensis* Cronin and Khalifa.

Occurrence.- The Qarara Formation; the middle Eocene of section 5 of the study area.

Stratigraphical range.- middle Eocene (late Lutetian).

Subfamily Trachyleberidinae Sylvester-Bradley 1948

Genus *Trachyleberis* Brady 1898
Trachyleberis nodosus Bassiouni 1969b
(Fig. 7-4, 9; Fig. 8-1)

Trachyleberis nodosus nodosus Bassiouni 1969b, p. 385, pl. 24, figs. 1-3; Zalat 1987, pl. 8, figs. 1-3; Allam et al. 1991, p. 457, pl. 2, fig. 18; Elewa 1994, p. 164, pl. 5, figs. 8-9; Elewa 1998b, p. 12, fig. 4: 16.
Trachyleberis nodosa - Cronin and Khalifa 1979, p. 401, pl. 2, figs. 9-10; Guernet et al. 1991, p. 302, pl. 2, figs. 1, 4; Boukhary et al. 1993, p. 196, pl. 1, figs. 1-4.

Dimensions.-	L	H
	1.125	0.575

Remarks.- Bassiouni (1969a) recognized three subspecies (*T. nodosus nodosus*; *T. nodosus nodosulcatus*; *T. nodosus reticulatus*) from the Eocene of Gebel Mokattam, east of Cairo city. On the other hand, Elewa (2002a) concluded that the change from heavily tubercular pattern in *T. nodosus nodosus* to more reticulate pattern in *T. nodosus reticulatus* through an intermediate form in *T. nodosus nodosulcatus* is due to the availability of the calcium carbonates in the water which changes with fluctuations in water depth and other environmental factors. Therefore, he considered these three subspecies as three morphs of a single species, *Trachyleberis nodosus*. The above dimensions are for *T. nodosus* morphotype A (MTA).

Occurrence.- The Qarara Formation; the middle Eocene of section 5 of the study area.

Stratigraphical range.- middle Eocene (late Lutetian).

Genus *Acanthocythereis* Howe 1963
Acanthocythereis projecta Bassiouni 1969b

Acanthocythereis projecta Bassiouni 1969b, p. 391, pl. 25, figs. 5-6; Boukhary et al. 1993, p. 198, pl. 1, figs. 7-8; Elewa 1998b, p. 13, fig. 4: 18.

Dimensions.-	L	H
	0.750	0.400

Occurrence.- The Qarara Formation; the middle Eocene of sections 5 of the study area.

Stratigraphical range.- middle Eocene (late Lutetian).

Acanthocythereis? sp.

Dimensions.-	L	H
	0.875	0.425

Remarks.- Specimens of this form are questionably assigned to the genus *Acanthocythereis* on the basis of shape and reticulation, however, the internal features are required for a definite identification of this form. This form resembles *Acanthocythereis*? sp. of Cronin and Khalifa (1979) but it differs in having larger size and relatively coarser pattern of reticulation. It also resembles *Acanthocythereis salahii* Bassiouni but it differs in having less spinose surface and coarser pattern of reticulation.

Occurrence.- The Maghagha Formation; the middle Eocene of sections 3, 4 of the study area.

Stratigraphical range.- middle Eocene (early to late Lutetian; P10-14).

Genus *Cativella* Coryell and Fields 1937

Cativella qurnensis Bassiouni 1969b

Cativella qurnensis Bassiouni 1969b, p. 398, pl. 27, figs. 8a-c; El Khoudary and Helmdach 1981, pl. 7, figs. 3-5; Shamah 1981, pl. 21, figs. 11a-b; Bassiouni, Boukhary, Shamah and Blondeau 1984, p. 183, pl. 1, figs. 11a-b; Allam, Shamah and Zalat 1991, p. 457, pl. 2, fig. 5; Elewa 1994, p. 176, pl. 6, figs. 1, 2; Elewa 1998b, p. 13, fig. 4: 10.
Costa qurnensis - Cronin and Khalifa 1979, p. 403, pl. 1, figs. 20-21.
Cativella? *qurnensis* - Boukhary et al. 1993, p. 198, pl. 2, figs. 1-2, 4.

Dimensions.-	L	H
	0.725	0.400

Remarks.- The present form resembles that described by Bassiouni (1969a) from the Eocene deposits of Helwan.

Occurrence.- The Qarara Formation; the middle Eocene of section 5 of the study area.

Stratigraphical range.- middle Eocene (late Lutetian).

Genus *Paracosta* Siddiqui 1971
Paracosta mokattamensis (Bassiouni 1969b)
(Fig. 9-8)

Costa mokattamensis Bassiouni 1969b, p. 399, pl. 27, figs. 3-4; Boukhary et al. 1982, p. 57, pl. 2, figs. 5a-c; Aref 1982, p. 172, pl. 22, figs. 5a-c.
Costa berggreni - Khalifa and Cronin 1979, p. 175, pl. 2, figs. 3-5.

Dimensions.-	L	W	H
	1.000	0.375	0.500 M
	0.875	-------	0.475 F

Remarks.- Elewa (this volume) concluded, based on geometric morphometrics of carapace shape, that *Paracosta mokattamensis* (Bassiouni), *Paracosta humboldti* (Bassiouni) and *Paracosta ducassae* (Bassiouni et al.) should be considered as three morphs of one species (*Paracosta mokattamensis* (Bassiouni)), MTA, MTB and MTC, respectively. The present specimens belong to MTA. On the other hand, Morsi in Morsi et al. (2003) assigned the previously named *P. ducassae* to *Reymenticosta aliena* (Bassiouni). He based his conclusion on the typical resemblance, in his opinion, of the males of *Reymenticosta aliena* with the original illustrations for *P. ducassae* (Bassiouni et al.). However, the smaller size of *P. ducassae* (MTC of Elewa, this volume), the doubt about the type species of the genus *Reymenticosta* (see Keen et al. 1994), the morphometric study of Elewa (this volume) and neglecting the comparison for females in both forms indicate that *P. ducassae* is nothing more than a morph of *P. mokattamensis* as stated by Elewa (this volume).

Occurrence.- The Maghagha Formation; the middle Eocene of sections 3, 4 of the study area.

Stratigraphical range.- middle Eocene (early to late Lutetian; P10-14).

Paracosta mokattamensis praemokattamensis (Bassiouni 1969a)
(Fig. 9-2)

Costa mokattamensis praemokattamensis Bassiouni 1969a, p. 10, pl. 1, figs. 6-7; Bassiouni and Luger 1990, p. 777, pl. 1, figs. 1-2, 4-5; Elewa 1998b, p. 9, fig. 4: 1.

Dimensions.-	L	W	H
	0.450	0.275	0.275

Remarks.- Bassiouni and Elewa (1999) stated that *Paracosta mokattamensis praemokattamensis* (Bassiouni) seems to be a good indicator of the lower/middle Eocene boundary.

Occurrence.- The Minia Formation; the early Eocene of section 1 of the study area.

Stratigraphical range.- late early Eocene (Ypresian; P8-9).

Paracosta praetricostata praetricostata (Bassiouni 1969b)
(Fig. 9-9)

Costa praetricostata praetricostata Bassiouni 1969b, p. 406, pl. 25, figs. 7-9; Zalat 1987, pl. 6, fig. 20, pl. 7, fig. 11; Helal 1990, p. 220, pl. 11, figs. 1-2; Allam et al. 1991, p. 457, pl. 2, fig. 7; Elewa 1994, p. 170, pl. 6, figs. 9-12.
Costa praetricostata – Boukhary et al. 1993, p. 199, pl. 3, figs. 1-3, 5-6.

Dimensions.-	L	W	H
	0.675	0.325	0.400

Occurrence.- The Maghagha Formation; the middle Eocene of section 3 of the present study.

Stratigraphical range.- middle Eocene (middle Lutetian; P11).

Chrysocythere Ruggieri 1962
Chrysocythere bahnasaensis Elewa 1997
(Fig. 9-4)

Costa sp. 1 Khalifa and Cronin 1979, p. 176, pl. 1, figs. 11-12.
Chrysocythere sp. Boukhary et al. 1993, p. 200, pl. 3, fig. 4.
Chrysocythere bahnasaensis Elewa 1997, p. 373, fig. 20: 14, 17, 18.
Paleocosta sp. aff. *mokattamensis* Guernet et al. 1991, p. 308, pl. 2, fig. 5.

Dimensions.-	L	H
	0.750	0.425

Remarks.- Elewa (1997) in his description of the holotype of this species mentioned that the surface is ornamented by reticulation and three longitudinal ribs, however, it shows a fourth longitudinal rib in the lateral view. The figured specimen of Guernet et al. (1991) as *Paleocosta* sp. aff. *mokattamensis* seems to represent *Chrysocythere bahnasaensis* Elewa.

Occurrence.- The Maghagha Formation; the middle Eocene of sections 3, 4 of the study area.

Stratigraphical range.- middle Eocene (early to late Lutetian; P10-14).

Genus *Reticulina* Bassiouni 1969a
Reticulina cameroni (Khalifa and Cronin 1979)
(Fig. 8-12)

Costa? *cameroni* Khalifa and Cronin 1979, p. 175, pl. 1, figs. 10, 13-15.

Dimensions.-	L	W	H
	0.600	0.250	0.300

Remarks.- This species is attributed to the genus *Reticulina* Bassiouni based on the undeveloped longitudinal ridges and the ornamentation that is fundamentally reticulation.

Occurrence.- The Maghagha Formation; the middle Eocene of sections 3, 4 of the study area.

Stratigraphical range.- middle Eocene (early to late Lutetian; P10-14).

Reticulina heluanensis (Bassiouni 1969b)
(Fig. 8-5)

Carinocythereis (*Reticulina*) *heluanensis* Bassiouni 1969b, p. 394-396, pl. 26, figs. 5-10.
Carinocythereis (*Reticulina*) *ismaili* Bassiouni et al. 1984, p. 184, pl. 2, figs. 6-7.

Anticythereis cf. *heluanensis* - Cronin and Khalifa 1979, p. 405, pl. 2, fig. 13; Khalifa and Cronin 1979, p. 176, pl. 2, figs. 7-9.
Reticulina heluanensis - Morsi 1991, p. 110, pl. 4, figs. 9-10.
Reticulina ismaili - Boukhary et al. 1993, p. 198, pl. 1, fig. 11.

Dimensions.-	L	H
	0.875	0.375

Remarks.- Elewa (2002a) assigned the difference in outline between *Reticulina heluanensis* (Bassiouni) and *Reticulina ismaili* (Bassiouni et al.) to shape polymorphism described by Reyment (1988). Therefore, he considered these two species as two morphs of *Reticulina heluanensis* (Bassiouni), in which *Reticulina heluanensis* represents morphotype A (MTA) and *Reticulina ismaili* represents morphotype B (MTB).

Occurrence.- For *Reticulina heluanensis* MTA, the Maghagha Formation; the middle Eocene of section 3 of the study area. For *Reticulina heluanensis* MTB, the Qarara Formation; the middle Eocene of section 5 of the study area.

Stratigraphical range.- middle Eocene (middle Lutetian, for MTA, to late Lutetian, for MTB).

Subfamily Buntoniinae Apostolescu 1961
Genus *Buntonia* Howe 1935
Buntonia jordanica Bassiouni 1969c
(Fig. 9-11)

Buntonia jordanica Bassiouni 1969c, p. 206, pl. 24, figs. 5-6; pl. 25, figs. 1-4; Haggag 1979, pl. 11, fig. 19; Ismail 1992, pl. 2, fig. 3; Elewa 1994, p. 176, pl. 7, figs. 6-8; Elewa 1997, p. 374, fig. 20: 20.
Buntonia cf. *jordanica* - Honigstein et al. 1991, p. 102, pl. 1, fig. 13.

Dimensions.-	L	W	H
	0.500	0.250	0.325

Occurrence.- The Maghagha Formation; the middle Eocene of section 3 of the study area.

Stratigraphical range.- middle Eocene (middle Lutetian; P11).

Genus *Asymmetricythere* Bassiouni 1971
Asymmetricythere yousefi Bassiouni 1971
(Fig. 7-7)

Asymmetricythere yousefi Bassiouni 1971, p. 180, pl. 8, figs. 1-5; Cronin and Khalifa 1979, p. 406, pl. 1, figs. 13-15; Shamah 1981, pl. 21, figs. 9a-c; Bassiouni et al. 1984, p. 185, pl. 1, figs. 13a-c; Zalat 1987, pl. 9, figs. 3a-h; Helal 1990, p. 203, pl. 9, fig. 4; Allam et al. 1991, p. 457, pl. 2, fig. 1; Morsi 1991, p. 118, pl. 5, figs. 7-9; Boukhary et al. 1993, p. 204, pl. 3, figs. 13; Elewa 1994, p. 179, pl. 7, fig. 12; Elewa 1998b, p. 13, fig. 4: 12.

Dimensions.-	L	H
	0.900	0.500

Occurrence.- The Qarara Formation; the middle Eocene of section 5 of the study area.

Stratigraphical range.- middle Eocene (late Lutetian).

Genus *Digmocythere* Mandelstam 1958
Digmocythere cronini Elewa 2003
(Fig. 9-5)

Digmocythere cronini Elewa 2003, p. 8, fig. 16: 1-5.

Dimensions.-	L	W	H
	0.700	0.375	0.260

Remarks.- It is worth noting that the validity of the genus *Digmocythere* Mandelstam is questionable because of different factors in which the most important one is that the American species and the northern African species, all were apparently shallow water taxa, were separated by the Atlantic Ocean during the Eocene, therefore it is not clear how they are genetically related. Further studies for the internal features of this genus are needed to solve this serious problem.

Occurrence.- The Maghagha Formation; the middle Eocene of sections 3, 4 of the study area.

Stratigraphical range.- Middle Eocene (early to late Lutetian; P10-14).

Digmocythere ismaili (Bassiouni 1971)
(Fig. 7-8)

Brachycythere (*Digmocythere*) *ismaili* Bassiouni 1971, p. 170, pl. 7, figs. 5-6;Haggag 1979, pl. 11, fig. 11; Shamah 1981, pl. 23, figs. la-c; Bassiouni et al. 1984, p. 184, pl. 2, figs. 1a-c; Zalat 1987, pl. 10, figs. 5a-b; Helal 1990, p. 200, pl. 9, fig. 9; Allam et al. 1991, pl. 2, fig. 4; Morsi 1991, p. 87, pl. 2, figs. 1-3; Honigstein et al. 1991, p.104, pl. 2, fig. 6.
Brachycythere? *ismaili* – Boukhary et al. 1993, p. 202, pl. 4, figs. 1.

Dimensions.-	L	H
	0.800	0.450 M
	0.725	0.450 F

Remarks.- The greatest width is posterior in female carapaces while it is behind the middle in male carapaces. Moreover, the ventro-lateral part is more swollen in female carapaces than male ones. On the other hand, Elewa (2003) used the morphometric analyses to support the suggestion of Boukhary et al. (1982) in placing *Digmocythere omarai* Cronin and Khalifa as the ancestor of *D. ismaili* (Bassiouni).

Occurrence.- The Maghagha Formation; the middle Eocene of sections 3, 4 of the study area.

Stratigraphical range.- middle Eocene (early to late Lutetian; P10-14).

Subfamily Campylocytherinae Puri 1960
Genus *Anticythereis* Bold 1946
Anticythereis saitoi Khalifa and Cronin 1979
(Fig. 8-3)

Anticythereis saitoi Khalifa and Cronin 1979, p. 177, pl. 2, figs. 10-12; Elewa 1994, p. 186, pl. 8, figs. 13-15; Elewa 1997, p. 375, fig. 20: 25.
Reticulina cf. *saitoi* - Boukhary et al. 1993, p. 198, pl. 1, figs. 10.

Dimensions.-	L	W	H
	0.775	0.325	0.375

Occurrence.- The Maghagha Formation; the middle Eocene of sections 3, 4 of the study area.

Stratigraphical range.- middle Eocene (early to late Lutetian; P10-14).

Anticythereis seylingi Cronin and Khalifa 1979
(Fig. 9-10)

Anticythereis seylingi Cronin and Khalifa 1979, p. 405, pl. 2, figs. 19-21; Khalifa and Cronin 1979, p. 177, pl. 2, fig. 6; Elewa 1994, p. 187, pl. 9, figs. 1-3; Elewa 1997, p. 375, fig. 20: 24.

Dimensions.-	L	W	H
	0.700	0.300	0.400

Occurrence.- The Maghagha Formation; the middle Eocene of sections 3, 4 of the study area.

Stratigraphical range.- middle Eocene (early to late Lutetian; P10-14).

Genus *Leguminocythereis* Howe and Law 1936
Leguminocythereis africana Bassiouni 1969d
(Fig. 8-9)

Leguminocythereis africana Bassiouni 1969d, p. 223, pl. 21, figs. 4-6; Cronin and Khalifa 1979, p. 404, pl. 1, fig. 25; El Khoudary and Helmdach 1981, pl. 7, fig. 6; Shamah 1981, pl. 21, fig. 6; Bassiouni et al. 1984, p. 186, pl. 1, fig. 6; Zalat 1987, pl. 7, figs. 1-2; Allam et al. 1991, pl. 2, fig. 12; Morsi 1991, p. 93, pl. 2, figs. 9-10; Bassiouni et al. 1994, p. 262, pl. 2, figs. 1-2, Elewa 1994, p. 188, pl. 9, figs. 4-6; Elewa 1997, p. 375, fig. 20: 23.

Dimensions.-	L	W	H
	0.800	0.275	0.400 M
	0.775	0.300	0.450 F

Remarks.- Female carapaces are shorter, wider and higher than males.

Occurrence.- The Maghagha Formation; the middle Eocene of sections 3, 4 of the study area.

Stratigraphical range.- middle Eocene (early to late Lutetian; P10-14).

Leguminocythereis sadeki Bassiouni 1969d
(Fig. 8-11, 13)

Leguminocythereis praesadeki Boukhary et al. 1982, p. 60, pl. 2, figs. 3a-b; Aref 1982, p. 176, pl. 22, fig. 3; Elewa 1994, p. 190, pl. 9, figs. 8-10; Elewa 1997, p. 376, fig. 20: 28.

Dimensions.-	L	W	H
	0.800	0.450	0.475

Remarks.- Bassiouni and Elewa (1999) clued that *Leguminocythereis sadeki* Bassiouni, *Leguminocythereis praesadeki* Boukhary et al. and *Leguminocythereis hassani* Khalifa and Cronin are representatives of environmentally cued polymorphism and they are nothing more than three ecomorphs of *Leguminocythereis sadeki* Bassiouni, MTA, MTB and MTC, respectively. The present specimens belong to MTB.

Occurrence.- The Maghagha Formation; the middle Eocene of section 3, 4 of the study area.

Stratigraphical range.- middle Eocene (early to late Lutetian; P10-14).

Family Xestoleberididae Sars 1928

Genus *Xestoleberis* Sars 1866

Xestoleberis kenawyi Khalifa and Cronin 1979

(Fig. 8-7, 10)

Xestoleberis? *kenawyi* Khalifa and Cronin 1979, p. 181, pl. 1, figs. 7-8; Honigstein et al. 1991, p. 100, pl. 1, figs. 7-8; Elewa 1994, p. 192, pl. 10, figs. 1-3; Elewa 1997, p. 376, fig. 20: 27; Elewa 1998b, p. 14, fig. 4: 19. *Xestoleberis kenawyi* Honigstein and Rosenfeld in Honigstein et al. 2002, p. 380, pl. 4, fig. 12.

Dimensions.-	L	W	H
	0.500	0.325	0.350

Occurrence.- The Maghagha Formation; the middle Eocene of section3 , 4 of the study area.

Stratigraphical range.- middle Eocene (early to late Lutetian; P10-14).

Xestoleberis kiseibaensis Bassiouni and Luger 1990

(Fig. 8-4)

Xestoleberis kiseibaensis Bassiouni and Luger 1990, p. 848, pl. 25, figs. 8-12 Elewa 1994, p. 193, pl. 10, figs. 4-6; Elewa 1998b, p. 14, fig. 4: 20.

Dimensions.-	L	W	H
	0.450	0.275	0.275

Occurrence.- The Minia Formation; the early Eocene of section 1 of the study area.

Stratigraphical range.- late early Eocene (Ypresian; P8-9).

Genus *Uroleberis* Triebel 1958
Uroleberis triebeli Bassiouni and Luger 1990

Uroleberis triebeli Bassiouni and Luger 1990, p. 848, pl. 25, figs. 13-17.

Dimensions.-	L	H
	0.470	0.320

Remarks.- The present form compares well with the specimens of Bassiouni and Luger (1990), however, their specimens have larger carapaces.

Occurrence.- The Minia Formation; the early Eocene of section 1 of the study area.

Stratigraphical range.- late early Eocene (Ypresian; P8).

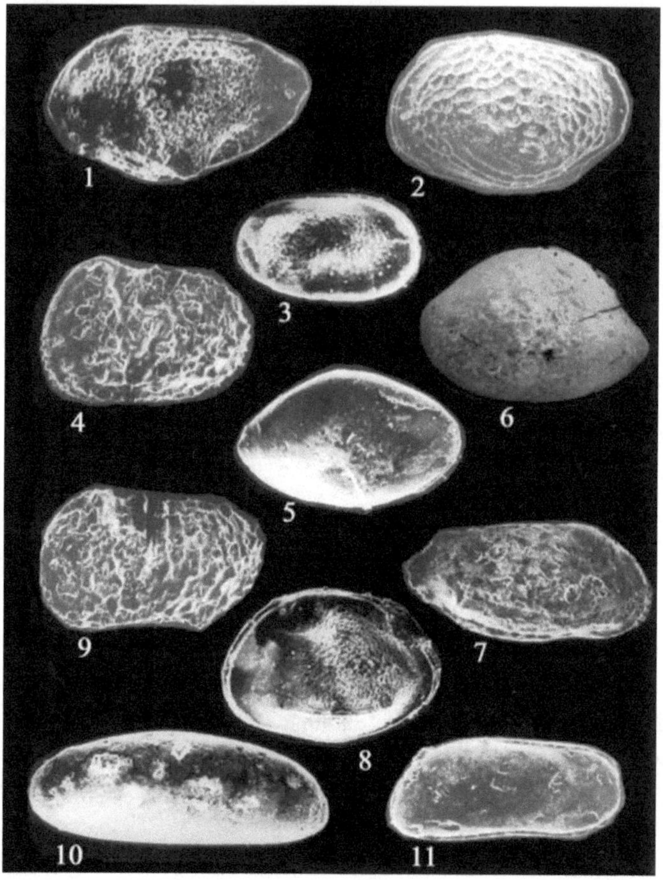

Fig. 7. 1. *Cytheropteron boukharyi* Khalifa and Cronin; left side view of carapace, x 200; 2. *Loxoconcha mataiensis* Khalifa and Cronin; left side view of carapace, x 200. 3. *Cytherella alii* Elewa; left side view of carapace, x 100. 4, 9. *Trachyleberis nodosus* MTA Bassiouni; 4. left side view of female carapace, x 100; 9. left side view of female carapace, x 100. 5. *Bairdoppilata crebra* (Deltel); right side view of carapace, x 100. 6. *Bairdia ilaroensis* Reyment and Reyment; left side view of carapace, x 100. 7. *Asymmetricythere yousefi* Bassiouni; right side view of carapace, x 150. 8. *Digmocythere ismaili* (Bassiouni); left side view of carapace, x 100. 10. *Novocypris eocenana* Ducasse; left side view of carapace, x 100. 11. *Krithe bartonensis* (Jones); right side view of carapace, x 200

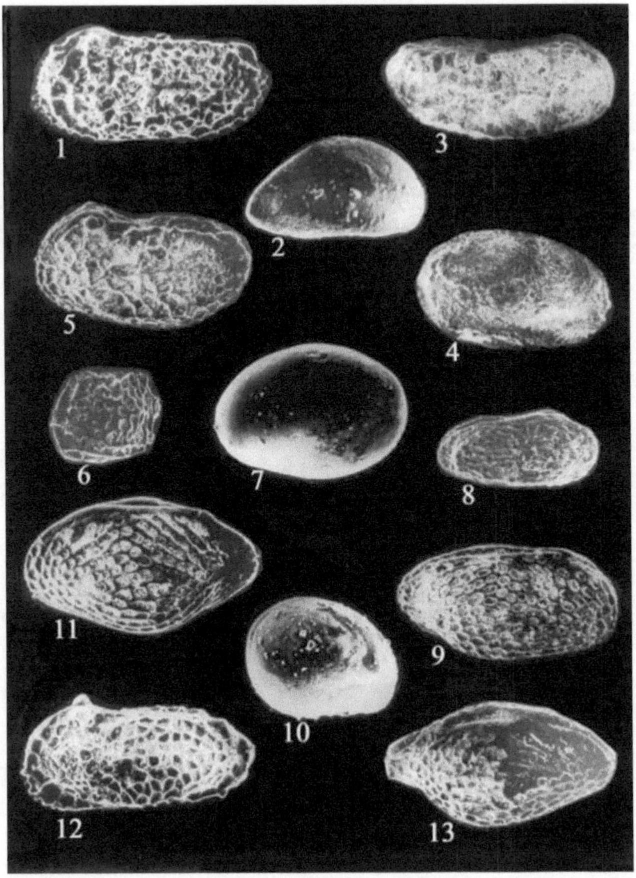

Fig. 8. 1. *Trachyleberis nodosus* MTA Bassiouni; left side view of male carapace, x 100. 2. *Paracypris maghaghaensis* Khalifa and Cronin; right side view of carapace, x 100. 3. *Anticythereis saitoi* Khalifa and Cronin; right side view of carapace, x 100. 4. *Xestoleberis kiseibaensis* Bassiouni and Luger; right side view of carapace, x 150. 5. *Reticulina heluanensis* MTA (Bassiouni); left side view of female carapace, x 100. 6. *Schizocythere fadlensis* Cronin and Khalifa; left side view of carapace, x 150. 7, 10. *Xestoleberis kenawyi* Khalifa and Cronin; 7. left side view of carapace, x 150; 10. right side view of carapace, x 100. 8. *Microcosmia gibberosa* Bassiouni and Luger; right side view of female carapace, x 200. 9. *Leguminocythereis africana* Bassiouni; right side view of female carapace, x 40. 11, 13. *Leguminocythereis sadeki* MTB Bassiouni; 11. left side view of carapace, x 100; 13. right side view of carapace, x 100. 12. *Reticulina cameroni* (Khalifa and Cronin); left side view of carapace, x 100

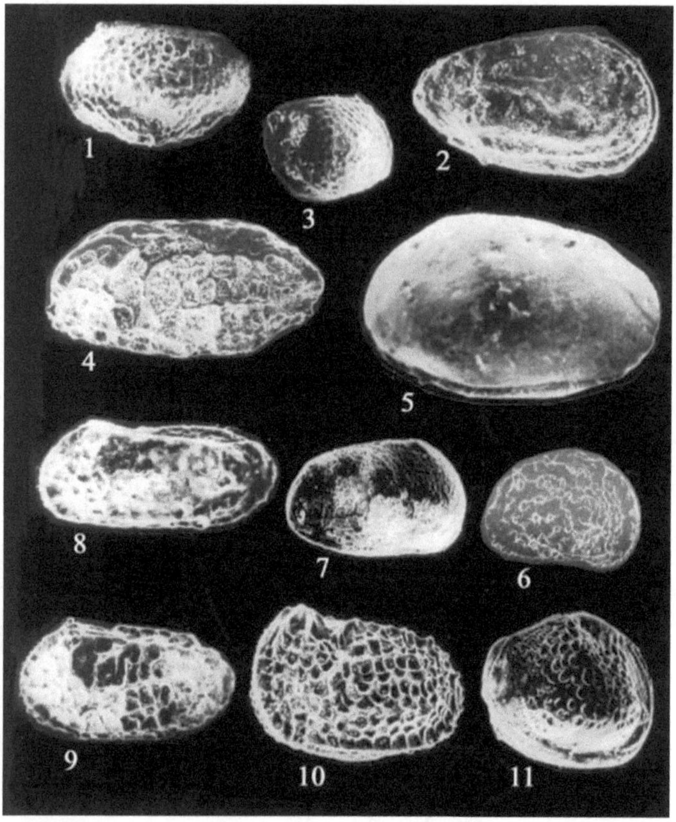

Fig. 9. 1. *Loxoconcha mataiensis* Khalifa and Cronin; right side view of female carapace, x 200. 2. *Paracosta mokattamensis praemokattamensis* (Bassiouni); right side view of female carapace, x 200. 3. *Loxoconcha blanckenhorni* Bassiouni and Luger; right side view of female carapace, x 100. 4. *Chrysocythere bahnasaensis*; left side view of female carapace, x 150. 5. *Digmocythere cronini* Elewa; left side view of female carapace, x 350. 6. *Martinicythere samalutensis samalutensis* Bassiouni; right side view of carapace, x 150. 7. *Clithrocytheridea*? *tarfaensis* Khalifa and Cronin; right side view of carapace, x 150. 8. *Paracosta mokattamensis* MTA (Bassiouni); left side view of male carapace, x 100. 9. *Paracosta praetricostata praetricostata* (Bassiouni); left side view of carapace, x 100. 10. *Anticythereis seylingi* Cronin and Khalifa; left side view of female carapace, x 100. 11. *Buntonia jordanica* Bassiouni; right side view of carapace, x 150

References

Allam A, Shamah K, Zalat A (1991) Biostratigraphy of the middle Eocene succession at Gebel Mishgigah, Wadi Rayan, Libyan Desert, Egypt. Journal of African Earth Sciences 12 (3): 449-459

Apostolescu V (1961) Contribution à l'étude paléontologique (Ostracodes) et stratigraphique des bassins Crétacées et Tertiaires de l'Afrique Occidentale. Rév Inst Franç Pétrole, Pau 16: 779-867

Aref M (1982) Micropaleontology and biostratigraphy of Eocene rocks in the area between Assiut and Beni Suef of the Nile Valley, Egypt. Ph D Thesis, Assiut University, Egypt

Azab MM (1984) Biostratigraphy of the middle Eocene in Upper Egypt. Ph. D. Thesis, Ain Shams University, Egypt

Baird W (1845) Arrangement of the British Entomostraca, with a list of species, particularly noticing those which have as yet been discovered within the bounds of the Club. Hist Berwicksh Nat Club 2: 145-148

Baird W (1850) The natural history of the British Entomostraca. Roy Soc London 8: 364 pp

Barker MHS (1945) The stratigraphy and structure of the Eastern Desert between the latitude of Wasta and Assiut with data incorporated on additional sections. Unpublished AEO report

Bassiouni MA (1969a) Einige *Costa*- und *Carinocythereis* (*Reticulina*)- Arten aus dem Paläozän und Eozän von Jordanien (Ostracoda). Neues Jahrbuch für Geologie und Paläontologie Abhandlungen 134 (1): 1-16

Bassiouni MA (1969b) Ostracoden aus dem Eozän von Ägypten (1. Trachyleberidinae). Geologische Jahrbuch, Hannover 87 383-426

Bassiouni MA (1969c) Einige *Buntonia*- und *Soudanella*- Arten (Ostracoda, Crustacea) aus dem Eozän von Jordanien. Paleont Z 43 (3/4): 205-214

Bassiouni MA (1969d) Ostracoden aus dem Eozän von Ägypten (2. Die Unterfamilian Hemicytherinae, Thaerocytherinae und Campylocytherinae). Geologische Jahrbuch, Hannover 88: 203-234

Bassiouni MA (1971) Ostracoden aus dem Eozän von Agypten (3. Die Unterfamilian Brachycytherinae und Buntoniinae). Geologische Jahrbuch, Hannover 89: 169-192

Bassiouni MA, Elewa AMT (1999) Studies on some important ostracod groups from the Paleogene of Egypt. Geosound, Turkey 35: 13-27

Bassiouni MA, Luger P (1990) Maastrichtian to early Eocene Ostracoda from Southern Egypt (paleontology, paleoecology, paleobiogeography and biostratigraphy). Berliner Geowiss Abh (A) 120 (2): 755-928

Bassiouni MA, Luger P (1996) Middle Eocene Ostracoda from northern Somalia. Cour Forsch-Inst Senckenberg 192: 1-139

Bassiouni MA, Boukhary MA, Anan HS (1977) Ostracodes from Gebel Gurnah, Nile Valley. Egypt. Proc Egypt Acad Sci 30: 1-9

Bassiouni MA, Boukhary MA, Shamah, Blondeau A (1984) Middle Eocene ostracodes from Fayoum , Egypt . Geol Mediterraneenne 11 (2): 181-152

Bassiouni MA, Hamza FH, Morsi AM (1994) Ostracod and microfacies study of the Eocene sequence in the northwestern part of the Eastern Desert, Egypt. Revue de Paléobiologie, Genève 13 (1): 259-280

Berggren WA, Kent DV, Swishar CC III, Aubry M-P (1995) A revised Cenozoic geochronology and chronostratigraphy. In Berggren WA, Kent DV, Swishar CC III, Aubry M-P, Hardenbol J (eds) Geochronology, time scales and global stratigraphic correlation. Society for Sedimentary Geology, Special Publication pp 129-212

Bermúdez PJ (1961) Contribución al estudio de las Globigerinidea de la región Caribe-Antillana (Paleoceno-Reciente). Bol. Geología (Venezolano), Spec Publ 3 (Cong. Geol. Venezolano, 3rd, Caracas 1959, Mem 3) pp 1119-1393

Bishay Y (1961) Biostratigraphic study of the Eocene in the Eastern Desert between Samalut and Assiut by the larger foraminifera. 3rd Arab Petrol Congr, Alexandria 7 pp

Bishay Y (1966) Studies on the larger foraminifera of the Eocene of the Nile Valley between Assiut, Cairo and S.W. Sinai. Ph D Thesis, Alexandria University, Egypt

Blondeau, M. 1971. Contribution a l'etude des ostracodes Eocenes des Bassins de Campbon et de Saffre (Loire-Atlantique). Ph.D. diss. Nante University, France.

Bold WH van den (1946) Contribution to the study of Ostracoda with special reference to the Tertiary and Cretaceous microfauna of the Caribbean region. Reprint by Antiquariaat Junk: 167 pp, Amsterdam 1970

Bolli HM (1957) Planktonic foraminifera from the Eocene Navet and San Fernando formations of Trinidad. BWI Bull US Nat Mus 215: 155-172

Boukhary MA, Abdel Malik W (1983) Revision of the stratigraphy of the Eocene deposits in Egypt. Neues Jahrbuch für Geologie und Paläontologie Monatshefte 6: 321-337

Boukhary MA, Blondeau A, Ambroise D (1982) Etude sur les nummulites de la Région de Minia-Samalut, Vallée du Nil, Égypte. 1: Biometrie et biostratigraphie. 8 eme Colloque Africain de Micropaléontologie 3: 65-78

Boukhary MA, Toumarkine M, Khalifa H, Aref M (1982) Etude biostratigraphique a l'aide des foraminiferes planctoniques et des ostracodes de l'Éocène de Beni Mazar, Vallee du Nil, Egypte. 8 eme Colloque Africain de Micropaleontologie 3: 53-64

Boukhary MA, Guernet C, Strougo A, Bassiouni MA, Bignot G, Abdel Ghany O (1993) Eocene ostracodes of Mingar El Rayan (Fayoum district, Egypt); stratigraphic and paleogeographic implication. Revue de Micropaléontologie 36 (3): 191-211

Brady GS (1898) On new or imperfectly known species of Ostracoda, chiefly from New Zealand. Trans Zool Soc London 14 (8): 429-452

Brady GS, Crosskey HW, Robertson D (1874) A monograph of the post-Tertiary Entomostracea of Scotland including species from England and Ireland. Paleontograph Soc, London 1874: 274 pp

Brönnimann P, Bermúdez PJ (1953) *Truncorotaloides*, a new foraminiferal genus from the Eocene of Trinidad. BWI J Paleont 27: 817-820

Colom G (1954) Estudio de las biozonasa con foraminiferos del Terciario de Alicante. Bol Españ Inst Geol Y Minero 66: 1-279

Coryell HN, Fields S (1937) A Gatun ostracode fauna from Cativa, Panama. Am Museum Novitates 956: 1-18

Coryell HN, Sample CH, Jennings PH (1935) *Bairdoppilata*, a new genus of Ostracoda, with two new species. Am Museum Novitates 777: 1-5

Crane MJ (1965) Upper Cretaceous Ostracodes of the Gulf Coast area. Micropaleontology 11 (2): 191-254

Cronin TM, Khalifa H (1979) Middle and late Eocene Ostracoda from Gebel El Mereir, Nile Valley, Egypt. Micropaleontology 25 (4): 397-411

Cushman JA (1925) New foraminifera from the Upper Eocene of Mexico. Contrib Cushman Lab Foramin Res 1: 4-9

Cushman JA, Bermúdez PJ (1949) Some Cuban species of *Globorotalia*. Contrib Cushman Lab Foramin Res 25: 26-44

Cushman JA, Jarvis PW (1929) New foraminifera from Trinidad. Contrib Cushman Lab Foramin Res 5: 6-17

Deltel B (1964) Nouveaux ostracodes de l'Éocène et de l'Oligocène de l'Aquitaine Meridionale. Act Soc Linn Bordeaux 100: 127-221

Deroo G (1966) Cytheracea (Ostracodes) du Maastrichtien de Maastricht (Pays-Bas) et de réions voisines: résultats stratigraphiques et paléontologiques de leur étude. MGS C-V-2-2: 197 pp

Ducasse O (1967) Nouveaux ostracodes de l'Eocene Nord-Aquitain. Proc Verb Soc Sci Phys Nature, Bordeaux pp 23-89

Ducasse O (1969) Nouveaux ostracodes de l'Éocène Nord-Aquitanien. Proc Verb Soc Sci Phys Nature, Bordeaux 98 pp

El Dawoody AS (1971) Middle Eocene foraminifera from Samalut and Areg Oasis, Egypt Zb Geol Vied Rad Zk 14: 91-162

El Dawy MH (1997) Middle Eocene benthic foraminiferal biostratigraphy and paleoecology of east Beni Mazar area, Nile Valley, Egypt. Egyptian Journal of Geology 41 (1): 413-464

Elewa AMT (1994) Biostratigraphical studies of Eocene rocks of the El Sheikh Fadl-Ras Gharib stretch, Eastern Desert, Egypt. Ph D Thesis, Minia University, Egypt

Elewa AMT (1997) Ostracod assemblages from the Middle Eocene of the western bank of the Nile Valley between Samalut and Beni Mazar, Upper Egypt. Neues Jahrbuch für Geologie und Paläontologie Abhandlungen, Stuttgart 204 (3): 353-378

Elewa AMT (1998a) Fourier Biometrics: A case study on two species of the ostracode genus *Bairdoppilata* from the middle Eocene of Egypt. Neues Jahrbuch für Geologie und Paläontologie Mh, Stuttgart 1998 (4): 203-211

Elewa AMT (1998b) Ostracod assemblages at the Lower/Middle Eocene boundary in the Nile Valley, Egypt. Neues Jahrbuch für Geologie und Paläontologie Abhandlungen, Stuttgart 210 (1): 1-17

Elewa AMT (1999) The use of allochthonus microfossils in determining paleoenvironments: A case study using middle Eocene ostracods from Wadi

El Rayan, Fayoum, Egypt. Bulletin of Faculty of Science of Assiut University 28 (2): 33-52

Elewa AMT (2002a) Microevolution applied to ostracod biostratigraphy: the middle Cretaceous to middle Eocene ostracods of Egypt (with special reference to the genus *Paracosta* Siddiqui). Neues Jahrbuch für Geologie und Paläontologie Abhandlungen 226 (3): 289-318

Elewa AMT (2002b) Paleobiogeography of Maastrichtian to early Eocene Ostracoda of North and West Africa and the Middle East. Micropaleontology 48 (4): 391-398

Elewa AMT (2003) Morphometric studies on three ostracod species of the genus *Digmocythere* Mandelstam from the middle Eocene of Egypt. Palaeontologia Electronica 6 (2): 11 pp

Elewa AMT (2004) Quantitative analysis and paleoecology of Eocene Ostracoda and benthonic foraminifera from Gebel Mokattam, Cairo, Egypt. Palaeoclimatology Palaeogeography Palaeoecology, Elsevier 211 (3-4): 309-323

El Khoudary RH, Helmdach FF (1981) Biostratigraphic studies on the upper Eocene Apollonia Formation of NW Gebel AL Akhdar, NE Libya. Revista Española de Micropaleontologia 13 (1): 5-23

Guernet C, Bourdillon De Grissac C, Roger J (1991) Ostracodes Paleogenes du Dhofar (Oman). Interet stratigraphique et paleogeographique. Revue de Micropaléontologie 34 (4): 297-311

Haggag, MAY (1979) Biostratigraphy of the west-central part of the Western Desert, Egypt. Ph D Thesis, Ain Shams University, Egypt

Haggag MAY (1989) Planktonic foraminiferal zonation and evolutionary groups of the middle Eocene in the Nile Valley, Egypt. Neues Jahrbuch für Geologie und Paläontologie Abhandlungen 178: 109-132

Haggag MAY, Anan HS (1987) Middle Eocene planktonic foraminifera from the Maghagha area, Egypt. Ain Shams Science Bulletin (B) 25: 149-172

Hanai T (1957) Studies on the Ostracoda from Japan, II. Subfamily Pectocytherinae n subfam. Journal of Faculty of Science of University of Tokyo 10: 469-482

Haskins CW (1970) Tertiary Ostracoda from the Isle of Wight and Barton, Hampshire, England, Pt. 5. Revue de Micropaléontologie 13 (1): 13-29

Helal SA (1990) Stratigraphic and paleontologic studies on the Eocene sediments at Gebel Shabrawet area, Eastern Desert, Egypt. M.Sc Thesis, Cairo University, Egypt

Honigstein A, Rosenfeld A, Benjamini C (1991) Ostracodes and foraminifera from the early-middle Eocene of Qeren Sartaba, Jordan Valley. Journal of Micropalaeontology 10 (1): 95-107

Honigstein A, Rosenfeld A, Benjamini C (2002) Eocene ostracode faunas from the Negev, southern Israel: Taxonomy, stratigraphy and paleobiogeography. Micropaleontology 48 (4) 365-389

Howe H (1935) Ostracoda of the Arca Zone of the Choctawhatchee Miocene of Florida. Geol Bull Fla13: 1-37

Howe H, Law J (1936) Louisiana Vicksburg Oligocene Ostracoda. Geol Bull, State of Louisiana Dept Conserv, Louisiana Geol Surv 7: 1-96

Howe R (1963) Type saline Bayou Ostracoda of Louisiana. Geol Bull, State of Louisiana Dept Conserv, Louisiana Geol Surv 40: 1-62

Hussein AIM (1992) *Bulimina jacksonensis* Cushman 1925 from the Mokattamian Stage (middle and upper Eocene) of Egypt and its biostratigraphic significance Middle East Research Center, Ain Shams University, Earth Science 107-113

Hussein AIM (1994) Remarks on some uvigerines from the middle Eocene rocks in the Nile Valley. Egyptian Journal of Geology, Egypt 38 (1): 21-34

Irizuki T (1989) Fossil ostracod assemblages from the Pliocene Sasaoka Formation, Akita city, Japan- with reference to sedimentological aspects. Trans Proc Paleont Soc Japan, NS 296: 318

Ismail AA (1992) Late Campanian to early Eocene Ostracoda from Esh El Mallaha area, Eastern Desert, Egypt. Revue de Micropaléontologie 35 (1): 39-52

Jones TR (1849) A monograph of the Entomostraca of the Cretaceous Formation of England. Palaeontograph Soc, London 1849: 40 pp

Jones TR (1857) Monograph of the Tertiary Entomostraca of England. Paleontogr Soc, London 1-68

Kaufmann A (1900) Zur systematik der Cypriden. Naturf Gesell Bern pp 103-109

Keen MC, Al Sheikhly SSJ, Elsogher A, Gammudi AM (1994) Tertiary ostracods of North Africa and the Middle East. In Simmons MD (ed) Micropaleontology and hydrocarbon exploration in the Middle East, Chapman and Hall Inc, London pp 371-400

Keij AJ (1957) Eocene and Oligocene Ostracoda of Belgium. Inst Roy Sc Nat Belgique 139: 1-210

Khalifa H, Cronin TM (1979) Ostracodes de l'Eocene moyen de El Sheikh Fadl, est de Beni Mazar, Haute Egypte. Revue de Micropaléontologie 22 (3): 172-185

Khalilov DM (1956) Pelagic foraminifers of the Paleogene deposits of Azerbaydjhan, Akad Nauk Azerbaydjhan, SSR Geol Inst Baku 17: 234-261 (in Russian)

Latreille PA (1806) Genera crustaceorum et insectorum. Paris 1: 1-303

Mandelstam MI (1958) Trudy Vsesoyuznogo Neftyanogo Nauchno-Issledovatelskogo Geologo. Razvedochnogo Instituta (VNIGRI), Leningrad 115 (9): 232-299

Mandelstam MI (1960) Trilobitomorpha and Crustaceomorpha. In Orlov JA (ed) Fundamentals in Paleontology (the Arthropoda), Moscow pp 264-421 (in Russian)

Marliére R (1958) Ostracodes du Montien des Mons et résultats de leur étude. Mém Soc Belge Géol, Paléont, hydrol 5: 1-53

McCoy F (1844) A synopsis of the characters of the Carboniferous fossils of Ireland. Dublin Univ Press 207 pp

Moore RC (ed) (1961) Treatise on Invertebrate Paleontology, Part Q, Arthropoda 3. Geol Soc Amer and Kansas Univ Press 442 pp

Moos B (1965) Die Ostrakoden-Fauna des Unteroligozäns von Bünde (Bl. Herford-West, 3817) und einiger verwandte jüngere Arten (Ostr, Crust). I. *Quadracythere* (*Hornibrookella*) n subg, *Pokornyella*, *Hemicythere*, *Hermanites*. Geologische Jahrbuch, Hannover 82: 593-630

Morsi A (1991) Stratigraphic and ecologic evaluation of the Eocene rocks in the area southwest of the Cairo-Suez road. M Sc Thesis, Ain Shams University, Egypt

Morsi A, Boukhary M, Strougo A (2003) Middle-upper Eocene ostracods and nummulites from Gebel Na'alun, southeastern Fayoum, Egypt. Revue de Micropaléontologie 46: 143-160

Müller GW (1894) Die Ostrakoden des Golfes von Neapel und der angrenzenden Meeres-Abschnite. Fauna und Flora des Golfes von Neapel und der angrenzenden Meeres-Abschnite, herausgegeben von der zoologischen Station zu Neapel, Berlin 21(I-VIII): 1-404

Nuttall WLF (1930) Eocene foraminifera from Mexico. J Paleont 4: 271-293

Oertli HJ (ed) (1985) Atlas des ostracodes des France. Bull Centres Rech Explor-Prod Elf-Aquitaine 9: 396 pp

Omara S, Mansour HH, Youssef M, Khalifa H (1977) Stratigraphy, paleoenvironment and structural features of the area east of Beni Mazar, Upper Egypt. Bulletin of Faculty of Science, Assiut University 6 (3): 171-197

Puri H (1953) The ostracod genus *Hemicythere* and its allies. J Washington Acad Sci 43: 169-179

Puri H (1960) Recent Ostracoda from the west coast of Florida. Trans Gulf Cst Ass Geol Soc 10: 107-149

Reyment RA (1988) Evolutionary significant polymorphism in marine ostracods. In Hanai T, Ikeya N, Ishizaki K (eds) Evolutionary biology of Ostracoda (its fundamentals and applications). Proceedings of the 9th International Symposium on Ostracoda, Shizuoka pp 987-1001

Reyment RA, Reyment ER (1959) *Bairdia ilaroensis* sp nov aus dem Paläozän Nigeriens und die Gueltigkeit der Gattung *Bairdoppilata* (Ostr, Crust). Stockholm Contributions in Geology 3: 59-70

Ruggieri G (1962) Alcuni Ostracodi Quaternari e Recenti pertinenti al genere *Costa* Neviani. Boll Soc Paleont Ital 1: 3-9

Said R (1960) Planktonic foraminifera from the Thebes Formation, Luxor, Egypt. Micropaleontology 6: 277-286

Said R (1990) Cenozoic. In Said R (ed) The Geology of Egypt. A A Balkema Publishers pp 451-486

Saito T (1962) Eocene planktonic foraminifera from Hahajima (Hillsborough Island). Trans Proc Paleont Soc Japan 45: 209-225

Salahi D (1966) Ostracodes du Crétacé supérieur et du Tertiaire en provenanced'un Sondage de la Région de Zelten (Lybie). Inst Fran. Pétr 21 (1): 3-43

Sars GO (1866) Oversigt af Norges marine Ostracoder. Norske Vidensk-Akad Forhandl pp 1-130

Sars GO (1922-1928) An account of the Crustacea of Norway. Crustacea: Bergen Mus, pt. XV-XVI 9: 277 pp

Shamah K (1981) Le Paleogene de la province du Fayoum, Egypte. Mem Sci de la Terre, Paris 1: 383 p

Shutskaya EK (1958) Izmenchivosti nekotorich nignepaleogenovikh planktonnikh foraminifer severnogo Kavkaza. Akad Nauk SSSR Voproso Mikropaleont 2: 84-90

Siddiqui QA (1971) Early Tertiary Ostracoda of the Family Trachyleberididae from West Pakistan. Bull Brit Mus Nat Hist Geol, London 9: 98 pp

Stephenson MB (1936) Shell structure of the ostracode genus Cytheridea. J Paleont 10 (8): 695-703

Strougo A, Boukhary MA (1987) The middle Eocene-upper Eocene boundary in Egypt: present state of problem. Revue de Micropaléontologie 30: 122-127

Subbotina NN (1947) Foraminifers of the Danian and Paleogene deposits of the northern Caucasus. Microfauna of the Caucasus Emba region, and central Asia. Trudy (VNIGRI) pp 39-160 (in Russian)

Subbotina NN (1953) Fossil foraminifers of the USSR: Globigerinidae, Hantkeninidae, and Globorotaliidae. Trudy (VNIGRI) 76: 296 pp (in Russian), Translated

Sylvester-Bradley PC (1948) The ostracod genus *Cythereis*. J Paleont 22: 792-797

Triebel E (1950) Homöomorphe Ostracoden-Gattungen. Senckenbergiana 31 (5-6): 313-330

Triebel E (1958) Zwei neue ostracoden-Gattungen aus dem lutet des Pariser Beckens. Senckenbergiana Lethaea 39 (1/2): 105-117

Toumarkine M, Bolli HM (1970) Evolution de *Globorotalia cerroazulensis* (Cole) dans l'Éocène moyen et supérieur de Possagno (Italie). Revue de Micropaléontologie 13: 131-145

Whatley RC (1983) The application of Ostracoda to paleoenvironmental analysis. In Maddocks RF (ed) Application of Ostracoda. University of Houston Geosciences pp 51-77

Whatley RC (1988) Ostracoda and paleogeography. In De Deckker P, Colin JP, Peypouquet JP (eds) Ostracoda in the Earth Sciences, Elsevier Publ Co, Amsterdam, Oxford, New York, Tokyo pp 103-123

Whatley, R.C. 1991. The platycopid signal: a means of detecting kenoxic events using Ostracoda; J Micropaleont 10 (2), 181-185

Wehrli H (ed) (1969) Palaontologische Abhandlungen, Abteilung A, Palaozoologie. Akademie Verlag, Berlin, Germany 162 pp

Yassini I (1979) The littoral system ostracodes from the Bay of Bou Ismail, Algiers, Algeria. Revista Española de Micropaleontologia 11 (3): 353-416

Zalat A (1987) Paleontologic and stratigraphic studies on the Eocene limestones at Gebel Mokattam and Mishgigah, Egypt. M Sc Thesis, Tanta University, Egypt

4 Morphological variability and adaptability in Egyptian Eocene ostracod species *Paracosta mokattamensis* (Bassiouni)

Ashraf M. T. Elewa

Geology Department, Faculty of Science, Minia University, Egypt, aelewa@link.net

4.1 Abstract

Microscopic observations as well as geometric morphometric analyses of specimens belonging to three previously identified species of the genus *Paracosta* Siddiqui (*P. mokattamensis*, *P. humboldti* and *P. ducassae*), from the middle Eocene of Egypt, led to conclude that these three species should be considered as three morphs of one species (*Paracosta mokattamensis* (Bassiouni)). The results reveal that the use of quantitative shape analyses is the most powerful tool for differentiating forms within the species exhibiting shape polymorphism in their life cycles. Nonetheless, the microscopic observations of the different characteristics of the organisms should be a helpful tool in identifying the ornamental polymorphism. The study of the environmental and geographical distribution of *Paracosta mokattamensis* MTA and its two related morphs (MTB, MTC) could interpret the lack of economic phosphate deposits in the Paleogene of Egypt, the Middle East and North Africa. This study exemplifies adaptability as a parallel way to migration for organisms to survive.

Keywords: Adaptability, polymorphism, geometric morphometrics, *Paracosta mokattamensis,* Eocene, Egypt.

4.2 Introduction

The relationship between *Paracosta mokattamensis* (Bassiouni) and *Paracosta humboldti* (Bassiouni) and its related form: *Paracosta ducassae* (Bassiouni et al.) was unknown for a long time since 1969 until Boukhary et al. (1993) tried to find out a clear relationship between the last two forms. These authors, in their study on the middle Eocene ostracods of Mingar El Rayan in Fayoum, arrived at the conclusion that *P. ducassae* is nothing more than a morph of *P. humboldti*. However, for no reason, they mentioned their specimens as a subspecies of *P. humboldti* (e.g. *P. humboldti ducassae*) in their explanation of the photographed specimens. In 2002, Elewa stated that *P. ducassae* (see Bassiouni et al. 1984 for the systematic description) is a synonym of *P. mokattamensis* and the later species together with *P. humboldti* are two morphs of the same species. He attributed this type of polymorphism to the ornamental polymorphism of Reyment (1988) in which the irregularly ornamented morph (*P. humboldti*; MTB in the present study) possesses a more bluntly rounded posterior end than the other morph (*P. mokattamensis*; MTA in the present study). It is worth mentioning that *P. mokattamensis* was recognized from the Nile Valley, however the other two forms were from Fayoum. In general, all the above mentioned studies did not spot the light on the quantitative shape differences of the forms under scope of the present work.

The aim of the present study is to differentiate between shapes of these related forms by means of geometric morphometrics to establish whether there is an adequate shape variation between these forms leading to separate them into three different species, as it was believed earlier, or there is a type of polymorphism that leads to assign the studied forms to one species with different morphs. Moreover, the environmental and geographical distribution of these forms will be discussed hereafter.

To achieve the aim of this study the thin plate spline and relative warps analyses were applied. The thin plate spline method is based on analogy of a 2D morphological object to a thin homogenous deformable metallic plate (Bookstein 1989, 1991); thus one specimen is fit to another by its stretching, and the numerical estimate of degree of such a smooth deformation is the *bending energy* coefficient.

The shape variation encompasses two components, an affine (uniform) part and non affine (non uniform) part (Bookstein 1991). In the affine change, the orthogonality of principal axes is preserved, and parallel lines remain parallel, like the deformation of a square into a parallelogram or a circle into an ellipse. The non affine change is represented by the residual of size-free change that remains after the difference due to any affine change has been subtracted from the total change in shape, for example, when an initially flat object is twisted or warped. For some examples on the method I refer the reader to Reyment (1993, 1995a, b and 1997), Reyment and Bookstein (1993) and Reyment and Elewa (2002).

4.3 Material and measures

The studied material was collected by the author from two different areas (Fig. 1): one section opposite to Beni Mazar city on the eastern side of the Nile Valley, about 60 km north of Minia city (for *P. mokattamensis*); and two sections from Wadi El Rayan, about 100 km south of Fayoum in the Western Desert (for *P. humboldti* and *P. ducassae*). Table 1 shows the lithology of the three sections from which the studied material was collected.

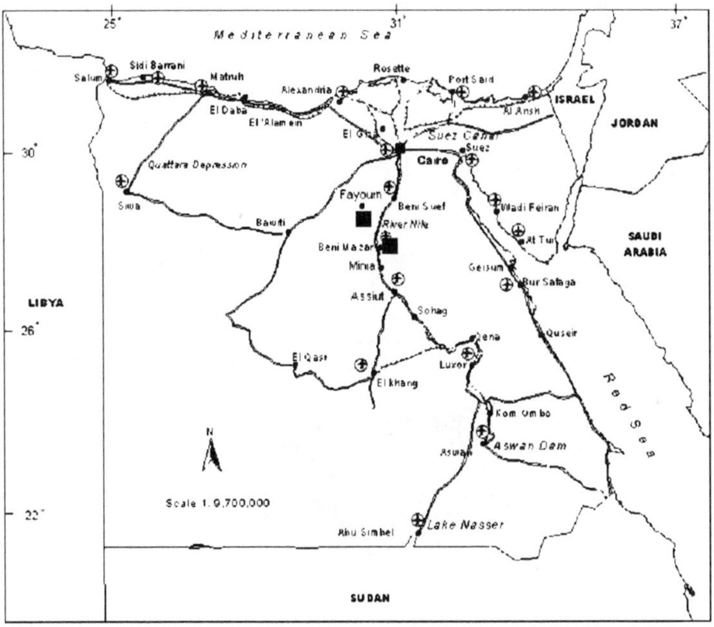

Fig. 1. Location map of the studied forms. Big black squares show the locations from which the author collected the studied forms

Eighteen well-preserved, adult specimens from the studied ostracod species (10 males and 8 females; 10 specimens of *Paracosta mokattamensis*; 4 specimens of *P. humboldti*; and 4 specimens of *P. ducassae*) were selected for the morphometric analyses. Six landmarks were selected on the left views of carapaces, or otherwise the right views of carapaces were rotated by 180 degree to be compatible with the left views of carapaces, for the thin plate spline and relative warps analyses (Fig. 2): the mid-point along the anterior margin (no. 1), the location of the eye tubercle (no. 2), the location of the adductorial area (no. 3), the contact

Table 1. The lithology of the three sections from which the studied material was collected

Morph	Age	Section and lithology
MTA (Paracosta mokattamensis)	Middle Eocene (Lutetian)	El Sheikh Fadl section, on the eastern bank of the Nile Valley, opposite to Beni Mazar city, about 60 km north of Minia city; chalk-marl complex
MTB (Paracosta humboldti)	Middle Eocene (Lutetian)	Two sections at Wadi El Rayan, about 100 km south of Fayoum, Western Desert; sandy marl to sandy shale facies
MTC (Paracosta humboldti ducassae)	Middle Eocene (Lutetian)	Two sections at Wadi El Rayan, about 100 km south of Fayoum, Western Desert; sandy marl to sandy shale facies

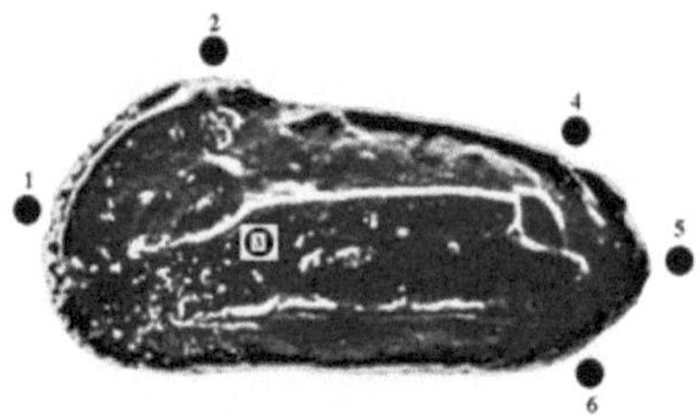

Fig. 2. Sketch showing the locations of the six landmarks used in the present study. Figure is based on MTC (photomicrograph after Boukhary et al. 1993)

4.4 Geometric morphometry

4.4.1 The relative warps analysis

The program of the relative warps analysis was written by F. J. Rohlf (1998b), version 1.2.

Non affine shape differentiation in the three forms

The ordination for the 1st and 2nd warps (Fig. 3) is the representation of the non affine (= non uniform) shape variation. The first singular value for these data constitutes 39.60% of the variance, the second one comprises 28.13% of the variance, thus, more than 67% of the variance is included within the first two relative warps which are sufficient for interpreting the shape variation within the studied data. Figure 4 indicates that the 1st relative warp has a tendency to separate most specimens with subrectangular shape (at the upper section; like 2, 3, 5, 15) from those with ovate shape (at the lower section; like 6, 12, 13, 14, 18). On the other hand, the 2nd relative warp separates most specimens with symmetrical anterior margin (at the left; like 5, 8, 9, 11, 13) from those with asymmetrical anterior margin (at the right; like 3, 16, 17). It seems that the 2nd relative warp exhibits a polymorphic character in the studied forms.

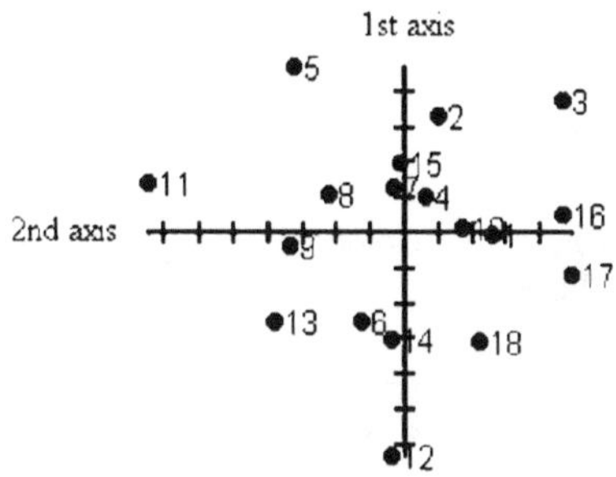

Fig. 3. Ordination by the non-affine shape components for the three studied forms. 1-10: *Paracosta mokattamensis*, 11-16: *P. humboldti*, 17-18: *P. humboldti ducassae*

Affine shape differentiation in the three forms

Figure 4 shows the plot of the 1st vs. 2nd uniform axes. From this figure, It is clear that the 1st uniform axis could separate between the forms exhibiting acutely triangular posterior margin (at the left; like 3, 12 15, 18) from the forms with obtusely triangular posterior margin (at the right; like 2, 4, 8, 11, 16). The 2nd uniform axis differentiates most females (at the upper section; like 1, 3, 6, 9, 11, 13, 16) from most males (at the lower section; like 2, 5, 7, 8, 10, 12, 14, 15, 17, 18). As it was expected, the affine projection could provide additional information about the relationship between the studied forms.

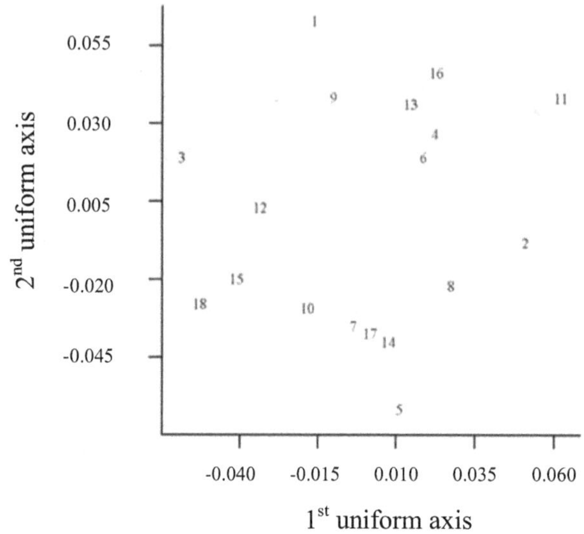

Fig. 4. Ordination by the affine shape components for the three studied forms. Numbers are same like Fig. 3

4.4.2 The thin plate spline analysis

The program used for the thin plate spline analysis is that of F. J. Rohlf (1997), version 2.13. This technique is a graphical representation of mapping from one shape to another. Left views of three males of *P. mokattamensis*, *P. humboldti* and *P. ducassae* (one specimen for each) were considered as typical (reference) specimens for the analysis. Figure 5 portrays the mapping of the non affine case of *Paracosta mokattamensis* (MTA in the present study) into *P. humboldti* (MTB in the present study). The deformation is not so strong and the bending energy is 0.04684. Both the dorsal and ventral zones are affected. Figure 6 displays the warp of MTA into *P. ducassae* (MTC in the present study). The bending energy is weaker than the previous case and equals 0.03706. The figure shows that the ventral margin is more affected than other sides. From Fig. 7, showing the warp of MTB into MTC, it can be noticed that the deformation is weaker than the previous two cases with bending energy equal 0.01625. The dorsal zone is more affected than other sides.

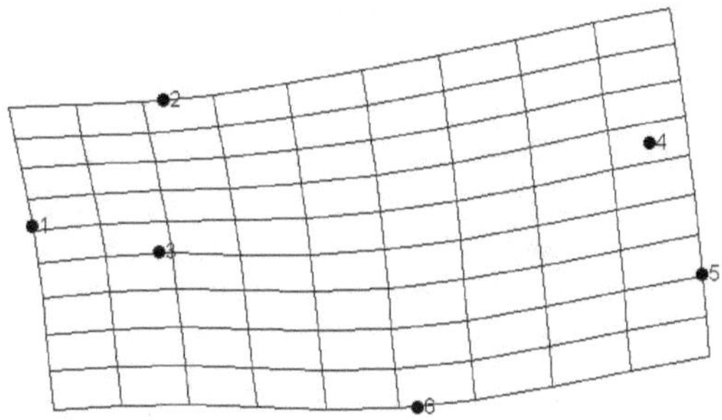

Fig. 5. The non affine deformation for the comparison between MTA and MTB

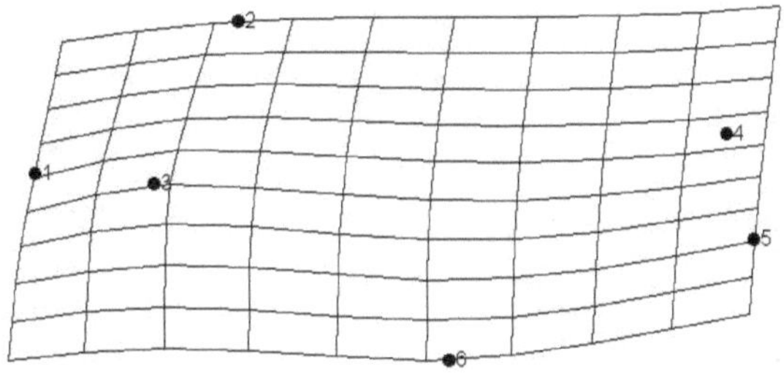

Fig. 6. The non affine deformation for the comparison between MTA and MTC

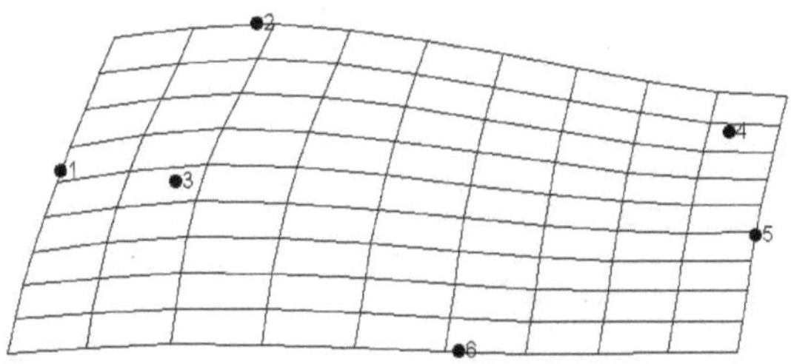

Fig. 7. The non affine deformation for the comparison between MTB and MTC

The three comparisons indicate that the similarity of MTB to MTC is stronger than that of MTA to MTC. In the mean time, the similarity of MTA to MTC is stronger than that of MTA to MTB. In other words, MTC could be a transitional form between MTA and MTB.

4.5 Discussion and conclusions

The genus *Paracosta* was introduced to science by Siddiqui (1971) from the Eocene of West Pakistan to represent those forms with normally developed four longitudinal ridges and reticulation on the lateral surface of the carapace. This genus is perhaps the most characteristic taxon of the epicontinental seas of the southern Tethys (Keen et al. 1994). The importance of this genus in Egypt is due to its occurrence from the Maastrichtian to middle Eocene representing some distinct phylogenetic trends, and the short stratigraphical ranges of some of its species, hence, they could be used as zonal markers. For detailed discussion on the biostratigraphical, paleogeographical and evolutionary significances of this genus, see Elewa (2002).

For the present work, the author examined several specimens of the studied forms of this genus, using the microscopic observations as well as the morphometric analyses, and could separate these forms into three different forms within a single species that is *Paracosta mokattamensis* (the first recognized form by Bassiouni 1969b from Gebel Mokattam, at the east of Cairo).

According to the microscopic observations (Fig. 8), MTA is ovate to subrectangular in outline, however, MTB is subrectangular. The median rib in MTB is branching and gives rise to a small riblet, but in MTA it ends at the muscle node. However, in some specimens of MTB the median rib is stopped at the muscle node indicating a transitional form between these two forms. Moreover, MTB has a curved eye rib connecting the eye tubercle to the median rib while it is absent in MTA. On the other hand, the median rib in MTB possesses a depression at the anterior end which is lacking in MTA and MTC. Also, the surface is smooth in MTA and MTC, but it shows transverse riblets in MTB. It is clear that the microscopic observations could successfully define the ornamental polymorphism of Reyment (1988) in the studied forms.

In the mean time, the geometric morphometry led to some important results:

* The 1st non affine relative warp axis could separate most specimens with subrectangular shape from those with ovate shape. In fact, not all specimens of MTA show an ovate outline but some show subrectangular outline like MTB. This leads to a somewhat overlap between the two morphs.

* The 2nd relative warp axis separates most specimens with symmetrical anterior margin from those with asymmetrical anterior margin. This could be attributed to the shape polymorphism (type 1) of Reyment (1985) in

which the swung rounded anterior is opposed to the regularly rounded anterior margin.

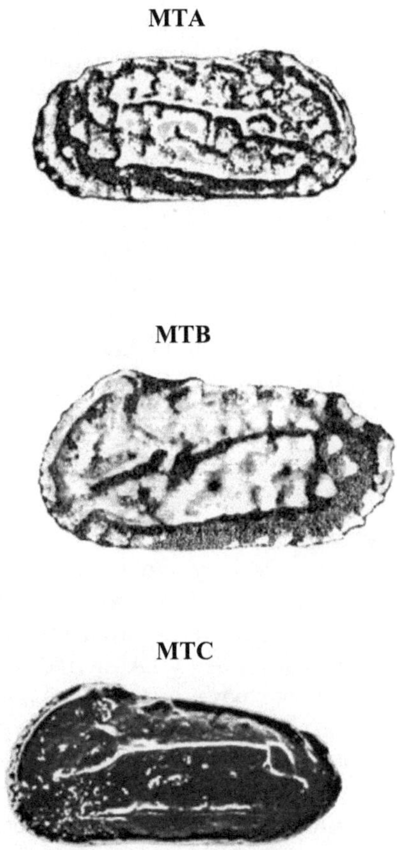

Fig. 8. Photomicrographs of the three studied forms (after Bassiouni 1969b for MTA and MTB; after Boukhary et al. 1993 for MTC)

* The 1st uniform axis could separate between the forms showing acutely triangular posterior margin from the forms with obtusely triangular posterior margin.

* The 2nd uniform axis could differentiate most females from most males, or in other words it could detect the sexual dimorphism within the studied forms.

* The thin plate spline analysis led to the conclusion that the similarity of MTB to MTC is stronger than that of MTA to MTC. Meanwhile, the similarity of MTA to MTC is stronger than that of MTA to MTB. In conclusion, MTC is presumed in the present study to be a transitional form between MTA and MTB.

Paleoenvironmentally, as I mentioned above, *P. mokattamensis* was recognized from the Nile Valley, however the other two forms were from Fayoum. By tracing the environmental conditions under which these forms were distributed, it was found that *P. mokattamensis* is a member of ostracod association representing shallow water conditions in the Maghagha Formation of the Nile Valley. On the other hand, during the deposition of the Qarara Formation in the Nile Valley (= Midawara Formation in Fayoum), the area was subjected to shallow water conditions during the interval of *Acarinina aspensis* Zone to *Morozovella lehneri* Zone of early middle Eocene, and saw a transient shift toward outer shelf or deeper environments in the latest period, in the interval of *Truncorotaloides rohri* Zone of late middle Eocene (see Elewa and Ishizaki 1994; Nishi et al. 1994; Elewa et al. 1995). Therefore, the gradual change in reticulation expressed by the increasing prominence of the longitudinal ridges (from degradation to agradation according to Peypouquet et al. 1986; in other words, from weak ornament to strong ornament) from *P. mokattamensis* (MTA in the present study) to *P. ducassae* (MTC in the present study) to *P. humboldti* (MTB in the present study) may be related to a gradual increase in depth from shallow to deeper and in salinity (the saturation of $CaCO_3$ in the water) from reduced to normal (environmentally induced polymorphism).

Paleogeographically, the species under consideration represents a group of ostracods that is the most characteristic taxon of the epicontinental seas of the southern Tethys (Keen et al. 1994). Al Sheikhly (1981) believes that there is a close relationship between the geographical distribution of *Paracosta* and the Paleogene phosphate belt running from the Middle East through North Africa. At the same time, Peypouquet et al. (1986) are convinced that the agradation-degradation phenomenon should be related to physico-chemical conditions prevailing at the water/sediment interface immediately preceding phosphate deposition, and in general to be related to upwelling of deeper water on to a shelf with protected bays. In contradiction, the studied *Paracosta* species and its morphs either inhabited a shallow water environment as a whole, during their life cycles, or show a distinct adaptability with the vertical change in depth from shallow to deeper environments. This may interpret the lack of economic phosphate deposits in the Paleogene of Egypt as well as the same interval of the surrounding areas of the Middle East and North Africa. In general, it seems that the studied *Paracosta* species is pandemic and tends to adapt to changes in depth and saturation of $CaCO_3$ in the water through the ability to exhibit polymorphism during its life cycle. On the other hand, the ancestor of this species (*Paracosta mokattamensis praemokattamensis* Bassiouni, according to Bassiouni and Elewa 1999) is possibly an endemic species and was recorded by Bassiouni (1969a) from the Eocene of Jordan. Nevertheless, auxiliary studies of the geographical distribution of the latter species is needed to demonstrate its routes of migration.

4.6 Acknowledgement

I appreciate the critical reviewing of this manuscript by Prof. Dr. Rashad Abdel Kireem of the Geology Department, Faculty of Science, Alexandria University, Egypt, as well as another reviewer who preferred to be anonymous.

References

Al Sheikhly SAJ (1981) Maastrichtian-Upper Eocene Ostracoda of the subfamily Trachyleberidinae from Iraq, Jordan and Syria. Unpublished Ph D Thesis, University of Glasgow 229 pp

Bassiouni MA (1969a) Einige *Costa*- und *Carinocythereis* (*Reticulina*)- Arten aus dem Paläozän und Eozän von Jordanien (Ostracoda). Neues Jahrbuch für Geologie und Paläontologie Abhandlungen 134 (1): 1-16

Bassiouni MA (1969b) Ostracoden aus dem Eozän von Ägypten (1. Trachyleberidinae). Geologische Jahrbuch, Hannover, 87: 383-426

Bassiouni MA, Boukhary MA, Shamah K, Blondeau A (1984) Middle Eocene ostracods from Fayoum, Egypt. Géologie Méditerranéenne 11 (2): 181-192

Bookstein FL (1989) Principal warps: thin-plate spline and the decomposition of deformations. IEEE Transactions on Pattern Analysis and Machine Intelligence 11: 567-585

Bookstein FL (1991) Morphometric tools for landmark data: geometry and biology. Cambridge Univ. Press, New York, 435 pp

Boukhary MA, Guernet C, Strougo A, Bassiouni MA, Bignot G, Abdel Ghany O (1993) Eocene ostracodes of Mingar El Rayan (Fayoum district, Egypt). Stratigraphic and paleogeographic implication. Revue de Micropaléontologie 36 (3): 191-212

Elewa AMT (2002) Microevolution applied to ostracod biostratigraphy: the middle Cretaceous to middle Eocene ostracods of Egypt (with special reference to the genus *Paracosta* Sddiqui). Neues Jahrbuch für Geologie und Paläontologie Abhandlungen 226 (3): 289-318

Elewa AMT, Ishizaki K (1994) Ostracodes from Eocene rocks of the El Sheikh Fadl-Ras Gharib stretch, the Eastern Desert, Egypt (Biostratigraphy and paleoenvironments). Earth Science (Chikyu Kagaku), Tokyo, 48 (2): 143-157

Elewa AMT, Ishizaki, K, Nishi H (1995) Ostracoda from the El Sheikh Fadl-Ras Gharib stretch, the Eastern Desert, Egypt -With reference to distinguishing sedimentary environments-. In Riha J (ed) Ostracoda and Biostratigraphy (The 12th International Symposium on Ostracoda, Prague), AA Balkema, Rotterdam pp 203-213

Keen MC, Al Sheikhly SSJ, Elsogher A, Gammoudi AM (1994) Tertiary ostracods of North Africa and the Middle East. In Simmons MD (ed) Micropalaeontology and hydrocarbon exploration in the Middle East. Chapman and Hall Inc pp 371-400

Nishi H, Elewa, AMT, Ishizaki K (1994) Planktonic foraminiferal biostratigraphy of upper Paleocene to middle Eocene sequences in the Eastern Desert area, Egypt, Trans Proc Palaeont Soc Japan, NS, Tokyo 175: 521-552

Peypouquet JP, Grousset F, Mourguiart P (1986) Paleooceanography of the Mesagean Sea based on ostracods of the northern Tunisian continental shelf between the Late Cretaceous and Early Paleogene. Geologische Rundschau 75: 159-174

Reyment RA (1985) Phenotypic evolution in a lineage of the Eocene ostracod *Echinocythereis*. Paleobiology 11(2): 174-194

Reyment RA (1988) Evolutionary significant polymorphism in marine ostracods. In Hanai T, Ikeya N, Ishizaki K (eds) Evolutionary biology of Ostracoda (its fundamentals and applications). Proceedings of the 9th International Symposium on Ostracoda, Shizuoka pp 637-657

Reyment RA (1993) Ornamental and shape variation in *Hemicytherura fulva* McKenzie, Reyment, Reyment (Ostracoda; Eocene, Australia). Revista Española de Paleontologia 8 (2): 125-131

Reyment RA (1995a) On multivariate morphometrics applied to Ostracoda. In Ostracoda and Biostratigraphy, Proceedings of the 12th International Symposium on Ostracoda, Prague (1994). Balkema, Rotterdam pp 203-213

Reyment RA (1995b) Analysis of geographic shape-variation in *Vargula hilgendorfii* (Ostracoda, Crustacea). Revista Española de Paleontologia, Homenaje al Dr. Guillermo Colom pp 26-30

Reyment RA (1997) Evolution of shape in Oligocene and Miocene *Notocarinovalva* (Ostracoda, Crustacea): A multivariate statistical study. Bulletin of Mathematical Biology 59 (1): 63-87

Reyment RA, Bookstein FL (1993) Infraspecific varaiability in shape in *Neobuntonia airella*: an exposition of geometric morphometry. In: Ostracoda in the Earth and life sciences, Proceedings of the 11th International Symposium on Ostracoda, Warrnambool (1991), Balkema, Rotterdam pp 291-314

Reyment RA, Elewa AMT (2002) Size and shape variation in Egyptian Eocene *Loxoconcha* (Ostracoda) studied by morphometric methods. In Thiergärtner H (ed) Mathematical methods and data bank application in paleontology, Mathematische Geologie, CPress Verlag, Germany 6: 3-14

Rohlf FJ (1997) TPSsplin: thin-plate spline, version 1.15. New York, State University at Stony Brook. (program)

Rohlf FJ (1998a) TPSdig, version 1.20. New York, State University at Stony Brook. (program)

Rohlf FJ (1998b) TPSrelw: relative warps, version 1.20. New York: State University at Stony Brook. (program)

Siddiqui QA (1971) Early Tertiary Ostracoda of the family Trachyleberididae from West Pakistan. Bull Brit Mus Nat Hist Geol 9: 98 pp

5 Insect migration and dispersal with emphasis on Mediterranean ecosystems

Panos V. Petrakis[1] and Anastasios Legakis[2]

[1] National Agricultural Research Foundation, Institute for Mediterranean Forest Research, Laboratory of Entomology, Terma Alkmanos, 115 28 Athens, Greece, pvpetrakis@fria.gr, [2] University of Athens, Dept. of Biology, Zoological Museum, 157 84 Athens, Greece, alegakis@biol.uoa.gr

5.1 Abstract

Insects migrate in many different scales ranging from a few metres to many hundreds of kilometres. The *holistic conceptual model* hosts a suite of widely accepted definition of insect migration. The predisposition of some insects for migration is called *migratory syndrome* and involves the condition of the flight muscles, the oviposition period and timing, the fecundity, the energy uptake and wing polymorphism. Pheromones are another useful trait of migratory insects since they are used to gather the scattered migrants in a restricted space. The cost of a migration event was compared to the cost of other solutions to the problems of overcrowding or habitat destruction. In many cases migration is the less costly solution apart from other benefits that may incur to the insect such as natural selection, since migration is a safe estimator of the abilities of the insect and its health status. To estimate the cost of insect migration several authors have proposed formulas with hardly estimable parameters. Moreover, it is very difficult for the insect to manipulate parameters such as the migration time and the available energy for migration. It is speculated that in evolutionary time, the manipulation in evolutionary time of some parameters is not possible in the physiological context of an insect, and for this, a migration event may last more than two or three generations. The migration of insects is now studied using, apart from biological methods, new approaches such as specifically constructed entomological radars. Climate change in general and global warming in particular affect insect migration in several ways. It causes phenological shifts to the host plants, destruction of habitat, extensive fragmentation of landscapes, marine incursions and faunal reshuffling. Also, many insects and host plants are responding in different ways to global warming and greenhouse gases. In Mediterranean insects such as *Euplagia quadripunctaria* there is a short scale migration primarily caused by searching for ideal climatic conditions.

Keywords: global warming, migration syndrome, population trajectory, altitudinal migration, oogenesis flight syndrome, planetary boundary layer, Mediterranean ecosystems, entomological vertical looking radar.

5.2 What is insect migration and how it can be predicted?

In general the distinction between migration and other movements of insects is artificial (Baker 1981; Tipping 1995). The constant parameter to all types of insect movements is the change of habitat. As Dingle (1978) pointed out, the change of habitat is not only the escape from old ones, which have possibly deteriorated, but also the colonization of new ones and the concomitant potential for better reproduction. This idea emerged quite early as Southwood (1960) did not separate migratory movements from dispersion, while Andrewartha and Birch (1954) discussed migration under a chapter headed "Dispersion". Other authors such as Rainey (1960) considered migration as a seasonal displacement of a population while Kennedy (1961) considered it as a persistent and straightened–out insect movement with a simultaneous inhibition of factors that arrest it. Williams (1958) stated that migration is a continuous insect movement to a more or less definite direction, which is under the control of the animal. Schneider (1962) reviewed the old works and in general talked about migration as a type of dispersion; in particular when dealing with the flight of winged aphids he used the term "*dispersive migration*". Many authors have recognised that these definitions are acceptable generalisations only within specific taxonomic groups since they involve ecological and ethological aspects, which differ among groups.

Instead of trying to define exactly what migration is, it is better to set the framework of the discussion about its components and the associated parameters of movement, thus providing a less strict definition. In insect migration one may have to define first a temporal component. Usually the term *insect migration* is used to specify insect movements that take less than one year per direction and involves at least two generations. Some altitudinal migrations such as the hill–topping of several insects, which are frequent in the Mediterranean, are an exception to this rule in the sense that they involve only one generation. Within this type of movement we cannot exclude the embarkation on a vehicle – e.g. the butterflies moving across large water bodies using ships for transportation (Darlow 1951; Vanholder 1996)- or the assistance of wind (Brower 1995). A second component is purely spatial and involves the distance at which insects move. This distance can be everything from a few meters to several hundreds of kilometres – e.g. *Danaus plexippus* (Linnaeus) [Lepidoptera, Danaidae] and *Schistocerca gregaria* (Forskål) [Orthoptera, Acrididae] (Tipping 1995). Elaborating on the components needed to specify an insect movement as migration, Drake et al. (1995) have defined migration in the context of the "holistic conceptual model" (=HCM). This model involves four components: [1] the *migration arena,* which is the vertical and horizontal space in which a specific migration takes place; [2] the *population trajectory*, which involves the demographic parameters in space and time of the migrant and the receptive or the newly established populations; [3] the *migration syndrome,* which involves all the physiological, behavioral and morpho–anatomical traits that are associated with migrants; [4] the *genetic complex*, which is actually the ecology of the genes that are responsible for the traits comprising the migration syndrome. It is believed that the HCM can serve as an organizing framework for all those studies involved with insect migration. For instance, it has been stated (Mendel 2000) that the pine processionary moth *Thaumetopoea pityocampa* Den. & Schif. [Lepidoptera, Thaumetopoeidae] migrated in Europe from Africa by following the spread of Mediterranean pines – *sensu* Klaus (1989) – through land bridges that joined Europe and Africa 20 Ma BP. It has been reported that the Sahara at those times was occupied by Mediterranean-type vegetation. Although the insect may exhibit local migratory behavior, this type of evolutionary migration cannot be included in the term *insect migration* given here. In other cases, the migration cannot be exactly identified if the direction of the movement is unknown. For instance, the high numbers of ladybirds in the east coast of Britain in 1976 was considered as a migration of Coccinellidae (Coleoptera) from the European continent. However, it was proved that these predatory beetles simply looked for food and populated the coast because they were stopped by the sea, which they were unable to cross (Majerus and Kearns 1989).

In the Mediterranean countries the altitudinal migrations and the hill topping of butterflies are more or less known and predictable events. In these countries, especially those that have a rich topographic relief, it is a common practice for farmers and gardeners to collect ladybirds that occupy the summits of small mountains in large numbers (Petrakis and

Legakis *pers. obs.*). All these activities have only a calendar basis in the sense that they are not deduced from the examination of insect parameters.

Environmental parameters are among the earliest factors that have been implicated to control the timing of insect migration. Kennedy (1961) stated that the migratory desert locust *S. gregaria* enters a developmental phase named phase *gregaria* when its population densities are high. The inhibition of the release of the juvenile hormone is responsible for this development. When in high concentration, the juvenile hormone forces the insect to enter phase *solitaria*. These changes coincide with the onset and the end of the rains; the coupling of the physiological and environmental effects on desert locusts seems to be tight.

Several authors have stated that in some insects there is a predisposition for migration because they possess the traits necessary for movement. Dingle (1996, see also 1985) has called this set of traits "m*igratory syndrome*". The syndrome is manifested by traits such as well-developed flight musculature; relatively long pre–oviposition period; high fecundity for the establishment in the new habitats by means of eventually a small number of individuals which survived migration; propensity for flight; no histolysis of main and secondary flight muscles; high polymorphism and plasticity within and among insect populations; and a relatively high supply of rapidly built up energy in order to resume their migration.

The polymorphism of the flight apparatus is usually maintained at the expense of energy stored in both male and female insects (Zera and Denno 1977). Usually the polymorphism is of two types: [1] wing polymorphism and [2] flight muscle polymorphism. Both types involve degradations in the number and length of wings and muscles. Juvenile hormone (JH) and ecdysone (ED) have been extracted from insects and analysis of titres has been found responsible for both types of polymorphism. In a recent work, Dingle and Winchel (1997) found that the soapberry bug *Jadera haematoloma* (Heteroptera, Rhopalidae) maintains a sophisticated polymorphism in wing length that involves four different morphs ranging from macropters (long-winged) to brachypters (short-winged) and apters (no wings are present). In the co-existence of the four morphs on the Florida keys (mainly on Key Largo Isle) JH plays an important role by genetically governing not only the production of morphs but also the calling behavior of females, and it directly influences the migratory flight of the bug and of some aphids. Interestingly, the existence of morphs is regulated also by the availability of food (Dingle and Winchell 1997). In the neotropical congener bug *Jadera aeola* Dallas it was found that morphs are maintained as a response to environmental cues such as food availability and the inset of the wet season (Tanaka and Wolda 1987). These facts imply that in these insects, migration is regulated by external

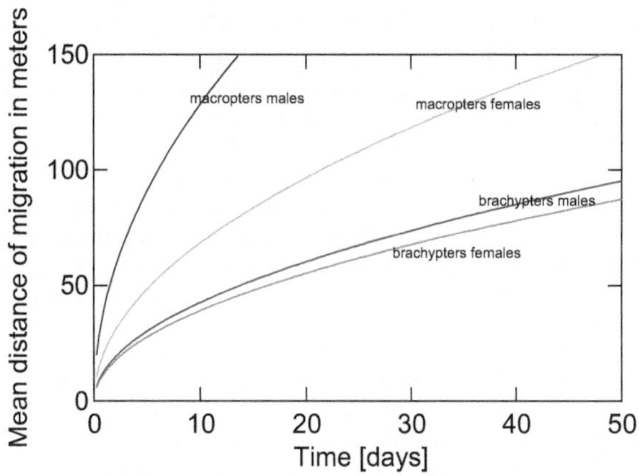

Fig. 1. Mean dispersal distance of macropters and brachypters males and females for the first fifty days of their lives. The calculations were done according to Socha and Zemek (2003) by using their formulas and estimated values

factors such as food availability through a JH mechanism. Internally, the JH system seems to function by the speeding up of the biogenetic pathway of JH and the breakdown of JH by JH esterase according to the Fairbairn (1994) model for the regulation of JH.

With regard to flight musculature the polymorphism is more flexible since the flight muscles can undergo histolysis (Zera and Denno 1997). This is also true for the soapberry bug (Dingle and Winchell 1997). Andersen (1993) has found that the proportion of macropters varies in all seasons in bivoltine Scandinavian pond skaters (Heteroptera, Gerridae). This failure to find a correlation between wing polymorphism and migration was attributed to the various degrees of histolysis of the flight muscles that masked the effect of macropters in the flight behavior of Gerridae. If macropters have a higher propensity for flight than brachypters then what is the meaning of keeping wing polymorphism in walking insects? Socha and Zemek (2003) addressed this question and found that macropters is an adaptation of *Pyrrhocoris apterus* (L.) (Heteroptera, Pyrrhocoridae), which [1] have higher dispersal activities than brachypters, [2] they spread and establish better in new areas, [3] have a higher mean and angular velocity and [4] move at higher distances Fig. 1. All these happen in a walking mode of movement. Moreover, Socha et al. (1998) found that there is a fasting period that always precedes the long distance dispersal of *P. apterus*. In this way the insect enters a non–diapause inhibition of reproduction with very low sexual activity, which is a prerequisite to channel all resources to migration. These inhibitions, except for giving the time to the insects to disperse, happen when the brachypters are feeding and are reproductively active. Flightlessness of *P. apterus* is believed to have evolved secondarily and did not result in the loss of the physiological mechanisms that mobilizes the energy necessary for flight. Only the mode of dispersal has changed to the less costly walking that does not need the maintenance of flight muscles.

Table 1. Mean values of dispersal distances (in meters ± standard deviations) experiments designed for the estimation of macropters and brachypters in laboratory populations of *P. apterus*

Sex / Wing morph	Males	Females
Macropter	156.0 ± 81.1	83.1 ± 43.2
Brachypter	51.8 ± 26.9	47.5 ± 24.7

5.3 How migration is measured

An important activity of many insects must be studied with the deserved accuracy and extent that the imagination of entomologists and the availability of technology permit. The methods listed in this unit are not confined to Mediterranean migrations. With the advent of cheap computing power a number of devices have appeared together with the associated algorithms that detect insect migrations and estimate parameters associated with them. The methodological framework in which a specific migration phenomenon is studied is very important in selecting the appropriate techniques. This account is concerned only with *long distance migration* (=LDD or LSM=Large Scale Migration) since *short distance migration* (=SDD or SSM=Small Scale Migration) has been reviewed in many papers and books (e.g. Cooter et al. 2000; Southwood and Henderson 2000). In all methods the most important concept is the *dispersal curve* (for LDD see Nathan et al. 2003), which is the frequency distribution of the dispersal distances of the individuals in a migrating group of insects. Any study of the migration pattern of insects must be designed to estimate the dispersal curve of the migration. Nathan et al. (2003) have used three main methodological groups.

5.3.1 Biogeographical methods

Biogeographic patterns are used to infer long distance migration of insects. Although this method has been used in pre–Darwinian work, Darwin and Wallace used it extensively. Recently Darlington (1957) gave many exemplary applications of the method and discussed the associated arguments. The method has recently become very powerful with the incorporation of phylogenetic methods using the large weaponry of molecular markers and the powerful computer systems where the associated phylogenetic algorithms are implemented. G. R. Coope (e.g. 1994) investigated the climatic changes through the distribution of beetles. It is believed that temperature changes as little as 1.5° C or sometimes even less can be deduced by beetles. Also, British Isles as more sensitive to climatic changes are globally important. On this basis Coope reconstructed temperature curves in Britain for the past 22,000 years. Also it has been found that some insects were distributed quite away from their current ranges implying that thee occurred extensive range shifts due to climatic changes. These extensive shifts could have been done through gradual short scale migrations as compensatory responses of the insect to climatic changes.

5.3.2 Movement/redistribution methods

These methods are using parameters estimated from movements of individuals in order to infer the population redistribution (Nathan et al. 2003). A further empirical subdivision of the methods includes the *Eulerian* and the *Lagrangian approaches*. Eulerian approaches are based on the monitoring of marked/unmarked individuals in space. Emphasis is given to the population and not the individual insects. Lagrangian approaches estimate the size and shape parameters of the movement of individuals, such as velocity, spread and direction. They are more detailed than Eulerian though their application is more costly.

5.3.3 Genetic methods

By using genetic markers these methods are estimating the effective rather than the actual dispersal of organisms. Moreover, if the employed markers are cytoplasmic (and not nuclear) such as mitochondria and chloroplasts of the insects' food plants, they have a uniparental inheritance. In insect groups where one of the sexes is heterogametic – e.g. Lepidoptera – the comparison of mitochondrial DNA to that taken from the Y–chromosome gives important information for the differential dispersal of males and females. The method has been extensively used in the work on mammals (Petit et al. 2001). In insects mitochondrial DNA (COI gene) was extensively used to find the differentiation of populations of sparsely distributed insects over vast areas on the globe (Andersen et al. 2000) and in a restricted area (i.e. Finland, Roslin 2001). On the basis of the COI gene Roslin stated that the Finnish dung beetle *Aphodius fossor* (L.) (Coleoptera, Scarabaeidae) is rather a patchily distributed large population than many small populations, a fact that is possibly due to the high dispersal mobility of the insect. This last was expected as showed by direct observations.

The data collected through genetic techniques were used to estimate gene flow (*Nm*) between pairs of populations, usually between source and sink sites, or to assess the diversity of insect populations. Important statistics for these are Wright's F_{ST}, Nei' s G_{ST}, and Weir' s θ (Bilton et al. 2001).Usually the estimates of *Nm* deviate extensively, a fact attributed to the violation of the various assumptions (e.g. divergence of the population from Hardy–Weinberg equilibrium and complete lack of linkage disequilibrium. The employment of different molecular markers and the various mutation patterns result also in dramatically different estimates of *Nm*. Bilton et al. (2001) emphasize that a common feature of all these approaches is that all sampled individuals are considered as members of a population regardless of the population of origin.

5.3.4 Other techniques

The techniques that can be applied in most methodologies are practically innumerable. However, some techniques are more frequently applied and have greater success. Radar techniques are the most advanced (Nieminen et al. 2000) and many systems are constructed, together with the necessary routines for their function. This type of remote sensing technology is entomologically extremely important since the volume of the air sampled and the radius of detection are high (*c*. 1 Km) while the device functions almost continuously unaffected by sunlight or moonlight (Chapman et al. 2003). Existing GIS systems were also used in conservation studies. Chefaoui et al. (2005) used also remote sensing in combination with the multivariate technique *ecological-niche factor analysis* (ENFA) in a dung beetle conservation program in Madrid, Spain area. The method was used to model the conservation status of *Copris hispanus* and *C. lunaris* (Coleoptera, Scarabaeidae) and characterise their niches.

Entomological Vertical Looking Radar (=VLR)

The first radars used in entomology were non-autonomous X-band scanning radars, which required a lot of labour. VLR's were constructed to overcome such difficulties but the earlier constructions lacked the necessary resolution and identification capabilities of newly constructed VLR's (Chapman et al. 2003). In Europe the most extensive network is that of Rothamsted Radar Entomology Unit, which employs VLR systems. Basically the function of such a VLR system involves the emission of a vertical beam, which is linearly polarized, while the plane of polarization is continuously rotated by mechanical means in addition to an angular motion around a vertical axis forming with it a very small angle (*c.* 0.18°). Returned signals are captured for a period of 5 min every 15 min. These images are analysed by means of a complex Fourier transformation. If this procedure converges to a solution it gives seven parameters (see Chapman et al. 2003 for details). The signals having a high correlation coefficient (> 0.9) are meant to represent the same object, which according to the theoretical model of the constructors this object is of biological origin. In this way raindrops or other materials of non-biological origin are excluded. Although VLR's units are not commercially available they have been used to carry out several investigations (Chapman et al. 2002 for the diamondback moth *Plutella xylostella*).

Mark-recapture technique (=MR)

The safest, though laborious, technique is the mark-recapture method. To avoid redundancy the reader is directed to the respective chapter of the book written by Southwood and Henderson (2000). Recently have appeared many computer programs that achieve a more or less "good" estimate of population densities. The reader may have to use an efficient search engine and use the words "mark", "capture" and "recapture". Schneider (2003) gives examples for searching in the BIOSIS database in order to find works on butterflies that use mark-recapture. The same author stated also that the dispersal distance of many butterflies and the meadow brown butterfly in particular (*Maniola jurtina* L., Lepidoptera, Satyridae). On the basis of five mark-recapture studies she found a linear relationship between the size of the study area and the distance of dispersal. The same relationship was also found in other 21 works on 21 other species of Lepidoptera. She concluded that the dispersal ability of butterflies can only be estimated in large scale studies involving mark-recapture techniques. Moreover, if the species of butterfly shows a sedentary behavior in small scale studies, this can be misleading since it can be the outcome of an "unfavorable matrix rather than a lack of movement ability".

Marking of individuals

The direct marking of individual insects has been recently reviewed by Haggler and Jackson (2001) from whom further information can be obtained. The only caution should be that since marking is expected to confer a decrease in the flying abilities of insects, especially in LDD, special care should be taken.

The marking of individual migrating organisms could be very helpful in studying the energetic balance of migrations. This is only achieved in birds, mammals and in a much smaller degree in fish. In insects, while the marking of individuals is rather common for the drawing of migrating trajectories especially, in *D. plexippus* individuals as a rule are not monitored for energetic examination. What we know on the subject comes from bird and mammal studies. The formula proposed by Hedenström and Alerstam (1997) is quite helpful in discussions of the cost of insect migrations and the necessary adaptations towards its minimisation.

Another parameter of high importance for insects living in seasonal environments is the duration of the migration, which can be also measured by direct marking. Again the formula proposed by Hedenström and Alerstam (1997) can be helpful. It was found that adaptations in insects appeared to regulate the duration of migration in order to synchronise the hatching of larvae with the phenological stage of the host plant as is the case of the summer Mediterranean red admiral butterfly *Vanessa atalanta* (L.) (Stefanescu 2001). The manipulation of the Hedenström – Alerstam equation is not always possible in the physiological system of constraints of an insect. In this way the cost of migration is distributed in two or more generations such as monarchs (Urquhart and Urquhart 1978) and hawker dragonflies (*Anax junius* (Drury) [Odonata, Anisoptera, Aeshnidae] (Freeland et al. 2003). In cases where the time or the cost of migration is distributed over more than one generation only the marking with radioactive isotopes can be somewhat useful.

Electrophoresis

This technique is actually used in any case we want to know the genetic variation of a population and origin of an organism. The introductory book of Brewer (1970) is suggested for the details of the method. Several individuals of the suspected migrants are grounded and placed on a gel (usually made of starch) on which an electric field is applied. The proteins are thus arranged according to the distance they run on the gel, which is roughly inversely proportional to their molecular weight. The gel together with the proteins is processed with a stain, which makes them visible. If several individuals are analysed in this way it is possible not only to find the origin of an insect individual but also the genetic variation in the migrant population. With the explosion of molecular techniques it is also possible to identify the origin and the relationships of a migrant population of insects with other populations. A study that used both electrophoretic techniques and DNA markers is that of Roslin (2001). Although the study was not designed to show the superiority of one technique over the other both methods gave similar results on the spatial population structure and the heterogeneity of the Finnish dung beetle *A. fossor*.

5.4 Why insects migrate?

It is a common part in any folk wisdom that in a certain places there are good and bad times for insect collecting without regard to the season or the number of generations of the respective insects. Voltinism could be the effect of the insect biology of which three main characteristics are subjected to natural selection. First, the ability of the insects to discover the suitable habitat quickly. Second, the reproductive potential of the species as expressed by the intrinsic rate of increase [*r*] (Stearns 1992). Third, the ability of the insects to exhaustively exploit the available resources well before the arrival of other insects. Migration and dispersal are simple mechanisms which are manifestations of the way the insect is adapted to maximise either the *R* (i.e. the net reproductive rate) or the *r* (Kozlowski 1999).

Mediterranean biota are believed to be invasive in other ecosystems since they are better adapted to disturbance than the indigenous taxa (Di Casti 1991). Their geographic distribution is highly fragmented since Mediterranean habitats are shaped by anthropogenic impact (Balleto and Casale 1991). For these insects the migration at a short distance, relatively equal to the distance among patches, in the sense of favorable biotope act as a communicative mechanism among the various populations as it is predicted by theory of metapopulations (Hanski 1999). All documented Mediterranean taxa are not very active dispersers or migrants and their introduction was either accidental such as the Argentine ant *Iridomyrmex humilis* (Mayr), (Hymenoptea, Formicidae) and the *Drosophila subobscura* Collin (Diptera, Drosophilidae) in the homoclimatic Chile or deliberate importation such as various species of dung beetles (Coleoptera,

Scarabaeidae) (Kirk and Lumaret 1991; see page 421 for list of imported dung beetles except *Aphodius* spp.).

Around the Mediterranean basin, all areas are considered to have very long summers – a summer of six to nine months is quite common – with a clear shaping effect on the local insect faunas. Multivoltine and univoltine insect species are almost equal in numbers (for Heteroptera and Auchenorrhyncha in east Mediterranean insect assemblages, see Petrakis et al. 1988, 1994). Insect phenologies exhibit a rather common pattern shared between all regions having a Mediterranean-type climate (Shapiro 1975) although it has been predicted that a climate change may cause phenological asynchrony (e.g. between herbivorous insects and host plants; Dixon 2003). For instance, in upland biotopes of Attica in Greece (elevation between 400 and 900 masl at Mt Poikilo, Mt Aigaleo and southern slopes of Mt Parnis) two peaks appear in an average year. One rather flat peak appears in spring including March, April and May; a second peak appears typically in October. The phenological pattern in spring is formed because univoltines emerge either in early or late spring. The peak in autumn is rather abrupt and it is composed [1] of multivoltine species that exhibit their last generation in the year or [2] of univoltines that migrated from their usual feeding and breeding sites in the lowlands but can only feed here. In unproductive biotopes with entirely different vegetation cover – e.g. forests dominated by *Pinus halepensis* Miller - the hemipteran and lepidopteran fauna consists mainly of two independent parts, those associated with the overstory trees and those related to the herbaceous ground vegetation. In either case there seems to be a migration on a local scale, which intensifies the autumnal peak in Attica.

A practical example of this fact is experienced on Crete Isl, Greece where the dense populations of the pine processionary moth – a serious pine defoliator in Mediterranean ecosystems - cannot be managed with preparations of *Bacillus thuringiensis* –BT- (Foray ®) since its endotoxins can kill every butterfly. In this way, in September and early October the second peak in autumn will disappear since the butterflies in the lowlands and especially those that breed here and migrate locally to nearby uplands are killed (P. V. Petrakis *unpublished data* from various places in Attica, Greece). To prevent severe reductions in the butterfly biodiversity of Crete (Chania prefecture) conservationists decided to manage PPC (=Pine Processionary Caterpillar) with a much more sophisticated application scheme involving pheromones, killing agents and BT.

It has been found that in Mediterranean conditions multivoltinism prevails among insects. This is also true for butterflies and true bugs. Added to this is the high species richness of Mediterranean-climate areas (Deacon 1983 for all Mediterranean-type ecosystems; Bond 1983 for

South African fynbos; Balleto and Casale 1991 for European Mediterranean-type ecosystems). In this scenery there seems to exist a close encounter (=exploitative) competition among organisms that exploit plants. This competition is relaxed by co-evolutionary associations between plant and insects (Petrakis 1994) and a thin, though extensive, partitioning or resources (Schoener 1986; Denno et al. 1991). The partitioning of resources is achieved by many ways, the most important being those related to limiting similarity, manipulations of ecological niche axes (Giller 1984; Schoener 1986) and small scale migrations of insects (Shapiro 1975). These relaxing ways make Mediterranean-type ecosystems less competitive for insects, especially those having temporal – both seasonal and diel - dimensions of insect niches not less important than food or biotope ones. That is, Mediterranean insect species prefer to perform SDD in order to relax competition instead of performing long distance flight in order to find suitable habitats as is done by monarch butterflies. In short-summer areas – e.g. high altitudes and small latitudes – the butterflies are largely univoltine. The butterfly fauna is peaked in spring and there is no competition for resources (Shapiro 1975). In regions where the snowfall and the cool winter winds are unpredictable univoltinism helps in reducing competition. Alternatively, the existence of the same species in predictable areas promotes the evolution of a migrating behavior.

Segregation of life stages may be an important ecological factor. For instance, in many lepidopteran species, larvae and adults use different food plants. The larvae are concentrated around *host plants* and adults around *nectar plants*. The segregation of the two is the rule in eventually all insects. The migration of the adults is an important mechanism to find a link between the two kinds of plants. *Parnassius apollo* (L.) does exactly this type of SSM (Brommer and Fred 1999). As is commonly invoked in historical human migrations can be a response of insects to density-dependence. This means that a migratory behavior is exhibited in those communities where there is not enough space. Because this issue has many conservational implications Fahrig (2001) investigated the parameters influencing the probability of extinction of a local population of a species, not necessarily an insect. It was found that the most important parameters were the reproductive rate of the organism and its emigration rate from the habitat. That is the extinction threshold increased with increasing emigration rate. On the other hand, the same author proved that the survival rate of the dispersing individuals increases as the optimal emigration rate increases. Insects that inhabit areas with high habitat cover evolve high emigration rates since the associated mortality is low. When the habitat cover is removed –e.g. by fire or climate change- then these insects increase their emigration mortality because of the placement of many individuals in an inappropriate habitat (Fahrig 2001).

5.5 How insects migrate?

Several reviews have recently appeared that relate insect migration to meteorological conditions (Drake and Farrow 1988). Two of them are most comprehensive or detailed. Rainey's review (Rainey 1974) is the most detailed and relates meteorological phenomena to insect flight particulars. All associations are judged on an energy basis for the benefit of the insect. While this is true, energy does not always reveal the way the insects exploit meteorological phenomena and therefore, several other aspects must be examined. The review of Drake and Farrow (1988) is the most comprehensive one and takes into account all recent advances in the study of insect migration such as radars.

It is known that most meteorological phenomena in the atmosphere of the earth take place in the troposphere (the first 10–15 Km) and the planetary boundary layer (=PBL) (the first 1–2 Km of the troposphere) is very important to insects. PBL is considered to be the troposphere layer that is most affected by the earth's surface. It exists in two forms, the vertically mixed one (*convective*) and the stratified form (*stable*). Insects exploit the structure of stable PBL for nocturnal migrations, which can last from a few minutes to several hours. Sometimes the migration may continue the next morning and noon. Nevertheless, radar observations showed that insect migrations are very rarely observed at sunsets. It is important to note that not only nocturnal insects undertake migrations during the night but also some diurnal insects such as grasshoppers, locusts and planthoppers and insects performing ballooning such as aphids and small spiders.

Insects, such as moths, that take advantage of the low–level jet currents in stable PBL' s have also been observed by Drake (1985). In other occasions moths have been observed to exploit all the atmospheric stratification which is associated with temperature inversions in the PBL and perform SSM's almost without spending energy and involving gliding on air currents. Insects always use vertical movements of the air to attain a certain height in the PBL. Apart from energetic reasons it is not known if the passive transfer in air currents gives insects any other advantage. It seems that the primary reason of migration is the abandoning of the habitat. The new habitat at the end of the migration process may be better, equal or worse than the previous one. This does not seem to affect the migration process, which is triggered solely by cues in the source biotope. An example of this could be the long distance migration of *D. plexippus*, which migrates southwards to the Mexican hills and mountains (but see Garland and Davis 2001) from northern territories of the USA and southern Canada. In regulating the time of the trip to the biotopes of migration the insect enters a diapause period until late August and early September. The feeding with old *Asclepias curassavica* plants was necessary to enter diapause on new plants while the congener of the food plant *A. syriaca* had no effect on the diapause status of the butterfly (Goehring and Oberhauser 2002). This indicates that the timing of the migratory trip is performed through feeding.

Another example of migratory cues is that related to the SSM of *P. apollo* butterfly, which abandons already occupied patches for other biotopes. However, these new biotopes are constrained either by food plants for larvae or dense adult populations (Brommer and Fred 1999) or even by the scale of the study (Schneider 2003). The cue needed for migration seems to be an adaptation towards finding food plants for larvae, and the source biotope provides the necessary phenological information for this. Stefanescu (2001) working in Iberian peninsula on red admirals (*V. atalanta*) has found that the entire phenology of the insect is directed towards finding acceptable larval food plants. The insect is migratory and in autumn it leaves mid and northern European biotopes for the Mediterranean basin. At the same time the larval food plant *Urtica dioica* L. is declining. The spring generation emerging in the Mediterranean migrates to summer biotopes when the nettle is fully available. Similarly, SSM at higher altitudes adds another generation to the red admiral and increases the biodiversity of mountainous biotopes in the Mediterranean.

To all these established movements of insects the atmosphere plays the role of the medium in which the process takes place. It also conveys to the candidate migrant insect the environmental messages that trigger the flight.

The ability of insects to migrate is not only dependent on the flight medium but also on the physiology and the anatomy of insects. Thomas et al. (2001) investigated the results from the expansion of ranges in two butterfly species the silver-spotted skipper (*Hesperia comma*, Lepidoptera, Hesperiidae) and the brown argus (*Aricia agestis* Den and Schif, Lepidoptera, Lycaenidae) and two species of bush crickets namely *Conocephalus discolor* and *Metrioptera roeselii*, (Orthoptera, Gryllacridoidae). Thomas and co-workers state that a range change may require an evolutionary stasis, which means no evident evolutionary changes take place in the butterfly but much more research is needed to detect changes in flight behavior and wing morphology such as those reported in the unit *What is insect migration*. For crickets it is known that environmental conditions determine whether larvae will develop to brachypters or macropters. Thomas's team have not established that wing changes are genetically governed but it is equally unlikely to be the result of response plasticity or a variation to latitude. For *H. comma* the same authors have found that the range expansion was led by increased habitat availability. For the *A. agestis* experiments showed that the range expansion of the northern half of the distribution can be completely explained by ecological processes. However, after the initiation of the expansion the frequencies of those phenotypes able to exploit the new habitats or the phenotypes that shows fast rates of expansion increase.

5.6 Cost of migration

In general there two ways to discuss the cost of migration. The first, is the direct way as those quoted by Schneider (1962) who reports that for Diptera the power required to maintain the flight ranges from 11 erg/sec/mg for small species to 38 erg/sec/mg for larger species. The second, is purely biological and takes into account all the necessary traits associated with the change of habitat. Since no Mediterranean insects have been studied in the first way and in general the energetic load associated with migration is only one aspect of the cost the second way will be followed. In what follows it s meant that migration is an escape strategy of the insect and the specific migratory strategy is largely controlled by the habitat (e.g. for butterflies Dennis et al. 2004; for planthoppers Denno et al. 2001).

If migration is more costly than other solutions – e.g. egg cannibalism in rock–pool corixids (Pajunen and Pajunen 1991) – to the problem of overcrowding and/or the shortage of food and mates, then there is no reason for the insect to adopt this solution (Rankin and Burchsted 1992). The methodological problem arises in instances where migration – usually a SSM – of an insect takes place even if it is energetically costly (Harada 1998). Harada studied two skater species (Heteroptera, Gerridae) that inhabit temporary and permanent water bodies in Japan. The first species *Aquarius paludum* (Fabricius), when presented with a dried biotope, stayed on site, histolysed the secondary flight muscles, diapaused and lived longer. The second species *G. latiabdominis* Miyamoto, when presented with the same environmental conditions with *A. paludum*, migrated to new biotopes by retaining the flight capability (Harada 1998). It is important to note that the first species is dimorphic and both macropters and brachypters co–exist in the same population (Andersen 1993). Probably the strategy of not–to–fly is energetically more feasible than undertaking the challenge to migrate to new water habitats. The hypothesis first stated by Lack (1968) that migration needs already existing but extensively modified traits and not novelties, seems to hold true in these two insects (Alerstam et al. 2003). A slight modification of the enzymatic apparatus associated with the histolysis forces *A. paludum* to stay on the same site with many evolutionary opportunities for the development of other traits such as diapause.

In a recent co-authored treatment of the subject (Woiwod et al. 2001) various authors examined the biomechanics, physiology and behavior of the migrating insects. However, the work is not restricted to insect migration but includes many types of insect movements not necessarily by means of wings. With regard to the active insect movements it is stated that the success of the insects is mainly attributed to the evolution of wings while the loss of wings has evolved secondarily (Dudley 2001). Many other physiological traits also evolved to make flight possible. The metabolism of lipids in locusts has been modified so as to involve the amino acid proline, which is believed to be the fuel for the flight of the insect (Zera et al. 1994; Zera and Mole 1994).

Other modifications or evolution of new structures occurred in the organs associated with behavior. Sensitive receptors of chemicals, temperature and proprioceptors (Klein et al. 1988; D' Urso and Ippolito 1988) are usually well developed in Rhynchota (or Hemiptera for North American entomologists) to detect the right time for migration – light, temperature, humidity –, the appropriate vegetative substrate to land on and the odours associated with aggregations before migration and post alighting. Even in cases where specialised insect predators migrate to find a patchily distributed prey, the evolution of an effective sensory system is vital. Through these sensors the predators detect the potential damage incurred on plants by prey arthropods and decide to land or not. Such a situation was proved to be the case for two specialised insect predators. One predator is *Oligota kashmirica benefica* Naomi (Coleoptera, Staphylinidae) and the other is *Scolothrips takahashii* (Thysanoptera, Thripidae) and their prey *Tetranychus urticae* (Acari, Tetranychidae). Shimoda and Takabayashi (2001) were able to prove through experimentation that the two predators migrate towards patches having more prey spider mites. The location of such prey-rich patches is done through volatiles emanating from damaged tissues of the plant. Because this cue can be deceptive since it might be that the damage on the plant is still fresh when there are no longer prey insects on the plants, the predators usually combine these odours with volatiles associated with living spider mites. In some cases, it seems that migration plays a role of natural selection analogous to the plumage of bird feathers or the courtship of wolves and domesticated dogs. It is an estimator of the abilities of the partner and its health status. It has been shown (Altizer et al. 2000) that in the best known and the most well studied migration example, which is the monarch butterflies in North America, migrants of long distances have very low levels of parasitism by the protozoan *Ophryocystis elektroscirrha* whereas non-migratory populations showed the greatest levels of parasitism. It seems that the seasonal migration of *D. plexippus* together with the parasite exerts a selection pressure on the butterfly or it mitigates the parasitism levels of the host (Leather 2002). It is not known whether the parasite prevents migration in non-migratory populations or it cannot develop in long distance migrants. In the case of the large and small milkweed bugs *Oncopeltus fasciatus* and *Lygaeus kalmii* (Het., Lygaeidae) it has been found that the colonization of new habitats is done by means of the *oogenesis flight syndrome* (Rankin et al. 1986). This syndrome states that females migrate to new habitats and establish a new population at those places where they lay eggs. However, Aldrich et al. (1999) have shown that this migration is done by means of males that spread their pheromones in those places where they want to guide potential mates. This is helped by the pheromone production peculiarities of *O. fasciatus* (Staddon 1995) in which adult specific glands in males produce and store the necessary compounds, while in the predacious asopin *Podisus maculiventris* (Say) (Heteroptera, Pentatomidae) the production of pheromone per unit body weight is much higher since predators have an augmented search space (Aldrich et al. 1999). Also the smaller body size of males and presumably their faster developmental time ensures that males are the first insects to be born from a clutch of eggs. Since generalists are more likely to be scattered in the environment it is feasible to hypothesize that the generalist *L. kalmii* would use more pheromone than the specialist *O. fasciatus*. Pheromones seem to be an effective way to gather the scattered migrants in the new habitat.

An insect corroborating the proposed production scheme is the predacious soldier bug *P. maculiventris* (Aldrich 1995). The generalist predatory habits of this insect force it to produce and spread much more pheromone per unit weight than the lygaeids, in new habitats where it migrated. It is not known what exactly happens in other Asopinae pentatomids (predatory). In Mediterranean countries there are more than twelve species distributed in eight genera. Despite this quite high number of potential biological control agents, Mediterranean farmers prefer to use *P. maculiventris* probably because of the ease of finding it in the market and of the reluctance of local entomologists to study the control potential of other Asopinae.

In studies involving cost versus type of movement it was found that among the many ways of migration the passive one is not the lowest in cost since it involves the risk of landing at inappropriate habitats. Locomotion by leg movements is the most costly while energetic migration by using the wings is intermediate. Maybe this is the reason why migrations in insects are usually performed by winged forms.

Another parameter of high importance for insects living in seasonal environments is the duration of the migration. Adaptations in insects appear to regulate the duration of migration in order to synchronise the hatching of larvae with the phenological stage of the host plant as is the case of the summer Mediterranean red admiral butterfly *V. atalanta* (L.) (Stefanescu 2001). A term proposed by Schmit (2003) to describe in a broader sense this situation is *"climatic compensation"*. In this way the cost of migration may be distributed in two or more generations such as in monarchs (Urquhart and Urquhart 1978) and hawker dragonflies (*Anax junius* (Drury) [Odonata, Anisoptera, Aeshnidae] (Freeland et al. 2003).

In a migrating organism the body size has been found to have a direct relation to the migrating distance and velocity. Combining the energy that can be stored in the body of the organism and the maximum speed that can be attained during the migration, Hedenström and Alerstam (1997 1998) have found that at least in birds the maximum migration distance is constrained by the amount of food that can be stored in the body. The same authors explain a comparison of the migration distances of two European swan species. The small swan *Cygnus bewickii* migrates at longer distances than the large swan *C. olor*. In insects there seems to be no such relation. In an important paper Hanski et al. (2002) studied the migration rate of the Glanville fritillary butterfly (*Melitaea cinxia* (L.), Lepidoptera, Nymphalidae) in northern Europe (Åland islands and Paldiski, Estonia). In this work Glanville fritillaries from four regions were released in a natural environment. The influence of population and life history parameters on the migration rate of the butterflies was examined. While there were differences in body sizes they were unable to find any significant relation between body size and migration rate in both between -and within- regions comparisons. Instead, they found that migration rate is influenced by the host plant rather than by body size. In all influences male fritillaries were absent. Only females seem to be affected by the parameters examined.

The fact that only females have to change or evolve a novel trait in order to cope with migration events and new habitats is also supported by the study of the stink bug *Nezara viridula* (L.) (Rhy., Heteroptera, Pentatomidae) (Musolin and Numata 2003). The migration of the stinkbug from southern localities to the new habitats in north Japan (Osaka) is not without cost to the insect. The insect produced an autumnal generation the individuals of which are destined to die in the coming winter because of the northern habitat. The authors predict that the insect will probably evolve an elongation of the photoperiod that induces diapause. In this way, the insect will diapause earlier and is expected to have the last generation earlier in the year in order to avoid the winter loss of progeny. This concerns females while the cost incurred to males is the one associated with the defence metathoracic glands and the recruitment of adults in newly discovered places (Aldrich et al. 1999). The Mediterranean area is characterized by a marked seasonality and, at least in the north, by an intensely sculptured and patchy terrain. Overcrowding, shortage of food and adverse climatic conditions, especially during summer, are significant factors that shape the behavior of insects. Long distance migrations are not common in this area as their energetic cost can be quite heavy. The insects have to spend a lot of time moving up and down in different altitudes and the patchiness of the environment does not ensure that a suitable site will be found for settlement.

5.7 Evolutionary traits associated with insect migrations

A trait related to migration and dispersal must be sufficiently variable and inheritable in order to serve as an evolutionary trait. Also the fixing of this trait can be done only from the way that the advantages that may confer to the insect outweigh the disadvantages such as migratory predation and failure to locate a mate or new resources (Bilton et al. 2001).

Apart from the obvious benefits that migration may have to the insects as a habitat selecting strategy, migration may be an adaptation to fragmented or small populations, a situation that is very frequent in Mediterranean-type ecosystems. The effect of mobility of insects from declining small populations has been reported for British populations of swallowtails *Papilio machaon* L. (Lepidoptera, Papilionidae) and large blue butterflies *Maculinea arion* (L.). Dempster (1991) found that there is a significant difference between morphometric measurements (ratio of thoracic width to length) in resident and migrant populations of the butterflies. He showed that butterflies with low ratio of thoracic width to thoracic length are weaker fliers. This was observed in the Wicken Fen population of *P. machaon*, which declined until it became extinct probably at 1950. This lack of migration in British butterflies seems to be the rule since their majority are weak fliers and furthermore they live in small closed populations (Thomas 1984). This fact explains well why British butterflies are declining and need specifically designed reserves for their conservation so that the reserves are within easy reach of one another (Dempster 1991). However, as Dempster stated these results are only tentative since there is no way to know the connection between morphometric changes and the behavior of the insects.

Warren and Key (1991) recognize that insect ranges are dynamic, especially in the post-glacial period, and very vulnerable to climatic changes in extreme habitats at the edges of their range. For forest insect species it is conceived that any fragmentation or isolation of their ranges is very likely to cause their extinction. Migration is profoundly absent from many forest insects and this could be adaptive since their habitat used to be extensive and continuous. The fragmentation of present woodlands possibly poses a risk to these insects since they cannot migrate to find forest habitats and more importantly they are unable to take advantage of favorable climatic conditions. Bark beetles (Coleoptera, Scolytidae) may be an exception to this since although they are heavy, they actively fly to find vulnerable trees for building their galleries where they will attract females for copulation and egg laying.

An adaptive role can also be found in the short scale migratory behavior presented by a Mediterranean population of *Lasiommata megera (L.)* (Lep., Satyridae) (Dennis 1987). Dennis observed on Corfu town on the island of Kerkyra, Greece that the butterfly exhibited hill-topping behavior, which is quite common in Mediterranean. He interpreted this as a mate finding behavior. This mechanism has been already examined by Scott (1968) and was found to be a common in low density species as is usually the case of Mediterranean taxa.

Migration may be an adaptation of species to surpass bottlenecks due to the existence of small populations. Menéndez et al. (2002) elaborated on the issue that migration increases the persistence of small populations. They thought that the phenomenon is a manifestation of the *rescue effect* (Brown and Kodric–Brown 1977) and believe that migration prevents small populations from becoming extinct. While existing theory predicts that migration should take place in large and dense populations, a decrease in the number of resident insects, of the insect *Zygaena filipendulae* (L.) (Lepidoptera, Zygaenidae) was found.

In all studies of this type the problem is how to define and measure connectivity among different patches. Using Hanski's (1999) formula:

$$S_i = \sum e^{(-\alpha d_{ij})} N_j \qquad (5.1)$$

where d_{ij} := is the distance between the patches i and j ; N_j := is the estimated population size in the patch j; a := an exponent representing the extent to which immigration probability declines with distance and it is empirically estimated (the selected value was 3 from Menéndez et al. (2002) to reflect the similarity of dispersal distances of *Z. filipendulae* and the already known values of *Plebejus argus* (L.) (Lepidoptera, Lycaenidae)), and regression analysis of the area of the patch, connectivity, population size, population density and host plant density on the number of residents, the number of emigrated and the number of immigrated individuals in each patch, they found that insects were attracted to habitats of high quality that have potentially high population densities. The behavioral sequence of mate finding and the avoidance of inbreeding were observed in these insects, as this behavior is widespread in all animals. It was also found that patches not having the food plant showed an increased emigration rate. Expectedly, the migration rates for *Z. filipendulae* were shown to be higher for small- and low-density sites. It must be noted that this insect is aposematic in all life stages. All these indicate that migration is of adaptive significance to many insects as happens to many other animal groups such as birds. Menéndez et al. (2002) agree that Allee effects may be important as rescue effects for the spatial dynamics of insects.

In the two very important rice pests (Wilson and Claridge 1991) *Nilaparvata lugens* (Stål) and *Sogatella furcifera* (Horváth) (Auchenorrhyncha, Delphacidae) laboratory experiments performed by Matsumura and Suzuki (2003) showed that the emergence of macropters was in habitats either crowded by mainly intraspecific insects or in plants already bearing feeding spots by other sucking insects (Fig. 2). The authors stated that the form suitable for migration – e.g. the macropters – was produced in crowded conditions or when many intraspecific insects coexisted on food plants. The underlying mechanism for the production of macropters is based on the alteration of the food plant chemistry and is certainly an adaptive mechanism in order to avoid crowding.

In the past, several authors have searched for a mechanism that underpins the production of macropters. Raatikainen (1967) who studied *Javesella pelucida* F. (Auchenorrhyncha, Delphacidae) stated that whatever the mechanism, it must not be sophisticated. Vepsäläinen (1974) and Järvinen and Vepsäläinen (1976) proposed that Gerridae evolved a developmental switch, which is affected by several environmental cues and in unstable habitats it triggers the production of macropters thus promoting migration. Denno and Grissell (1979) in a detailed study of *Prokelisia marginata* (Auchenorrhyncha, Delphacidae) proposed that the proximate mechanism for the production of macropters is the same developmental switch as the one proposed by Järvinen and Vepsäläinen. This switch permits the wing dimorphism to be present even though it is not locally the optimal strategy (Bilton et al. 2001).

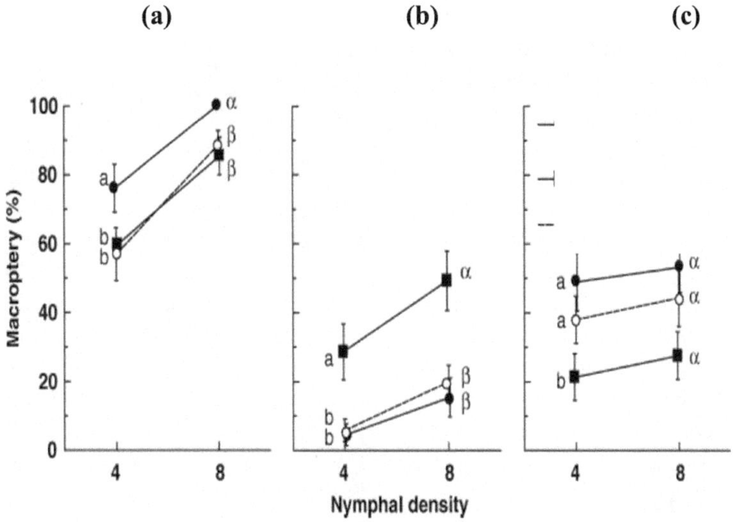

Fig. 2. Diagram showing the percentage macroptery in the two rice pests *Sogatella furcifera* and *Nilaparvata lugens* at two nymphal densities (4 and 8). The filled symbols denote the existence of competition in the form of a previous feeding, while the open symbols the lack of it. The circles denote intraspecific and the squares interspecific competition. (a) *S. furcifera* females. (b) *N. lugens* females and (c) *N. lugens* males

As it happens in many biological systems, insect migration may be a sophisticated behavior triggered by non-sophisticated mechanisms. The warming of the climate plays a part in these mechanisms but we do not know the details. Nevertheless, the expansion and contraction of the geographic ranges of various insect species or the observation of migrating population could give us an estimate of their impact in the ecology of insects.

5.8 How insects face global climate change?

It is well documented in many different ways that the mean temperature of the planet has increased (Barber et al. 2000) and will continue to increase the next 50–100 years (Parmesan et al. 1999). With this warming insects are expected to migrate to more suitable habitats in order to compensate for climate change.

The impact of climate change on insects fall in two broad categories. First, some insects shift their ranges as a compensatory response to global warming (Parmesan 1996; Parmesan et al. 1999; Schmitt 2003). Second, changes in the biology of particular species is a compensatory response to climate change, mainly global warming. The natural history collections (NHC) in museums and private collections provide the necessary data to compare present-day data with historical records (Suarez and Tsutsui 2004; Graham et al. 2004). The efficient use of NHC data is expected to provide the necessary information background for conservation planning in the face of a changing global climate (Cabeza 2003; Araujo et al. 2004). For a variety of historical reasons, NHC usually have no organised and comprehensive data on insects inhabiting Mediterranean areas and it is expected that any improvement in museum informatics will leave unaffected our knowledge on insect migrations in these areas. However, as Graham et al. (2004) emphasized NHC data can be of a great value in studies where the physiology and the natural history of a species is connected to statistical modelling. A purely mechanistic understanding of the involved processes would help the extrapolation of these findings to novel situations such as climate change.

With regard to the biology change of insects many plant phenological parameters have been significantly shifted while many herbivorous insect populations seem to have become extinct as a result of these shifts that cause asynchrony (Dixon 2003). These phenological shifts were found not only as a result of global warming but also as an effect of greenhouse gases though a separation of the two is very difficult.

As a result of the shifts in plant phenologies many specialist herbivorous insects of specific plant tissues or organs such as pollinators are expected to become extinct or be significantly reduced in population densities (Bond 1995; Thomas et al. 2004). Insects that closely follow their host plant phenological phases are also expected to follow the route of the plant towards higher altitudes or latitudes, though many authors believe that it is the temperature rise that causes this shift in the northern boundary of their geographical range (Parmesan et al. 1999). This type of shifts of the northern boundary of the geographic range of insect species has been extensively observed in Mediterranean insects (Schmitt 2003). It is believed that the movement of the species boundaries is not restricted to peripheral populations. Reemer et al. (2003) showed for Dutch hoverflies (Diptera, Syrphidae) that the migratory behavior may be affected by the availability of suitable substrates that is recently killed trees. Since herbivorous insects respond fast – i.e. in a few generations – to the changes of their host plant (Claridge and Gillham 1992) the result could be a shaking up of insect–plant associations and a change in the specific composition of plant communities or the insect assemblages. In a few cases such a shaking of species has been observed in non-migratory species (butterflies in Parmesan et al. 1999; many vertebrates and insects in Jensen 2004) while migratory species are usually responding to climate changes by shifts in their migratory movements and several other aspects of their behavior such as gathering and shifting of their destination habitats. In general, the flexibility of insect responses and the intrinsic variety of their behavioral sequences make it difficult to separate the responses to those which are due to regular long term changes and those which are due to yearly weather or greenhouse emission variations. This variation is also difficult in every kind of biological system and not only the insects and their habitats (Jensen 2004).

Global warming is expected to adversely affect insects for a number of other reasons apart from food availability. The destruction of sensitive habitats such as water, the lowering of the quality of habitats, the loss of habitats as a consequence to the rise of the sea level will force the insects to disperse to new areas or cause their extinction. In most cases, this dispersion cannot be classified as migration as was stated in the beginning of this chapter. However, insects predisposed for migratory behavior will be at an advantage. In the Mediterranean area, one can expect that insects performing small scale migrations to restricted favorable areas such as the case of the Jersey tiger moth in Greek islands, will be unable to find new suitable habitats further north.

Apart from herbivores the majority of insects is affected through changes in the density or availability of their resources. Aquatic insects are an example of insects that are expected to be adversely affected by global warming. The destruction of many water habitats is one such way since increased evaporation is expected to empty several water bodies. Aukema (2003) refers to the case of *Aquarius najas* (De Geer) (Heteroptera, Gerridae). The insect was once very common in the Netherlands but now has disappeared from many parts most probably as a result of the destruction of its water habitats and the pollution received by the remaining ones. Aukema has observed that the insect is incapable of migration to compensate for habitat loss, since macropters are very rare in Dutch populations. However, the most important change that can be expected from aquatic and semiaquatic insects is the lowering of the quality of the affected water bodies, since the climatic and weather fluctuations is the cause of an increase in insect population densities of some species (Epstein 2000) as a result of the migratory behavior of insects that inhabit diminished habitats. Epstein anticipates that insect–borne disorders will become widespread. Several types of disorders that are believed to have disappeared are now appearing in a more aggressive mode (Suarez and Tsutsui 2004). For instance, a locally transmitted malaria epidemic was in outbreak in 1990 (the hottest decade of the century) in several North American states such as Texas, Florida, Georgia, Michigan, New Jersey and New York. Germane to this is the climbing up to the mountains of a previously lowland inhabitant mosquito *Aedes aegyptii* (Diptera, Culicidae). Malaria and dengue fever (a lethal virus disease) are now spreading in areas from where they disappeared. For both disorders there is no vaccine as of yet.

Since more aquatic biotopes are in lowlands, it is conceived that aquatic and lowland insect species are not threatened by extinction from long term global warming because they have room to expand at midlands and highlands. Nevertheless temperature rise and greenhouse gases cause deterioration of insect habitats and are expected to cause a decline in population densities of many insect species. Foster (1991) reports the insect *Emus hirtus* (L.) (Coleoptera, Staphylinidae) as an example of such a decline, which is not mentioned in the British Red Data Book. It is anticipated that all thermophilous insects are likely to expand their geographic range as a result of global warming. Petrakis (1991) has found in a coastal east Mediterranean maquis formation several Heteroptera species of eremic origin not previously reported – and presumably not present - from Greece (Drosopoulos, S. and Josifov, M. *pers. comm.* and Josifov 1986).

Apart from the indirect effects of global warming such as the loss of habitat, the impact of certain greenhouse gases such as CO_2 and ozone has been recently explored by several scientists (Whittaker 1999; Hamilton et al. 2005). It was found that the lowering of the nutritional quality of the plants, in the sense that herbivorous insects have to eat more to meet their nutritional needs, increased the susceptibility of soybean plants to herbivorous insects, while the ozone had no effect. Hamilton et al. (2005) found that the elevated CO_2 increased the total sugars in the leaves and acted as a phagostimulant to certain invasive insects such as the Japanese beetle *Popilia japonica* Newman (Coleoptera: Scarabaeidae). If this applies in a wide variety of plants and circumstances, whatever the increase in the agricultural production as a result of elevated CO_2, the herbivory will also increase and the relative proportion of the two has to be re–evaluated.

The effect that global warming may have on aquatic and semi aquatic insects makes them good bioindicators for the biological consequences of it. The sea skaters *Halobates* spp. (Heteroptera, Gerridae) are among the insects of which the migratory behavior and the similarities of distant populations have been studied in detail (Ikawa et al. 1998; Cheng 1989; Petrakis et al. 2003). Cheng has found that modern populations of *Halobates* spp. are occupying a band of 10 degrees around the equator. Their mobility and the unidirectional migration they usually perform (Ikawa et al. 1998) made the mixing of distant populations a serious obstacle in biogeographic studies (Andersen et al. 2000) despite the rich weaponry of molecular markers that are successful in some cases (for oceanic species of *Halobates*; see Petrakis et al. 2003).

The recent advances in the detection of plant responses to global warming and greenhouse gases gave the impression that a range expansion of a certain plant automatically causes the expansion of herbivorous insects that are specialising on its organs. An increase of the performance of chewers and in effect their population densities a elevated CO_2 may result from the increased palatability of the plant substrate (e.g. the common blue butterfly *Polyommatus icarus* (Rott.), Lep., Lycaenidae). This insect was found to lack adaptations to local plant communities in Sweden and this has been interpreted among others as a possibility of migration among sites (Bergstrom et al. 2004) unlike other lycaenids which are weak flyers. However, Janz et al. (2005) showed that possibly due to neural limitations the common blue prefers to oviposit on plants that promise sufficient nectar sources even if they do not give the best performance to their offspring. In this way the need to migrate in order to find nectar and larval sources is minimised.

At present, although insects do not have a clear and detectable response to global warming and greenhouse gases, CO_2 plants were found to possess traits that have been already used in higher order classifications (above genus) and are used as indices for environmental changes. Stomatal density is such an index. It has been used for the study of climatic effects on plants and prediction of future responses. In this context *Quercus* has been found to have wide fluctuations in stomatal densities during the last 10 million years (van der Burgh et al. 1993). Stomatal characters have been also studied in detail in the needles of a Miocene conifer formation in Idaho, USA (Huggins 1985) but the correlation between CO_2 concentration and the number of stomata in mm^2 is not strong (Fig. 3; Beerling and Woodward 1996) though it has been speculated that CO_2 may affect the activity of RuBP carboxylase, which is the most important carbon fixing enzyme in plants.

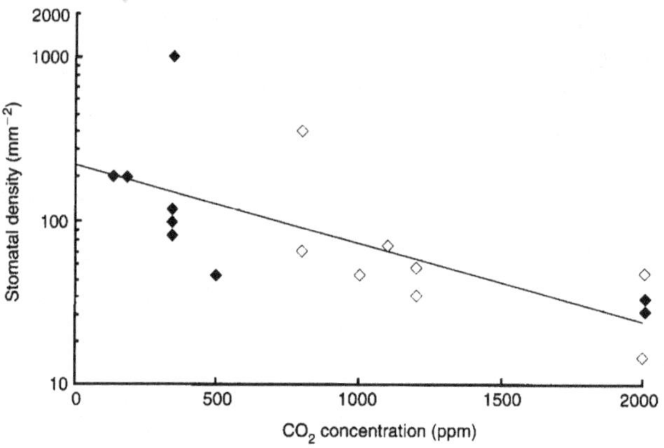

Fig. 3. Diagram showing the density of stomata in relation to the concentration of carbon dioxide in the atmosphere. Extant plants are open symbols and fossil taxa are filled symbols (after Beerling and Woodward, 1996)

On the other hand the rate of photosynthesis [A ($\mu mol^{-1} \cdot m^{-2} \cdot s^{-1}$)] is not severely affected by light [iradiance ≈ photon flux ($\mu mol^{-1} \cdot m^{-2} \cdot s^{-1}$)]. In the same work of Beerling and Woodward (1996) the authors state that apart from the pteridophytes, which do not increase their photosynthesis with irradiance, the influence of light is increased in plant groups where there are no specialist insects, such as grasses (Fig. 4). Therefore Beerling and Woodward state that it is not expected that past insect responses to environmental changes can be detected in the fossil plant record. However, there are cases in which the responses of insects to climate changes were accurately detected sometimes permitting the construction of temperature curves (Coope 1994). There is a complication added to the subject because the emergence of migration behavior in an insect does not leave direct traces even in extant taxa. A source of another complication came from the work of Arujo et al. (2004) who searched the selection methods of reserves in a climate change context. The assessment was done on 1200 plant species with European distributions and 6-11% of the modelled species are destined to become extinct in the next 50 years. In addition Dormann et al. (2004) performed phytotron experiments investigating the plant-plant interactions in the High Arctic for the plants *Salix polaris* and *Luzula confusa* and found that in warmer climates such as the CO2 induced global warming, the willow is able to exploit more efficiently the nutrients increasing thus its density in the field.

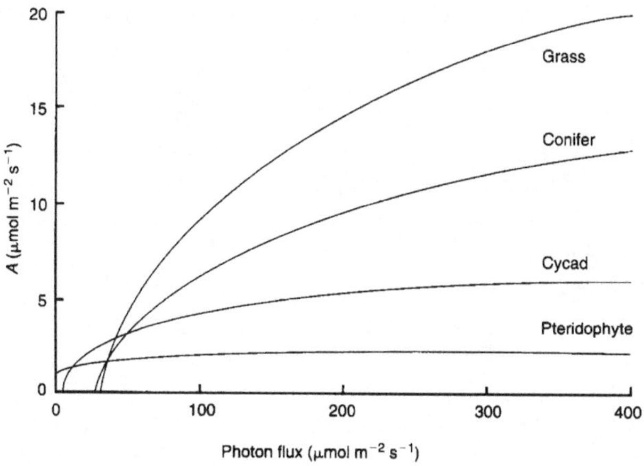

Fig. 4. Diagram showing the relationship between the photosynthetic rate and the photon flux (irradiance)

Both works have showed that the plant configuration of the habitats changes drastically with any global change and for this it is expected that herbivorous insects, especially the specialists, that have not the ability to disperse to more favorable habitats are destined to extinct.

5.9 Two example cases of insect migration

5.9.1 A case of insect migration in northern Mediterranean and central Europe

The cosmopolitan species *Vanessa cardui (L.) (*Lepidoptera, Nymphalidae*)* painted lady is an element of the Palaeotropical fauna in the Nearctic that migrated into Europe before the last glaciation (Schmitt 2003). In contemporary times it exhibits a large scale migration from northern Mediterranean and Southern Europe to the Middle and Northern Europe possibly as a demonstration of climatic compensation. The insect searches for habitats to which it is better adapted (Schmitt and Hewitt 2004). This is a general impression for the migratory behavior of the butterfly. In the northern territories where it arrives in spring or early summer it produces one or two generations and then by late summer or early autumn it returns to the south to hibernate. Some adults overwinter in northern Europe and it was thought that they were just temporary migrants (Scott 1992). The same, in general, migratory behavior is exhibited in the Nearctic populations and some adults have been observed to overwinter in Colorado. Scott believes that *V. cardui* shows the same seasonality in migration as *D. plexippus*. Locally, it can reach epidemic densities as a result of early fall migration to the European south. The responsible fuels for such a journey is provided by a neuro–11–peptide the adipokinetic hormone (Vanca AKH). Kollisch et al. (2003) have shown that this neuropeptide is produced by retrocerebral complex of *V. cardui* as in several other butterflies (e.g. *D. plexippus*). This hormone is responsible for the mobilization of the flight substrates in a variety of insects such as crickets, grasshoppers, locusts and butterflies where it mobilizes lipids. In cockroaches, flies and bees it mobilizes sugars, while in some coleopterans it mobilizes proline (Kollisch et al. 2003). The spring migratory flight of V. *cardui* is so massive and slower than the return in fall that allowed Scott (1992) to test the hypothesis of Baker' *s constant angle.* This hypothesis states that migrants keep a constant angle to the sun and for this they change direction when they fly in the day. Scott, in agreement with all the researchers of the migratory behavior of the insects, showed that the hypothesis is not true in the case of *V. cardui*, which keeps a constant direction during flight by means of neurological mechanism.

5.9.2 A case of insect migration in eastern Mediterranean

The Jersey tiger moth *Euplagia quadripunctaria* (Poda) (Lepidoptera, Arctiidae), is an example of a second order seasonal migrant, that is, species that at a specific time during the year leave their breeding ground and travel to areas in which they can survive i.e. in which they can over-winter or over-summer. After the end of the resting phase, the same individuals migrate back to the breeding areas to breed, i.e. breeding occurs in the original areas from which they come (Eitschberger et al. 1991). In the Mediterranean, the moth lives mainly in evergreen riparian forests and inside maquis valleys. Mountains or hills with steep sides border these valleys. At the bottom of the valleys run streams with water during the whole of the year. The valleys are fully covered with trees and shrubs. The vegetation together with the steep slopes, does not allow the sun to pass through during most of the day. All habitats are characterized by a lower temperature and higher humidity compared to the conditions outside the valleys. During summer, the insects migrate to the most humid and cool valleys of the area where they aestivate without feeding. Migration takes place during the night when temperatures are lowest and humidity is highest. The longest distance that has been recorded is 26 km (Elger 1969). All moths participate in the migrations, but each one flies as a single individual; there is no group flight. It seems that the direction of the migration is along temperature gradients. Experiments with marked individuals and antennal amputations showed that migration is mainly influenced by climatic factors. Thus the moths search for the cooler places (Lenau-Jürgens 1971). At these valleys, the moths are found at high densities (usually 50-100 ind/m^2 and reaching even up to 500-1000 ind/m^2 – Tauber & Tauber 1969) on the trunks, branches and leaves of trees and on humid rocks, migrating during the night towards higher positions in the canopy. During September they mate and the females disperse to their original valleys in order to oviposit. Oviposition occurs probably from late September until well into November at various suitable sites. After oviposition, the adults die. The new adults emerge around May and migrate until June to the most humid and cool sites of the region (Walker 1966). The insect is a typical example of a Mediterranean migrating insect and the study of its biology could be useful in the broader context of Mediterranean insect conservation (Balleto and Casale 1991).

5.10 Concluding remarks

It is well known to entomologists that the geographic distributions of many insect species are constrained by climate (Hill et al. 2001) by vicariance and dispersion-migration processes (Andrewartha and Birch 1954; Darlington 1957) and in several cases by extinctions that are either the initial or the result process (Thomas et al. 2004. The existing ample evidence of a modern global warming of the climate is also known to have reshuffled the boundaries of the ranges of many animals and plants worldwide. In this scenery, insect migration is directly connected to conservation purposes. However, it has been stated for plants (Bennett et al. 1991) that contemporary biogeographic patterns are the product of three processes, i.e. divergence, extinction and migration and throughout the climate oscillations of the Quaternary migration played a minor role in the maintenance of the biodiversity. It seams that insects do not conform to this prediction. There many reasons for this the most important being that of the speed of migrations and the associated maintenance of the heterogeneity, even in walking forms (Peterson et al. 2001). Peterson et al. found that the flightless salt-marsh planthopper *Tumidagena minuta* Mc

Dermott (Auchenorrhyncha, Delphacidae). These authors revealed a contradiction between fligtlessness and genetic structure of the insect. They studied allozyme variation in this insect and by employing G_{ST} (Nei 1973) as a measure of the genetic subdivision of populations. Indeed, the gene flow among *T. minuta* populations was very high reaching and sometimes surpassing that of highly mobile species (Fig. 5).

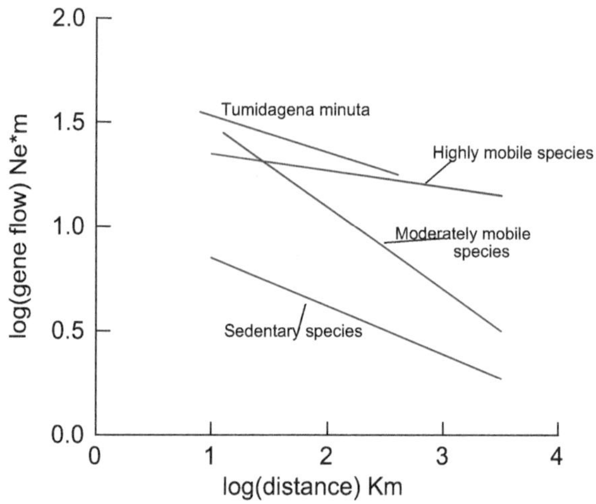

Fig. 5. Comparisons between *T. minuta* and other herbivorous insects. In this planthopper >99% of the individuals are flightless. However, the gene flow is similar to that of highly mobile insect species. Isolation by distance in this species is achieved at distances >20 Km, which are very similar to that of very mobile insects. This distance is very different from that corresponding to moderately mobile insects or sedentary ones in which isolation of populations (i.e. practically nil gene flow) is achieved 1 Km and 20 Km (after Peterson et al. 2001)

Insect populations respond in many ways to climate change the most important being range expansion, provided that there are suitable habitats to colonise (for butterflies, Hill et al. 2001). In the context of range expansion a lot of research has been devoted to the expansion of the polarward margin (e.g. Parmesan et al. 1999). Only a few studies have been devoted to the movement of the rear edge (Hampe and Petit 2005). It is believed that the populations of a certain species at the rear edge (low–latitude side) are much more important since they have peculiar ecological features and different conservation requirements from the species. Several conservation practices are directed towards the populations of the front margin or other parts of the range of a species are not equally productive for the rear edge populations. The rear edge of a species range has to be taken into account when reserves are designed while it is not certain that in designing reserve networks the inclusion of many parameters such as habitat quality, habitat heterogeneity or distance–dependent connectivity measures, will improve the selection of the reserves. In addition, in northern areas were lies the polarward limit of many butterfly ranges such as Britain, habitat degradation in many cases has obscured the northern limit expansion effect since habitats continuously disappear and deteriorate (Warren et al. 2001).

Migration and dispersal incur some cost to the insects, either directly or through population parameters (e.g. migration related mortality) insects are commonly performing it for population dilution, new resource finding, transportation to conventional encounter sites or in the recently as a compensatory response to global climate change. This may have many implications to pest management, disease spreading, reserve planning or biodiversity conservation. Only continuous research and the maintenance of extensive natural history collections are mechanisms expected to find answers and solutions to the newly emerged problems arising from global climate change.

5.11 Acknowledgements

Many persons contributed to this review. The discussions with Dr Vassilios Roussis are greatly acknowledged. Dr M. Gaetlich is thanked for sharing his observations on some migrant butterflies in the east Mediterranean. This work was greatly improved by the suggestions of Dr Joaquin Hortal. His critical reading of an early draft contributed substantially to the content and the structure of the manuscript. An anonymous reviewer is also thanked for suggestions and remarks. Mrs Athena Voulgaropoulou and Mrs Danae Panayotopoulou contributed to literature searching and collecting. This work was partially supported by the Greek secretariat of research and technology (grants to PVP; GSRT–KOR 017; SOUNIO–LAVRIO DEAL).

References

Aldrich JR (1995) Chemical communication in true bugs and exploitation by parasitoids and commensals In Carde R, Bell W (eds) Chemical Ecology of Insects, II. Chapman and Hall, New York, USA 318-363

Aldrich JR, Oliver JE, Taghizadeh T, Ferreira JTB, Liewehr D (1999) Pheromones and colonization: reassessment of the milkweed bug migration model (Heteroptera: Lygaeidae: Lygaeinae). Chemoecology 9: 63-71

Alerstam T, Hedenstrom A, Akesson S (2003) Long-distance migration: evolution and determinants. Oikos 103: 247-260

Altizer SM, Oberhauser KS, Broewer LP (2000) Associations between host migration and the prevalence of a protozoan parasite in natural populations of adult monarch butterflies. Ecological Entomology 25: 125-139

Andersen NM (1993) The evolution of wing polymorphism in water striders (Gerridae): a phylogenetic approach. Oikos 67: 433-443

Andersen NM, Cheng L, Damgaard J, Sperling FAH (2000) Mitochondrial DNA sequence and phylogeography of oceanic insects (Hemiptera: Gerridae: Halobates spp.). Marine Biology 136: 421-430

Andrewartha HG, Birch LC (1954) The distribution and abundance of animals. University of Chicago Press, Chicago

Araujo MB, Cabeza M, Thuiller W, Hannah L, Williams PH (2004) Would climate change drive species out of reserves? An assessment of existing reserve–selection methods. Global Change Biology 10: 1618–26

Aukema B (2003) Recent changes in the Dutch Heteroptera fauna (Insecta: Hemiptera). In Reemer M, van Helsdingen PJ, Kleukers RMJC (eds) Changes in ranges: invertebrates on the move. Proceedings of the 13th Inernational Colloquium of the European Invertebrate Survey, Leiden, 2-5 September 2001, Leiden: EIS-Nederland 39-52

Baker R (1981) The mystery of migration. Viking Press, New York

Balletto E, Casale A (1991) Mediterranean Insect Conservation. In Collins NM, Thomas JA (eds) Conservation of Insects. Academic press, San Diego, California 121 – 140

Barber VA, Juday GP, Finney BP (2000) Reduced growth of Alaskan white spruce in the twentieth century from temperature–induced drought stress. Nature 405: 668

Beerling DJ, Woodward FI (1996) Palaeo-ecophysiological perspectives on plant responses to global changes. Trends in Ecology and Evolution 11:20-23

Bennett KD, Tzedakis PC, Wilis KJ (1991) Quaternary refugia for north European trees. Journal of Biogeography 18: 103–115

Bergstrom A, Nylin S, Nygren GH (2004) Conservative resource utilization in the common blue butterfly – evidence for low costs of accepting absent host plants? 107: 345–351

Bilton DT, Freeland JR, Okamura B (2001)Dispersal in freshwater invertebrates. Annual Review of Ecology and Systematics 32: 159–181

Bond WJ (1995) Effects of global change on plant-animal synchrony: implications for pollination and seed dispersal in Mediterranean habitats. Ecological Studies 117: 181-202.

Brewer GJ (1970) An Introduction to Isozyme Techniques. Academic Press, New York

Brommer JE, Fred MS (1999) Movement of the Apollo butterfly *Parnassius apollo* related to host plant and nectar plant patches. Ecological Entomology 24: 125-131

Brown JH and Kodric–Brown A (1977) Turnover rates in insular biogeography: effect of immigration on extinction. Ecology 58:445–449

Brower LP (1995) Understanding and misunderstanding the migration of the monarch butterfly (Nymphalidae) in north America: 1857–1995. J Lepid Soc 49:304–385

Burgh van der J, Visscher H, Dilcher DL, Kurschner WM (1993) Paleoatmospheric signatures in Neogene fossil leaves. Science 260: 1788-90

Coope GR (1994) The response of insect faunas to glacial–interglacial climatic fluctuations. Philosophical Transaction of the Royal Society of London B 344: 19–26

Chapman JW, Reynolds DR, Smith AD (2003) Vertical - looking radar: A new tool for monitoring high-altitude insect migration. Bioscience 53:503-511

Chapman JW, Reynolds DR, Smith AD, Riley JR, Pedgley DE, Woiwod IP (2002) High-altitude migration of the diamondback moth, *Plutella xylostella*, to the UK: a study using radar, aerial netting and ground trapping. Ecological Entomology 27: 641-650

Chefaui RM, Hortal J, Lobo JM (2005) Potential distribution modelling, niche characterization and conservation status assessment using GIS tools: a case study of Iberial Copris species. Biological Conservation 122: 327-338

Cheng L (1989) Factors limiting the distribution of *Halobates* species. In Ryland JS, Tyler PA (eds) Reproduction, Genetics and Distributions of Marine Organisms. 23rd European Marine Biology Symposium, Olsen & Olsen, Fredensborg, Denmark 357-361

Claridge MF, Gillham MC (1992) In Sorensen JT, Foottit R (eds) Ordination in the Study of Morphology, Evolution and Systematics of Insects. Elsevier, Amsterdam, The Netherlands 241–260

Cabeza M (2000) Habitat loss and connectivity of reserve networks in probability approaches to reserve design. Ecology Letters 6: 665–672

Cooter RJ, Winder D, Chancellor TCB (2000) Tethered flight activity of *Nephotettix virescens* (Hemiptera: Cicadellidae) in the Philippines. Bulletin of Entomological Research 90: 49-55

D' Urso V, Ippolito S (1988). Proprioceptor organs in *Cicadella viridis* (L.) (Homoptera : Cicadellidae). Animalia 17: 381-391

Darlington PJ (1957) Zoogeography: the geographical distribution of animals. Wiley, New York

Darlow HM (1951) Observations on migrant Lepidoptera in the Mediterranean. Entomologist 84: 246–254

Deacon HJ (1983) The comparative evolution of Mediterranean-type ecosystems: A southern perspective In Kruger FJ, Mitchell DT, Jarvis JUM (eds) Mediterranean-type Ecosystems: The Role of Nutrients. Springer, Berlin, Germany 3-41

Dempster JP (1991) Fragmentation, isolation and mobility of insect populations In Collins NM, Thomas JA (eds) Conservation of Insects. Academic press, San Diego, California 143-153

Dennis RLH (1987) Hilltopping as a mate location strategy in a Mediterranean population of *Lasiommata megera* (L.) (Lepidoptea, Satyridae). Nota Lepidopterologica 10:65-70

Dennis RLH, Hodgson JG, Grenyer R, Shreeve TG, Roy DB (2004) Host plants and butterfly biology. Do host - plant strategies drive butterfly status? Ecological Entomology 29: 12-26

Denno RF, Gratton C, Langellotto GA (2001) Significance of habitat persistence and dimensionality in the evolution of insect migration strategies. In Woiwod I, Thomas C, Reynolds D (eds) Insect Moement: Mechanisms and Consequences. CABI, London, UK. 235–260

Denno RF, Grissel EE (1979) The adaptiveness of wing–dimorphism in the salt marsh–inhabiting planthopper, *Prokelisia marginata* (Homoptera: Delphacidae). Ecology 60: 221–236

Denno RF, Roderick GK, Olmstead KL, Dobel HG (1991) Density-related migration in planthoppers (Homoptera: Delphacidae): the role of habitat persistence. The American Naturalist 138:1513-41

Di Castri F (1991) The biogeography of mediterranean animal invasions. In Groves RH, Di Catsri F (eds) Biogeogaphy of Mediterranean invasions. Cambridge University Press, Cambridge, UK 439–452

Dingle H. (1978) Evolution of insect migration and diapause. Springer-Verlag, New York

Dingle H (1985) Migration. In Kerkut GA, Gilbert LI (eds) Comprehensive Insect Physiology, Biochemistry and Pharmacology. Pergamon, Oxford, UK 375-415

Dingle H (1996) Migration: the biology of life on the move. Oxford University Press, Oxford, UK

Dingle H, Winchell R (1997) Juvenile hormone as a mediator of plasticity in insect life histories. Archives of Insect Biochemistry and Physiology 35: 359-373

Dixon AFG (2003) Climate change and phenological asynchrony. Ecological Entomology 28: 380–381

Dormann CF, Van Der Wal R, Woodin SJ (2004) Neighbour identity modifies effects of elevated temperature on plant performance in the High Arctic. Global Change Biology 10:1587-1598

Drake VA (1985) Radar observations of moths migrating in a nocturnal low-level jet. Ecological Entomology 10: 259-265

Drake VA, Farrow RA (1988) The influence of atmospheric structure and motions on insect migration. Annual Review of Entomology 33: 183-210

Drake VA, Gatehouse AG, Farrow RA. 1995. Insect migration: A holistic conceptual model. In Drake VA, Gatehouse AG (eds) Insect Migration: Tracking Resources Through Time and Space. Cambridge University Press, Cambridge, 427-458

Dudley R (2001) The biomechanics and functional diversity of flight In Woiwod IP, Reynolds DR, Thomas CD (eds) Insect Movement: Mechanisms and Consequences. CAB International, London, UK 19-41

Easterling DR, Meehl GA, Parmesan C, Changnon SA, Karl TR, Mearns LO (2000) Climate extremes: observations, modelling, and impacts. Science 286:2068-2074

Eitschberger U, Reinhardt R, Steiniger H(1991) Wanderfalter in Europa (Lepidoptera). Appeal for international cooperation in the research on the migration of insects. Atalanta 22: 1-67

Elger R (1969) Freilandstudien zur Biologie und Ökologie von *Panaxia quadripunctaria* (Lepidoptera, Arctiidae) auf der Insel Rhodos. Oecologia (Berlin) 2: 162-197

Epstein PR (2000). Is global warming harmful to health? Scientific American 20: 19–23

Fairbairn DJ (1994) Wing dimorphism and the migratory syndrome: Correlated traits for migratory tendency in wing dimorphic insects. Res Pop Ecol 36: 157-163

Fahrig L (2001) How much habitat is enough? Biological Conservation 100: 65–74

Foster GN (1991) Conserving insects of aquatic and wetland habitats, with special reference to beetles In Collins NM, Thomas JA (eds) Conservation of Insects. Academic Press, San Diego, California 10:238-260

Freeland JR, May M, Lodge R, Conrad K (2003) Genetic diversity and widespread haplotypes in a migratory dragonfly, the common green darner *Anax junius*. Ecological Entomology 28: 413-421

Garland MS, Davies AK (2001) An examination of monarch butterfly (*Danaus plexippus*) autumn migration in Coastal Virginia. The American Midland Naturalist 147: 170-174

Giller PS (1984) Community Structure and the Niche. Chapman & Hall, London, UK

Goehring L, Oberhauser KS (2002) Effects of photoperiod, temperature, and host plant age on induction of reproductive diapause and development time in *Danaus plexippus*. Ecological Entomology 27: 674-685

Graham CH, Ferrier S, Huettman F, Moritz C, Townsend Peterson A (2004) New developments in museum–based informatics and applications in biodiversity analysis. 19: 497–503

Gurevitch J, Padilla DK (2004) Are invasive species a major cause of extinctions? Trends in Ecology and Evolution 19:470-474

Hagler JR, Jackson CG (2001) Methods for marking insects: Current techniques and future prospects. Annual Review of Entomology 46: 511-543

Hamilton JG, Dermody O, Aldea M, Zangerl AR, Rogers A, Berenbaum MR, DeLucia EH (2005) Anthropogenic changes in tropospheric composition increase susceptibility of soybean to insect herbivory. Environmental Entomology 34: 479–485

Hampe A, Petit RJ (2005) Conserving biodiversity under climate change: The rear edge matters. Ecology Letters 8 (online document)

Hanski I (1999) Metapopulation ecology. Oxford University Press, Oxford, UK

Hanski I, Breuker CJ, Schops K, Setchfield R, Nieminen M (2002) Population history and life history influence the migration rate of female Glanville fritillary butterflies. Oikos 98: 87-97

Harada T (1998) To fly or not to fly: Response of water striders to drying out of habitat. Ecological Entomology 23: 370-376

Hedenstrom A, Alerstam T (1997) Optimum fuel loads in migratory birds: distinguishing between time and energy minimization. Journal of Theoretical Biology 29: 424-432

Hedenstrom A, Alerstam T (1998) How fast can birds migrate? Journal of Avian Biology 29: 424-432

Hill JK, Collingham YC, Thomas CD, Blakeley DS, Fox R, Moss D, Huntley B (2001) Impacts of landscape structure on butterfly range expansion. Ecology Letters 4: 313–321

Huggins LM (1985) Cuticular analyses and preliminary comparisons of some Miocene conifers from *Clarkia*, Idaho In Smiley CJ (ed) Late Cenozoic History of the Pacific Northwest. American Association for the Advancement of Science 113-138

Ikawa T, Okubo A, Okabe H, Cheng L (1998) Oceanic dffusion and the pelagic insects *Halobates* spp. (Hemiptera: Gerridae) Marine Biology 131: 195-201

Janz N, Bergstrom A, Sjorgen A (2005) The role of nectar sources for oviposition decisions of the common blue butterfly *Polyommatus icarus*. Oikos 109: 535–538

Jarvinen O, Vepsalainen K (1976) Wing dimorphism as an adaptive strategy in water-striders (*Gerris*). Hereditas 84: 61-68

Jensen MN (2004) Climate warming shakes up species. Bioscience 54: 722-729

Josifov M (1986) Verzeichnis der von den Balkanhabinsel bekannten Heteropterenarten Faun. Abb. Mus. Tiet. Dresden 14: 61-93

Kennedy JS (1961) Phase transformation in locust biology. Biol. Rev., Cambridge Phil Soc 31:349–370

Kirk AA, Lumaret J.–P. (1991) The importation of Mediterranean–adapted dung beetles (Coleoptera: Scarabaeidae) from the northern hemisphere to other parts of the world. In Groves RH, Di Catsri F (eds) Biogeography of Mediterranean invasions. Cambridge University Press, Cambridge, UK 413–424

Klaus W (1989) Mediterranean pines and their history. Plant Systematics and Evolution 162:133-63

Klein U, Bock C, Kafka WA, Moore TE (1988) Olfactory sensilla on the antennae of *Magicicada cassini* (Fisher) (Homoptera: Cicadidae): structure, function and possible behavioral role. In Vidano C, Arzone A (eds) Proceedings of the

six-th Auchenorrhyncha meeting, Turin, Italy, September 1997. CNR-IPRA, Turin, Italy 93–98
Kollisch GV, Verhaert PD, Hoffmann KH (2003) Vanessa cardui adipokinetic hormone (Vanca–AKH) in butterflies and moth. Comparative Biochemistry & Physiology Part A 135: 303–308
Kozlowski J (1999) Adaptation: a life history perspective. Oikos 86: 185-194
Lack D (1968) Bird migration and natural selection. Oikos 19: 1-9
Leather S (2002) Excellence in ecological entomology - The Royal Entomological Society's Awards for Scientific Writing. Ecological Entomology 27: 513
Lenau-Jürgens H (1971). Untersuchungen zum Migrationsverhalten von *Panaxia quadripunctaria* Poda (Lepidoptera, Arctiidae) auf der Insel Rhodos. Forma et Functio 4: 1-45
Majerus M, Kearns P (1989) Ladybirds. Richmond Publishing Co., Slough, UK
Matsumura M, Suzuki Y (2003) Direct and feeding-induced interactions between two rice planthoppers, *Sogatella furcifera* and *Nilaparvata lugens*: effects on dispersal capability and performance. Ecological Entomology 28: 174-182
Mendel Z. 2000 The phytophagous insect fauna of Pinus halpensis and P. brutia in the Mediterannean. In Ne'eman G, Trabaud L (eds) Ecology, Biogeography and Management of *Pinus halepensis* and *P. brutia* Forest Ecosystems in the Mediterranean Basin. Backhuys Publishers, Leiden, The Netherlands 217-236
Menendez R, Gutierrez D, Thomas CD (2002) Migration and Allee effects in the six - spot burnet moth *Zygaena filipendulae*. Ecological Entomology 27: 317-325
Min KJ, Jones N, Borst DW, Rankin MA (2004) Increased juvenile hormone levels after long-duration flight in the grasshopper, *Melanoplus sanguinipes*. Insect Physiology 50: 531-537
Musolin DL, Numata H (2003) Timing of diapause induction and its life-history consequences in *Nezara viridula*: Is it costly to expand the distribution range? . Ecological Entomology 28: 694-703
Nathan R, Perry G, Cronin JT, Strand AE, Cain ML (2003) Methods for estimating long - distance dispersal. Oikos 103: 261- 273
Nei M (1973) Analysis of gene diversity in subdivided populations. Proceedings of the National Academy of Sciences (USA) 70: 3321-3323
Nieminen M, Leskinen M, Helenius J (2000) Doppler radar detection of exceptional mass-migration of aphids into Finland. Biometeorology 44: 172-181
Pajunen VI, Pajunen I (1991) Oviposition and egg cannibalism in rock-pool corixids (Hemiptera : Corixidae). Oikos 60: 83-90
Parmesan C (1986) Climate and species' range. Nature 382:765–766
Parmesan C, Ryrholm N, Stefanescu C, Hill JK, Thomas CD, Descimon H, Huntley B, Kaila L, Kullberg J, Tammaru T, Tennet W J, Thomas JA, Warren M (1999) Polarward shifts in geographical ranges of butterfly species associated with regional warming. Nature 399: 579– 583
Peterson MA, Denno RF, Robinson L (2001) Apparent widespread gene flow in the predominantly flightless planthopper *Tumidagena minuta*. Ecological Entomology 26: 629-637

Petit E, Balloux F, Goudet J (2001) Sex-biased dispersal in a migratory bat: a characterization using sex -specific demographic parameters. Evolution 55: 635-640

Petrakis PV (1991) Plant - Heteroptera Associations in an East -Mediterranean Ecosystem: An analysis of Structure, Specificity and Dynamics. PhD Dissertation, College of Cardiff, Wales, UK

Petrakis PV, Tselepatiotis V, Drosopoulos S (1988) Auchenorrhyncha - host plants associations in east Mediterranean communities: a multivariate approach on structure and specificity, In Vidano C, Arzone A (eds) Proceedings of the 6th International Auchenorrhynha Congress. CNR–IPRA, Turin, Italy 171-178

Petrakis PV, Tselepatioti V, Roussis V (1994) Plant taxonomy, Auchenorrhyncha specificity and niche seasonal dynamics in a maquis shrubland (Attiki, Greece). In Wilson MC, Stewart AJA (eds) Symposium on the Ecology of Planthoppers and Leafhoppers. Fifth European Congress of Entomology, University of York, York, UK

Petrakis PV, Tsoukatou M, Vagias C, Roussis V, Cheng L (2003) Evolution probing for semiochemicals based on secondary metabolites in the cuticles of three species of *Halobates* (Heteroptera; Gerridae). Biological Journal of the Linnean Society 80: 671-688

Raatikainen M (1967) Bionomics, enemies and population dynamics of *Javesella pellucida* (F.) (Hom., Delphacidae). Annales Agriculturae Fennicae 6: 1-149

Rainey (1960) The mechanisms of desert locusts swarm movements and the migration of insects. Proc of the 11th Intern Congr of Entomology. Vienna, Austria

Rainey RC (1974) Biometeorology and insect flight: Some aspects of energy exchange. Annual Review of Entomology 19: 407-439

Rankin MA, Burchsted JCA (1992) The cost of migration in insects. Annual Review of Entomology 37: 533-559

Rankin MA, McAnelly ML, Bodenhamer JE (1986) The oogenesis-flight syndrome revisited In Danthanarayana (ed.) Insect Flight: Dispersal and Migration. Springer-Verlag, Berlin 27 – 48

Reemer M, Smitt JT, van Steenis W (2003) Changes in ranges of hoverflies in the Netherlands in the 20th centrury (Diptera: Syrphidae) In Reemer M, van Helsdingen PJ, Kleukers RMJC (eds) Changes in ranges: invertebrates on the move. Proceedings of the 13th Inernational Colloquium of the European Invertebrate Survey, Leiden, 2-5 September 2001. EIS-Nederland, Leiden

Roslin T (2001) Spatial population structure in a patchily distributed beetle. Molecular Ecology 10: 823-837

Schmitt T (2003) Biogeography and ecology of southern Portuguese butterflies and burnets (Lepidoptera) In Reemer M, van Helsdingen PJ, Kleukers RMJC (eds) Changes in ranges: invertebrates on the move. Proceedings of the 13th Inernational Colloquium of the European Invertebrate Survey, Leiden, 2-5 September 2001. EIS-Nederland, Leiden 69-78

Schmitt T, Hewitt GM (2004) The genetic pattern of population threat and loss: A case study of butterflies. Molecular Ecology 13:21–31

Schneider C (2003) The influence of spatial scale on quantifying insect dispersal: an analysis of butterfly data. Ecological Entomology 28: 252-256

Schneider F (1962) Dispersal and migration. Annual Review of Entomology 7:223–242

Suarez AV, Tsutsui ND (2004) The value of museum collections for research and society. Bioscience 54: 66-74

Schoener T (1986) Resource partitioning In Kikkawa J, Anderson DJ (eds) Community Ecology: Pattern and Process. Blackwell, Melbourne, Australia 91-126

Scott JA (1968) Hilltopping as a mating mechanism to aid the survival of a low density species. Journal of Research on the Lepidoptera 7: 191-204

Scott JA (1992) Direction of spring migration of Vanesa cardui (Nympalidae) in Colorado. Journal of Research in Lepidoptera 31: 16–23

Shapiro AM (1975) The temporal component of butterfly species diversity In Cody ML, Diamond JM (eds) Ecology and Evolution of Communities. Belknap Press of Harvard University Press, Cambridge, Massachusetts 181-195

Shimoda T, Takabayashi J (2001) Migration of specialist insect predators to exploit patchily distributed spider mites. Population Ecology 43: 15 – 21

Socha R, Sula J, Zemek R (1998) Feeding behavior, digestive physiology and lipid content in macropterous females of *Pyrrhocoris apterus* (Heteroptera: Pyrrhocoridae). Physiological Entomology 23: 91-96

Socha R, Zemek R (2003) Wing morph-related differences in the walking pattern and dispersal in a flightless bug, *Pyrrhocoris apterus* (Heteroptera). Oikos 100: 35-42

Southwood TRE (1960) Migration– an evolutionary necessity for denizens of temporary habitats. Proc. of the 11th Intern. Congr. of Entomology. Vienna, Austria

Southwood TRE, Henderson PA (2000) Ecological Methods. Blackwell, Oxford, UK

Staddon BW (1995) Observations on growth, morphogenesis and programmed obsolescence of scent glands in the milkweed bug, *Oncopeltus fasciatus* (Heteroptera). Insect Physiology 41: 313-320

Stearns SC (1992) The Evolution of Life Histories. Oxford University Press, Oxford, UK

Stefanescu C (2001) The nature of migration in the red admiral butterfly *Vanessa atalanta*: evidence from the population ecology in its southern range. Ecological Entomology 26: 525-536

Tanaka S, Wolda H. 1987. Seasonal wing length dimorphism in a tropical seed bug: Ecological significance of the short winged form. Oecologia 73: 559-565

Tauber A, Tauber W (1969) Im Tal der tausend Schmetterlinge. Entomologische NachrichtenBl. Wien 16:23-26

Thomas CD, Bodsworth EJ, Wilson RJ, Simmons AD, Davies ZG, Musche M, Conradt L (2001) Ecological and evolutionary processes at expanding range margins. Nature 411: 577-581

Thomas CD, Cameron A, Green RE, Bakkenes M, Beaumont LJ, Collingham YC, Erasmus BFN, de Siqueira MF, Grainger A, Hannah L, Hughes L, Huntley B, van Jaarsveld AS, Midgley GF, Miles L, Ortega-Huerta MA, Peterson AT, Phillips OL, Williams SE (2004) Extinction risk from climate change. Nature 427: 145-148

Thomas JA (1984) The conservation of butterflies in temperate countries: past efforts and lessons for the future. In Vane–Wright RI, Ackery PR (eds) The biology of butterflies. Academic Press, London, UK 333 – 353

Tipping C (1995) The longest migration. In Walker TJ (ed) University of Florida book of insect records. University of Florida, Gainesville 23–25

Urquhart FA, Urquhart NR (1978) Autumnal migration routes of the eastern population of the monarch butterfly (*Danaus p. plexippus* L., Danaidae: Lepidoptera) in North America to the overwintering site in the Neovolcanic Plateau of Mexico. Canadian Journal of Zoology 56: 1759-1764

Vanholder B (1996) The migration of *Danaus plexippus* (Linnaeus, 1758) during October 1995 in the U.K. (Lepidoptera: Nymphalidae, Danainae). Nota Lepid 19: 129–140

Vepsäläinen K (1974) The life cycles and wing lengths of finnish *Gerris* Fabr. species (Heteroptera, Gerridae) Acta Zoologica Fennica 141: 1-73

Walker MF (1966) Some observations on the behavior and life-history of the Jersey tiger moth, *Euplagia quadripunctaria* Poda (Lep., Arctiidae) in the "Valley of the Butterflies", Rhodes. Entomologist 99(1): 1-24

Warren MS, Key RS (1991) Woodlands: Past, present and potential for insects In Collins NM, Thomas JA (eds) Conservation of Insects. Academic Press, San Diego, California 155-203

Warren MS, Hill JK, Thomas JA, Asher J, Fox R, Huntley B, Roy DB, Telfer MG, Jeffcoate S, Harding P, Jeffcoate G, Willis SG, Greatorex-Davies JN, Moss D, Thomas CD (2001) Rapid responses of British butterflies to opposing forces of climate and habitat change. Nature 414: 65-69

Whittaker JB (1999) Impacts and responses at population level of herbivorous insects to elevated CO_2. European Journal of Entomology 96: 149–156

Williams (1958) Insect migration. Collins, London, UK

Wilson MR, Claridge MF (1991) Handbook for the identification of leafhoppers and planthoppers of rice. C.A.B. International, International Institute of Entomology, London, UK

Woiwod P, Reynolds DR, Thomas CD (2001) Insect movement: Mechanisms and consequences. CABI Publishing, Wallingford, UK

Zera AJ, Denno RF (1997) Physiology and ecology of dispersal polymorphism in insects. Annual Review of Entomology 42: 207-231

Zera AJ, Mole S (1994) The physiological costs of flight capability in wing-dimorphic crickets. Res. Popul. Ecol. Kyoto 36: 151-156

Zera AJ, Mole S, Rokke K (1994) Lipid, carbohydrate and nitrogen content of long - winged and short- winged *G. firmus*: implications for the physiological cost of flight capability. Insect Physiology 40:1037-1044

6 Migration and adaptation of late Cenozoic cold-water molluscs in the North Pacific

Kazutaka Amano

Department of Geoscience, Joetsu University of Education, 1Yamayashiki, Joetsu, Niigata 943-8512, Japan, amano@juen.ac.jp

6.1 Abstract

Cold-water molluscan fauna originated in the North Pacific in accordance with the worldwide cooling events around the latest Eocene. The westward trans-Pacific migration of the cold-water molluscs occurred during the early to early middle Miocene, owing to shifting climatic belts and "ecological opportunity" rather than current direction. In contrast, the eastward migration of the cold-water molluscs occurred in cool climate ages from the early Oligocene to Holocene. As a result of the Plio-Pleistocene cooling, cold-water species spreaded to the Yellow and the East China Sea through the Japan Sea. Shifting climatic belts thus affected the zoogeographic range of cold-water species.

Keywords: Migration, Late Cenozoic, cold-water, Mollusca, North Pacific.

6.2 Introduction

The cold-water molluscan fauna in the North Pacific has a long and complex history involving species formation, migration and adaptation against a backdrop of changing climate and oceanographic conditions.

Migration and adaptation of molluscan species have been studied in relation to climate change and the opening of Bering Strait in the northern Pacific. Three types of migrations have been recognized among the North Pacific taxa (Amano 2000). The first type appeared in the northwestern Pacific and migrated to the northeastern Pacific. It includes *Mya* (MacNeil 1965), *Neptunea* (Strauch 1972), *Mizuhopecten* (Masuda 1986), Turritelloidea and Buccinoidea (Titova 1994). The species of the second type originated in the northeastern Pacific, then spread westward. The taxa include *Penitella* and *Platyodon* (Adegoke 1967), *Panomya* (Strauch 1972), *Littorina* (Reid 1996), *Liracassis* (Kanno 1973), *Nucella* (Amano et al. 1993), *Ceratostoma* (Amano and Vermeij 1998b), *Macoma* (*Rexithaerus*) (Amano et al. 1999), *Kaneharaia* (Amano and Hikida 1999) and *Lirabuccinum* (Amano and Vermeij 2003). The species of the third type such as *Cyrtodaria* (Strauch 1972) and Astartidae (Ogasawara 1986) invaded the North Pacific after the opening of the Bering Strait from the Arctic Sea region.

When Vermeij (2001) synthesized the history of North Pacific biota, he noted that the genera in the first two types appeared during the latest Eocene or the early Oligocene as a response to cooling. Based on Amano and Vermeij (1998b), he also emphasized the Oligocene to early Miocene origin of groups that subsequently migrated westward during the early middle Miocene. Vermeij (2001) also suggested that the turrid *Aforia* and *Yoldia* group might have been originated in the Southern Hemisphere. Based on a molecular study, Ozawa et al. (2001) noted that the buccinid *Kelletia* originated in Southern Hemisphere and reached the northwestern Pacific via the northeastern Pacific. However, a detailed scenario of the northward migration of the above mentioned genera remains speculative. Thus, these taxa are not included in this paper.

In consequence of the establishment of a new time table of the Pacific coast Cenozoic in North America based mainly on magnetostratigraphy (Prothero 2001), it is necessary to reexamine the timing of the molluscan migration events. Moreover, the bathymetric difference between the eastward and westward spreading molluscan groups has also not been discussed in detail.

The Pliocene trans-arctic interchange through the Bering Strait has been studied by many authors (e.g. Durham and MacNeil 1967; Vermeij 1991; Marincovich and Gladenkov 1999; Marincovich 2000; Marincovich et al. 2002; Gladenkov and Gladenkov 2004). When the Bering Strait opened, astartids and other bivalves invaded the Pacific from the Arctic region. Based on a drastic frequency change of calcareous nannofossil, *Coccolithus pelagicus* at 2.74Ma, Sato et al. (2004) concluded that this frequency change was correlated with the closure of the Central American Seaway and with onset of heavy glaciation in the Arctic Sea and the North Pacific Ocean. However, no detailed examination has been done on the route of southward expansion of the cold-water molluscs, including the astartids in the northwestern Pacific.

The purposes of this paper are to: (1) extend the list on the trans-Pacific migrants, (2) explore the bathymetric difference between different groups of migrants, and (3) specify routes of migration of cold-water species in the northwestern Pacific during the Plio-Pleistocene.

6.3 Material and Methods

Stratigraphic and geographic distributions of the North Pacific cold-water taxa listed by Amano (2000) and Vermeij (2001) are reexamined. Among them, the turbinid *Phanerolepida,* the muricid *Boreotrophon*, the buccinids *Japelion,* and *Sulcosipho* were not previously discussed in this context. In this paper, I reexamine their migration routes and ages in detail. Material stored at Joetsu University of Education (JUE) and Tohoku University (IGPS) was used in addition to a critical review of the literature.

I divided the North Pacific into western and eastern regions from the paleobiogeographic viewpoint. The northwestern Pacific includes Japan, Sakhalin, Kamchatka and Koryak Upland while the northeastern Pacific includes Alaska, Vancouver Island, Washington, Oregon and California. After checking the age of the oldest fossil record for each genus on both sides of the Pacific, I treat the older age as an appearance and the younger age as a migration age for each genus. It is possible for these ages to be younger than the real origin and migration timings mainly because of taphonomic reasons.

These Cenozoic ages are based mainly on the synthesized micropaleontological or magnetostratigraphic scheme by Kano et al. (1991) for Japan, Kurita et al. (2000) and Gladenkov et al. (2000) for Sakhalin, Gladenkov (1998) for Kamchatka, Marincovich (1990) for Alaska, and Prothero (2001) for Vancouver to California. Ecological data of Recent molluscs are mainly based on Higo et al. (1999), Coan et al. (2001) and McLean (1996).

For the Plio-Pleistocene, I examined the distribution of some cold–water molluscan species. In addition, the southern limits of some boreal species were investigated. As the Japan Sea was semi-enclosed and separated from the Pacific Ocean during the Plio-Pleistocene (e.g. Ogasawara 1994), the hydrographic condition of the Japan Sea differed from that of the Pacific. So, it is necessary to separately examine the southernmost limits of boreal species in the Pacific and Japan Sea. For this purpose, material housed at Joetsu University of Education (JUE), Tohoku University (IGPS) and University Museum of University of Tokyo (UMUT) was used.

6.4 Trans-Pacific migration

6.4.1 Case studies of some gastropods

Migration of Phanerolepida

The deep-water turbinid, *Phanerolepida transenna* (Watson 1879) is the only extant species of this genus and lives in warm-water habitats ranging from Sagami Bay, central Honshu to Southern Kyushu (200-1100 m in depth; Higo et al. 1999). However, the migration route is the same as that of some cold-water genera such as *Nucella, Ceratostoma* and *Lirabuccinum*.

Noda and Ogasawara (1976) and Majima and Murata (1992) compiled the fossil records of *Phanerolepida* species; *P. oregonensis* Hickman 1972, *P. expansilabrum* (Kuroda 1931), *P. pseudotransenna* Ozaki 1956, *P. rehderi* MacNeil 1960 (=*P. pseudotransenna* by Majima and Murata 1992), *P. transenna* (Watson 1879) and *P. okinawana* Noda 1988. *P. okinawana* resembles *Homalopoma granuliferum* Nomura and Hatai 1940 in having distinct spiral cords despite having coarser shagreen sculpture. In this paper, this species is excluded from the genus *Phanerolepida*.

According to Majima and Murata (1992) and Hickman (2003), the earliest species *P. oregonensis* was collected from the upper Eocene Keasy Formation (middle member) in Oregon and evolved from the middle Eocene *Nehalamia hieroglyphica* Hickman 1974 in Oregon (Fig. 1). The Pliocene to Recent *P. transenna* evolved by the way of *P. pseudotransenna* during the late Miocene in southwest Japan.

Two additional Miocene species were found in eastern Kamchatka and Hokkaido. Titova in Gladenkov et al. (2000) only illustrated *P. ozernensis* from the lower Miocene Ozernovsky Formation in eastern Kamchatka as a new species without any description. Judging from her illustration, this species is more similar to the Recent *P. transenna* than any fossil species especially in having fine shagreen sculpture. Fujii Environmental Geology Institute (1998) illustrated an ill-preserved specimen from the uppermost lower to lowermost middle Miocene Ponsubetsu Formation in Hokkaido as *Phanerolepida* sp.

The fossil record indicates the following history for this genus. The earliest deep-water species, *P. oregonensis* appeared in Oregon during the late Eocene. The genus reached eastern Kamchatka by the early Miocene. Then, it moved southward to central Honshu *via* Hokkaido in the latest early to earliest middle Miocene during the climatic warm interval.

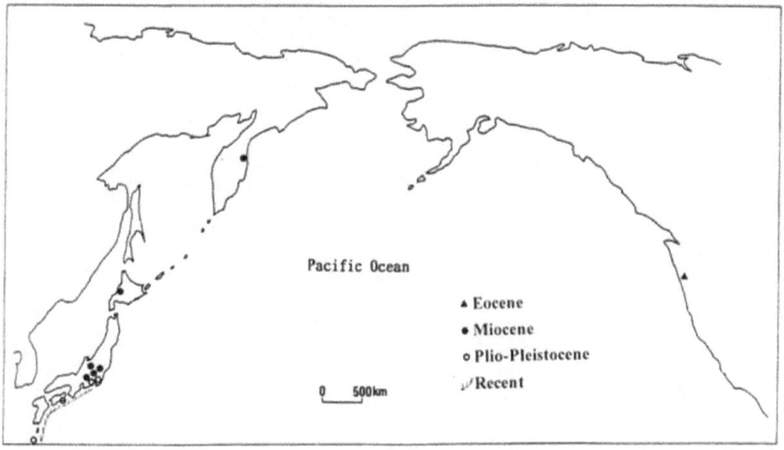

Fig. 1. Distribution of *Phanerolepida*

In the Pliocene, the genus adapted to the deep-water of warm area from central Honshu to Okinawa Islands and now lives there. This migration pattern is partly similar to the other shallow-water westward spreading group in its direction, but differs from the latter in that eastern Pacific members became extinct. No species now live in the North Pacific region and the northwestern part of America.

Migration of Boreotrophon

Many species of the muricid gastropod *Boreotrophon* occur in the sublittoral zone of the North Pacific, Arctic and North Atlantic (Tiba and Kosuge 1985; Egorov 1993; Houart 2001). The earliest fossil species are *Boreotrophon osawanoensis* (Tsuda 1959) from the lower Miocene Yamanouchi Formation in Gifu Prefecture, central Japan or *B.* sp. indet. described by Matsubara (1996) from the lower Miocene Yotsuyaku Formation in Iwate Prefecture, northeast Japan. Associated faunas of these species suggest that they might have lived in the upper sublittoral zone under the influence of warm water. I obtained a specimen of *B.* cf. *beringi* (Dall) from the middle Miocene Shibiutan Formation in the northern Hokkaido in association with *Anadara hokkaidoensis* Noda 1966, *Spisula onnechiuria* (Otuka 1940), *Megangulus protovenulosus* (Nomura 1935), *Kaneharaia ausiensis* (Ilyina 1954) and *Securella chitanana* (Yokoyama 1926). Thus, in the middle Miocene, the genus adapted to temperate-water in North Korea, Honshu and reached northern Hokkaido (Fig. 2). Sinelnikova in Gladenkov and Sinelnikova (1990) described *B. etolonica* as a new species from the middle Miocene Etolon Formation in western Kamchatka.

Fig. 2. Distribution of *Boreotrophon* in the early to middle Miocene. 1.Yamanouchi Formation in Gifu Prefecture, 2. Yotsuyaku Formation in Iwate Prefecture, 3. Shibiutan Formation in Hokkaido (see also the text)

However, as the type and only specimen lacks a basal part, it is difficult to determine the validness of this species within the genus *Boreotrophon*. There is no early to middle Miocene record of this genus in northeastern Pacific side. The oldest record is *Boreotrophon* from the upper Miocene to Pliocene Rio Dell Formation in California (Faustman 1964).

The genus originated in the warm water environment around Honshu during the early Miocene. Then, by the late Miocene, *Boreotrophon* spread to the northeastern Pacific. After the opening of the Bering Strait, the genus spread to the Arctic and North Atlantic regions in the Pliocene (Durham and MacNeil 1967). In the Pliocene, this genus diversified both on northwestern and northeastern sides of the Pacific Ocean (Grant and Gale 1931; Hatai and Nisiyama 1952; Masuda and Noda 1976).

Migration of Japelion

The Recent buccinid *Japelion* consists of *J. hirasei* (Pilsbry 1901), *J. pericochlion* (Schrenck 1862), *J. adelphicus* (Dall 1907) and *J. latus* (Dall 1918). The first two species live in the lower sublittoral to upper bathyal depths from central Honshu and northward to the Bering Sea. *J. latus* is an endemic species living mainly in the lower sublittoral of the Tsushima Strait, East China Sea and Yellow Sea.

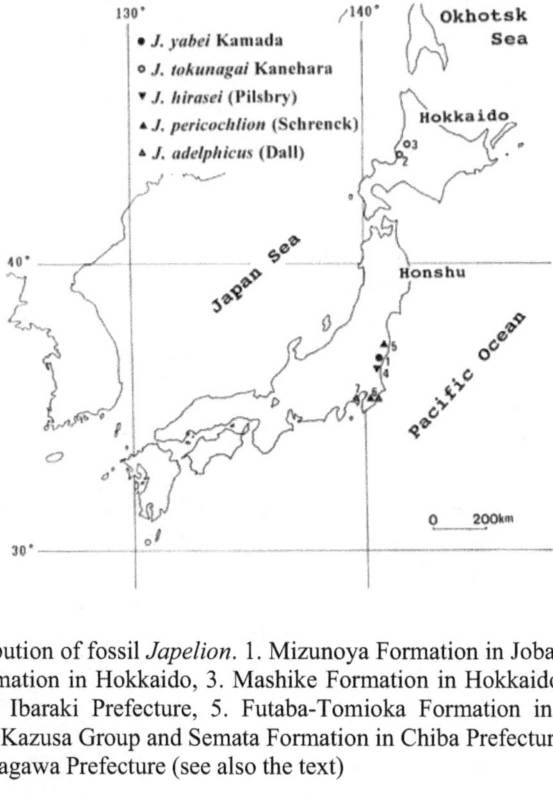

Fig. 3. Distribution of fossil *Japelion*. 1. Mizunoya Formation in Joban coal-field, 2. Morai Formation in Hokkaido, 3. Mashike Formation in Hokkaido, 4. Hitachi Formation in Ibaraki Prefecture, 5. Futaba-Tomioka Formation in Fukushima Prefecture, 6. Kazusa Group and Semata Formation in Chiba Prefecture, 7. Kazusa Group in Kanagawa Prefecture (see also the text)

J. adelphicus dwells in the lower sublittoral of warm-water area from central Honshu to southernmost Kyushu (Higo et al. 1999).

The fossil record of *Japelion* is confined to Japan. The oldest fossil species is *J. yabei* Kamada 1955 from the lower Miocene Mizunoya Formation in the Joban coal-field (Fig. 3). Concerning the upper Miocene fossil record, *J. tokunagai* Kanehara 1937 was collected from the Morai (Kanehara 1937) and Mashike (this study) Formations both in western Hokkaido.

In the Plio-Pleistocene, three Recent species appeared mainly in and around the Kanto Region, central Honshu. *J. hirasei* was described by Noda et al. (1995) from the Pliocene Hitachi Formation, Ibaraki Prefecture. *J. pericochlion* has been recorded from the Pliocene Futaba-Tomioka Formation in Fukushima Prefecture by Nemoto and O'Hara (1979) and from the Pleistocene Kazusa Group in Chiba Prefecture by Baba (1990). *J. adelphicus* has been known from the Pleistocene Kazusa Group and Semata Formation in Chiba and Kanagawa Prefectures (Okutani 1968; Baba 1990).

Based on the above fossil records, the genus originated in central Honshu during the early Miocene and may have spread recently to the Bering Sea, south Kyushu and East China Sea. However, it has never reached the northeastern part of the Pacific.

Migration of Sulcosipho

Sulcosipho was proposed by Dall (1916) as a new section of *Chrysodomus* Swainson 1840 (= *Neptunea* Roeding 1798), based on *Chrysodomus tabulatus* Baird 1863. *Sulcosipho tabulatus* now lives in 50-200 m depth of Petersburg, Alaska to Newport Bay, California (McLean 1996).

When he summarized the Neogene biogeography of *Neptunea*, Nelson (1978) included *N. "sachalinensis"* Khomenko 1938, *N. lamellosa* Golikov 1962 and *N. uwasoensis* Otuka 1935 (= *N. eos* (Kuroda 1931)) in *Sulcosipho*. However, as pointed by Goryachev (1987) and Amano (1997a), they should be classified as *Neptunea* (*Neptunea*) because the shell is not slender form and the shoulder is not tabulated. On the other hand, Goryachev (1987) treated *S. tabulatus*, *S. lawsoni* (Martin 1914) and *S. andersoni* (Martin 1914) as belonging to the Melangenidae (= Melongenidae) without any reasons. However, these species lack spines or nodes which are characteristic of the Melongenidae and the narrow siphonal canal resembles that of *Neptunea*. Dall (1919) stated that *S. tabulatus* was included in the same section of *Japelion adelphicus* by having the slender form and the tabulated shoulder. From these reasons, *Sulcosipho* should be treated as a genus of the Buccinidae rather than the Melongenidae.

The earliest fossil record of *Sulcosipho* is *Neptunea* (*Sulcosipho*) cf. *tabulatus* described by Kanno (1971) from the middle Miocene Yakataga Formation in Alaska. Then, this species migrated southward during the late Miocene. In contrast with *Japelion*, this genus did not reach the northwestern Pacific.

6.5 Pattern of origin, migration and adaptation

North Pacific cold-water genera or subgenera are classified into three groups as mentioned by Amano (2000). Among them, the group originating in the northwestern Pacific can be subdivided into two subgroups. The first subgroup (W1) migrated to the northeastern Pacific while the second one (W2) did not. The W1 subgroup consists of 21 genera or subgenera including *Boreotrophon* as mentioned above (Table 1).

On the origination age, the late Eocene (8 taxa) dominates the subgroup W1. Four taxa each originated in the early Oligocene and early Miocene. In contrast, there is no conspicuous age of migration ranging from the early Oligocene onwards.

On the other hand, only four genera are included in the W2 subgroup; *Ocinebrellus*, *Clinopegma*, *Japelion* and *Volutomitra*. Their appearances in the early to early middle Miocene are younger than those of the W1 subgroup. As pointed out by Vermeij (2001), most taxa of this group live on sandy mud or mud bottom at sublittoral to bathyal depths. It is noteworthy that 14 among 23 extant taxa are able to live in the bathyal zone.

Table 1. Origin and migration ages of W Group. *M, mud; sM, sandy mud; fS, fine-grained sand; cS, coarse-rgained sand; S, sand; R, rock

Genera or subgenera	NW Pacific	NE Pacific	Depth (m)	Bottom*	References
W1Group					
Turritella (Neohaustaor)	e.Miocene	e.m.Miocene	30-1500	sM	Titova, 1994
T. (Hataiella)	e.Eocene	e.m.Miocene	-	-	Titova, 1994; Oleinik and Marincovich, 2003
Tachyrhinchus	l. Eocene	l.Pliocene	12-1500	cS, sM	Titova, 1994
Boreotrophon	e.Miocene	l.Miocene	0-964	cS, sM	This study
Neptunea	l. Eocene	l.Oligocne	0-1380	sM	Strauch, 1972; Oleinik and Marincovich, 2003
Ancistrolepis	l. Eocene	e.Oligocene	100-690	M	Titova, 1993
Trominina	l. Eocene	e.Oligocene	-	-	Titova, 1994
Buccinum	e.Oligocene	l.Oligocene	0-1500	M	Titova, 1994
Arctomelon	m.Eocene	Holocene	100-200	M	Oleinik, 1996
Acila	Cretaceous	e.Oligocene	0-990	fS, M	Vermeij, 2001
Yoldia (Cnesterium)	m.Eocene	m.Miocene	9-400	sM	Vermeij, 2001
Portlandia (Megayoldia)	l. Eocene	Holocene	25-760	M	Vermeij, 2001
Mizuhopecten	e.Miocene	l.Miocene	2-80	cS, sM	Masuda, 1986
Cyclocardia	l. Eocene	e.Oligocene	10-400	sM	Popov, 1983
Clinocardium (Keenocardium)	e.Oligocene	l.Oligocene	10-200	sM	Kafanov, 1999, 2000
C. (Ciliatocardium)	l. Eocene	l.Oligocene	10-200	M	Kafanov, 2001
Serripes	e.Oligocene	l.Miocene	0-260	sM	Kafanov, 2003
Siliqua	l. Eocene	e.m.Miocene	0-100	S	Vermeij, 2001
Solen (Ensisolen)	e.m. Miocene	l.Miocene	0-50	S	Vermeij, 2001
Liocyma	e.Oligocene	l.Oligocene	0-1500	sM	Amano, 2000
Mya	m.Eocene	l.Oligocene	0-100	M	MacNeil, 1965
W2 Group					
Ocinebrellus	e.m. Miocene	-	0-200	R	Amano and Vermeij, 1998a; Matsubara and Amano, 2000
Clinopegma	e.m. Miocene	-	50-650	sM	Amano, 2000
Japelion	e.Miocene	-	50-500	M	This study
Volutomitra	e.Miocene	-	100-200	M	Amano,1997b

The group evolving in the northeastern Pacific also consists of two subgroups. One of them (E1) spread northwestward while the other (E2)

remained restricted to the Northeast Pacific. The E1 subgroup is composed of 22 genera or subgenera including *Phanerolepida* (Table 2).

Table 2. Origin and migration ages of E Group. *M, mud; sM, sandy mud; fS, fine-grained sand; cS, coarse-rgained sand; S, sand; R, rock

Genera or subgenera	NE Pacific	NW Pacific	Depth (m)	Bottom*	References
E1Group					
Puncturella	l.Eocene	e.m.Miocene	0-1200	R	Vermeij, 2001; Squires, 2003
Phanerolepida	l.Eocene	e.Miocene	200-1100	cS, M	This study
Littorina	m.Eocene	e. Miocene	0	R	Reid, 1996; Squires, 2003
Trichotropis	l. Eocene	e. Miocene	20-100	M	Amano, 2000; Squires, 2003
Euspira	Paleocene	m.Eocene	0-620	sM	Marincovich, 1977; Majima, 1989
Glossaulax	m. Eocene	e.m.Miocene	0-100	fS	Marincovich, 1977; Majima, 1989
Cryptonatica	e.Miocene	e.m.Miocene	20-300	sM	Marincovich, 1977; Majima, 1989
Liracassis	l.Eocene	e.Miocene	-	-	Kanno, 1973; Squires, 2003
Fusitriton	l.Eocene	e.m.Miocene	0-620	sM, fS	Smith, 1970; Squires, 2003
Nucella	e.Oligocene	e.m.Miocene	0-20	R	Amano et al., 1993
Ceratostoma	e.Miocene	e.m.Miocene	0-135	R	Amano and Vermeij, 1998b
Lirabuccinum	e.Oligocene	e.m.Miocene	0-270	R	Amano and Vermeij, 2003
Nuttalia	m. Eocene	e.Oligocene	0-20	sM, fS	Vermeij, 2001
Spisula (*Mactromeris*)	Paleocene	m.Eocene	0-200	fS, M	Vermeij, 2001
S. (*Pseudocardium*)	Paleocene	m.Eocene	0-30	fS	Vermeij, 2001
Macoma (*Rexithaerus*)	l.Oligocene	e.m.Miocene	0-100	S	Amano et al., 1999
Securella	e.Oligocene	e.m.Miocene	0-150	fS	Vermeij, 2001
Kaneharaia	e.Oligocene	e.m.Miocene	-	-	Amano and Hikida, 1999
Compsomyax	e.Oligocene	e.Miocene	2-500	M	Amano, 2000
Panomya	l.Oligocene	e.Miocene	0-300	sM, M	Strauch, 1972
Penitella	e.Oligocene	e.Miocene	0	R	Adegoke, 1967
Platyodon	e.Oligocene	e.Miocene	0-20	R	Adegoke, 1967
E2 Group					
Ocinebrina	e.Miocene	-	0-60	R	Marko and Vermeij, 1999
Sulcosipho	m.Miocene	-	54-392	sM	This study
Crassadoma	e.Miocene	-	0-80	R	Vermeij, 2001
Cumingia	Pliocene	-	0-65	R	Vermeij, 2001

Seven genera or subgenera first appeared in the early Oligocene, and four taxa each appeared in the middle and late Eocene. The migration age is also concentrated in the early to early middle Miocene (18 taxa). Only four genera belong to the E2 subgroup; *Ocinebrina, Sulcosipho, Crassadoma* and *Cumingia*. They originated after the early Miocene like did subgroup W2. As noted by Amano and Vermeij (1998b) and Vermeij (2001), this group includes many hard bottom- as well as soft bottom-dwellers mainly living in the intertidal to sublittoral zones. Fifteen taxa among 24 extant genera or subgenera are confined to these zones. *Phanerolepida* living in the bathyal zone is the single exception.

6.6 Causes of origin, migration and adaptation

During the Eocene-Oligocene transition, a global cooling affected both deep-water and shallow-sea molluscan faunas (Squires 2003; Oleinik and

Marincovich 2003; Nesbitt 2003; Hickman 2003). As noted by Vermeij (2001), the cold-water taxa appeared during the latest Eocene or the early Oligocene representing the first cooling interval of the Cenozoic. Most cold-water taxa evolved in the late Eocene and early Oligocene. It is noteworthy that no shallow-sea cold-water species evolved from the tropical Pacific biota in the latest Eocene to early Oligocene cooling. However, the reason for this is unknown.

The eastward migration took place at intervals from the early Oligocene to the Holocene, but there were few instances in the early to early middle Miocene. In contrast, the westward spreading occurred during the short interval of the early to early middle Miocene.

The deeper water W1 subgroup had opportunities to move northward and eastward in any cooling climate ages. As also noted by Amano and Vermeij (1998b) and Vermeij (2001), the northwestward migration of subgroup E1 occurred during the Neogene warm maximum. Two plausible causes for the westward migration are available.

For benthic organisms, current is usually considered as an important factor of larval transport (e.g. Cecca 2002). As noted by Kafanov (1982), many high latitude molluscs have a non-planktotrophic larval life; therefore planktonic dispersal would be unimportant. However, as the current sometimes transports the adult on algae, it may be important for the direction of migration. If there was a westward current during the Neogene warm maximum, it is easy to explain why many E1 taxa migrated westward but few W1 taxa spread eastward. However, it is difficult to explain the reason why circulation reversed only during the Neogene warm period.

Another plausible reason is the combination of a shifting climate belt and "ecological opportunity". In the Neogene warmest age, the northernmost Pacific lay in the temperate zone (Marincovich 1988; Ogasawara 1994; Amano et al. 1996). As the temperate zone extended northward, eastern Pacific taxa could move westward easily along the coast. The reason why the W1 subgroup could not migrate eastward is an "ecological opportunity" postulated as the most important reason for the trans-Arctic interchange by Vermeij (1991).

The E1 subgroup includes many rocky shore dwellers which are never found in the W1 subgroup. Only two muricid species and one boring bivalve species are known from the Paleogene of the northwestern Pacific region: *Ocinebra ashiyaensis* Nagao 1928 from the upper Oligocene Ashiya Group, in Fukuoka Prefecture, Kyushu, *Ocinebra tsuzurensis* Yokoyama 1924 from the lower Oligocene Iwaki Formation in the Joban coal-field, central Honshu, and *Parapholas satoi* Nagao 1928 from the upper Oligocene Meinohama Formation in Fukuoka Prefecture, Kyushu.

In other words, by the early Miocene, the hard bottom communities in shallow water were an open niche for species coming from the east.

In contrast, ten deep-water species among the twenty one species of W1 subgroup already migrated and adapted to the northeastern part by the end of Oligocene.

Moreover, similar niches of another species had been already occupied by the well established soft-bottom fauna. For example, the trophonines, volutids, yoldiids and pectinids appeared in the northeastern Pacific during the Paleocene and Eocene (Squires 2003; Moore 1983 1984). Thus, it was difficult for the W1 subgroup to adapt on the soft bottom in the northeastern Pacific.

The taxa restricted to either the northeastern or the northwestern Pacific (W2, E2) originated in the early Miocene or later. Five genera or subgenera first appeared in the early or early middle Miocene belong to the W1 subgroup.

As mentioned above, this interval is a timing of migration for most E1 and a few W1 taxa. The later appeared W2 and E2 taxa such as *Japelion* and *Sulcosipho* could not participate in the migration.

6.7 Plio-Pleistocene expansion of cold-water species

6.7.1 Recent cold-water species in Yellow Sea and East China Sea

A warm-water current flows at the surface of the East China Sea and the Yellow Sea. By contrast, there is a deep cold-water mass in the Yellow Sea (Tsuchida 1975). Seventeen cold-water species other than *Ancistrolepis vietnamensis* and *Arctomelon tamikoae* now live in the modern water of the sublittoral zone of the Yellow Sea (Yamashita 1978; Zhao et al. 1982; Qi et al. 1989; Table 3).

The southern limits of eight of these species in the Japan Sea are located at Tsushima or off Yamaguchi near the Yellow Sea. Populations of nine other species in the Yellow Sea are separated from the northern populations in the Japan Sea. Two Recent species of *Arctomelon* occur in Alaska and northwestern America (Weaver and du Pont 1970).

The distribution of *Arctomelon tamikoae* Kosuge 1970 living in 100-200 m depth of the East China Sea is isolated from the main distributional area of this genus.

Table 3. Cold-water species now living in the Yellow Sea, the East China Sea and the South China Sea

Species	Depth (m)	Bottom*	S.L.**	References
Acmaea pallida (Gould)	0-20	R	Tsushima	Qi et al., 1989
Homalopoma amussitatum (Gould)	0-300	cS, R	Niiigata	Zhao et al., 1982; Qi et al., 1989
Turritella (*Neohaustator*) *fortilirata* Sowerby	30-300	sM	Yamaguchi	Yamashita, 1978
Trichotropis bicarinata Sowerby	20-100	M	Hokkaido	Yamashita, 1978; Zhao et al., 1982
Boreotrophon beringi (Dall)	20-389	S	Hokkaido	Yamashita, 1978
Japelion latus (Dall)	50-200	sM	Tsushima	Qi et al., 1989
Ancistrolepis vietnamensis Sirenko and Goryachev	400-700	M	-	Sirenko and Goryachev, 1990
Plicifusus rhyssoides Dall	100-300	M	Tsugaru	Yamashita, 1978
Neptunea cumingi Crosse	0-200	R	Hyogo	Yamashita, 1978; Zhao et al., 1982
Buccinum plectrum Stimpson var.	-	-	Kuril Is.	Zhao et al., 1982; Qi et al., 1989
Volutharpa perryi (Jay)	10-70	sM	Yamaguchi	Qi et al., 1989
Arctomelon tamikoae Kosuge	100-200	M	-	Kosuge, 1970
Acila (*Truncacila*) *insignis* (Gould)	0-200	fS, M	Tsushima	Yamashita, 1978
Ennucula tenuis (Montagu)	0-2200	sM	Yamaguchi	Yamashita, 1978
Yoldia (*Cnesterium*) *johanni* Dall	10-100	sM	Hokkaido	Yamashita, 1978
Musculus nexus (Gould)	10-230	M	Primorie	Qi et al., 1989
Clinocardium (*Keenocardium*) *californiense* (Deshayes)	10-200	sM	Tsushima	Yamashita, 1978
Potamocorbula amurensis (Schrenck)	2-20	M	-	Qi et al., 1989
Cardiomya behringensis (Leche)	50-2900	sM	Hokkaido	Yamashita, 1978

Ancistrolepis vietnamensis Sirenko and Goryachev 1990, a descendant form of the cold-water genus *Ancistrolepis*, now lives in 400-700 m depth of the South China Sea. As proposed by Oleinik (1996), these species resulted from southward expansion in the Quaternary. However, there has been no discussion on the migration route of these species.

6.7.2 Plio-Pleistocene migration route

There are two ways for cold-water species to expand their distribution. One is the Japan Sea route and the other is the Pacific side route. However, the Pacific side of southwestern Japan lies in the subtropical zone affected by the warm Kuroshio Current during the late Cenozoic (e.g. Ogasawara 1994). Thus, it is difficult to assign the migration route to the Pacific side. *Tridonta borealis* and *T. alaskensis* are known from the Plio-Pleistocene strata in Hokkaido and the Japan Sea borderland (Fig. 4), but there is no Plio-Pleistocene record of this genus from the Pacific side of northeastern Honshu other than the Pliocene Togawa Formation and Pleistocene Hamada Formation. Ten cold-water species now living in Hokkaido and are known from the Japan Sea borderland at the horizon younger than the Datum A (Sato and Kameo 1996) which coincides in timing with the increasing abundance of *Coccolithus pelagicus* (Amano et al. 1987; Amano,1994 1996; Yoon 1988; Amano et al. 1996; Kase et al. 1996; Sato et al. 2003; Table 4).

In the late Pliocene, the southernmost occurrences of the following species are located at Akita Prefecture (Sasaoka Formation) and Niigata

Prefecture (Tanihama Formation): *Cyclocardia crassidens*, *C. isaotakii*, *Felaniella ohtai*, *Macoma middendorffi*, *M. lama meridionalis* and *Nuttallia commoda* (Fig. 5).

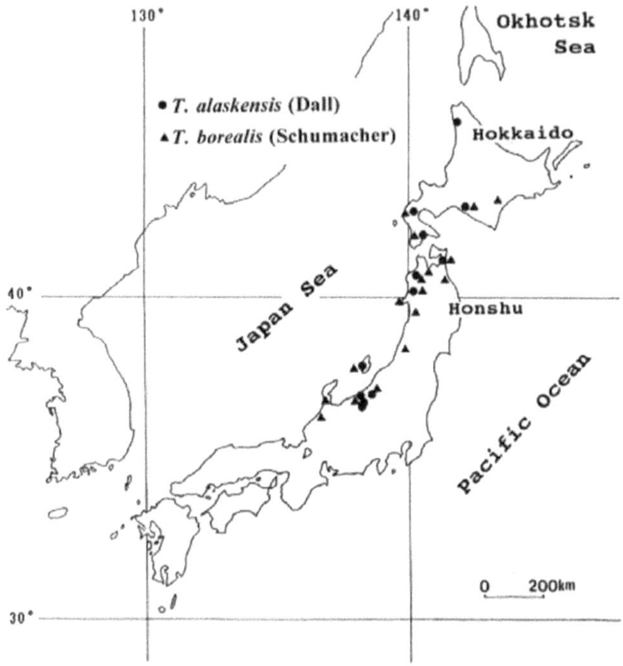

Fig. 4. Distribution of fossil *Tridonta alaskensis* (Dall) and *T. borealis* (Schumacher)

Later, *Siliqua alta* occurred from the lower Pleistocene Seoguipo Formation in Jeju Island near the Yellow Sea (Yoon 1988; Kang 1995). *Felaniella ohtai* was also found from the lower Pleistocene Sawane Formation in Sado Island, Niigata Prefecture. Four species have been recorded from the lower Pleistocene Omma Formation in Ishikawa Prefecture: *Ancistrolepis grammatus, Macoma middendorffi, M. lama meridionalis* and *Mya truncata*.

Table 4. Southernmost formation occurring the cold-water species during the late Pliocene to the middle Pleistocene

Species	l.Plio.	e.Pleist.	m.Pleist.	References
Ancistrolepis grammatus (Dall)	-	Omma F.	-	Amano et al., 1996
Chlamys (*Chlamys*) *straegus* (Dall)	-	-	Shibikawa F.	Amano, 1996
Cyclocardia crassidens (Broderip and Sowerby)	Sasaoka F.	-	Shibikawa F.	Amano, 1996, 2001
C. isaotakii (Tiba)	Tanihama F.	-	-	Amano et al., 1987
Felaniella ohtai Kase and Miyauchi	Sasaoka F.	Sawane F.	-	Kase et al., 1996
Siliqua alta (Broderip and Sowerby)	-	Seoguipo F.	-	Yoon, 1988
Macoma (*Macoma*) *middendorffi* Dall	Sasaoka F.	Omma F.	-	Amano, 1996
M. (*M.*) *lama meridionaris* Scarlato	Sasaoka F.	Omma F.	Shibikawa F.	Amano, 1996; Sato et al., 2003
Nuttallia commoda (Yokoyama)	Sasaoka F.	-	-	Sato et al., 2003
Mya (*Mya*) *truncata* Linnnaeus	-	Omma F.	-	Amano, 1996

Fig. 5. Plio-Pleistocene formations yielding some species living at present in Hokkaido and northwards. 1. Sasaoka Formation in Akita Prefecture, 2. Tanihama Formation in Niigata Prefecture, 3. Sawane Formation in Niigata Prefecture, 4. Omma Formation in Ishikawa Prefecture, 5. Seoguipo Formation in Korea, 6. Shibikawa Formation in Akita Prefecture (see also the text)

In the late middle Pleistocene, *Chlamys strategus*, *Cyclocardia crassidens* and *Macoma lama meridionalis* again reached Akita Prefecture (Shibikawa Formation).

Sato et al. (2002) showed that the cold–water calcareous nannofossil *Coccolithus pelagicus* assemblage migrated through the Japan Sea via the East China Sea to the South China Sea during 2.75-1.97Ma. Kitamura and

Kimoto (2004) showed that the southern strait to the Japan Sea existed in the glacial period during 1.45-1.32 Ma and during all interglacials as well. The horizon of the above cold-water species in the Sasaoka Formation is assigned in age to the late Pliocene just after 2.75 while that in the Omma Formation to the time interval of 1.45-1.32Ma.

Ancestral populations of some cold-water species migrated westward into the Yellow Sea and East China Sea through the Japan Sea during 2.75-1.97Ma and 1.45-1.32 Ma. Upper middle Pleistocene deposits are scarce in the Japan Sea coast. Although, some cold-water molluscs which live now in Hokkaido and northwards are known from the Shibikawa Formation, it is uncertain if these species could migrate to the Yellow Sea in the late midde Pleistocene. Because *Ancistrolepis vietnamensis* and *Arctomelon tamikoae* are known only from their type locality and have no fossil records, they may have arisen recently.

6.8 Conclusions

Cold-water taxa originated in the temperate zone of the North Pacific as a response to the worldwide cooling events around the latest Eocene. The taxa of northeastern origin migrated via the north to the northwestern Pacific mainly during the early to early middle Miocene, the warmest period of the Neogene. Such westward migration seems to be owing to the combination of shifting the climatic belts and the "ecological opportunity" rather than the current direction. On the other hand, species originating in the northwest migrated during the cooling climate ages from the early Oligocene to Holocene. Cold-water species of the North Pacific spread to the Arctic, North Atlantic regions through the Bering Strait and to the Yellow and East China Seas through the Japan Sea in the Pliocene to Pleistocene as climate cooled.

No shallow-sea cold-water species evolved from the tropical northwestern Pacific biota in the latest Eocene to early Oligocene cooling. In contrast, some species of cold-water taxa originated in the northeastern Pacific became adapted to warm waters in the western Pacific. For example, *Phanerolepida transenna* (Watson 1879), *Ceratosoma fournieri* (Crosse 1861) and *Securella chitaniana* (Yokoyama 1926) live or lived in the warm waters of southwestern Japan. The reason for the different histories of trans-Pacific species originated in the east and in the west remains unclear.

6.9 Acknowledgements

I am very grateful to Prof. Geerat J. Vermeij (University of California at Davis), Prof. Kazushige Tanabe (University of Tokyo) and Dr. Enrico Savazzi (Institute of Earth Sciences, Uppsala, Sweden) for their critical reading of this manuscript and comments. This study was partly supported by Grant-in-aid for Scientific Research of the Japan Society for the Promotion of Science (C, 15540450, 2003-2005).

References

Adegoke OS (1967) Earliest Tertiary west American species of Platyodon and Penitella. Proc California Acad Sci 35: 1-22

Amano K (1994) Astartidae (Bivalvia) from the Pliocene and Lower Pleistocene strata in the Shin'etsu district, Central Japan. Bull Joetsu Univ Educ 14: 241-248 (in Japanese with English abstract)

Amano K (1996) Taxonomy and distribution of boreal species of *Macoma* (Bivalvia) from the Plio-Pleistocene strata in the Japan Sea borderland. Bull Joetsu Univ Educ 16: 289-297 (in Japanese with English abstract)

Amano K (1997a) Biogeography of the genus *Neptunea* (Gastropoda: Buccinidae) from the Pliocene and the lower Pleistocene of the Japan Sea borderland. Paleont Res 1: 274-284

Amano K (1997b) Taxonomy and distribution of the fossil *Volutomitra* (Gastropoda) from Japan. Bull Joetsu Univ Educ 17: 377-383 (in Japanese with English abstract)

Amano K (2000) Origin and migration of boreal molluscs around the Japanese Islands. Month Mar Sci Extra 20: 18-24 (in Japanese, title translated)

Amano K. (2001) Pliocene molluscan fauna of Japan Sea borderland and the paleoceanographic conditions. Bio Sci (Tokyo) 53: 178-184 (in Japanese)

Amano K, Hikida Y (1999) Evolutionary history of the Cenozoic bivalve genus *Kaneharaia* (Veneridae). Paleont Res 3: 249-258

Amano K, Vermeij GJ (1998a) Taxonomy and evolution of the genus *Ocinebrellus* (Gastropoda: Muricidae) in Japan. Paleont Res 2: 199-212

Amano K, Vermeij GJ (1998b) Origin and biogeographic history of *Ceratostoma* (Gastropoda: Muricidae). Venus (Jap Jour Malac) 57: 209-223

Amano K, Vermeij GJ (2003) Evolutionary adaptation and geographic spread of the Cenozoic buccinid genus *Lirabuccinum* in the North Pacific. J Paleont 77: 863-872

Amano K, Kanno S, Ichikawa A, Yanagisawa Y (1987) Molluscan fauna from the Tasnihama Foirmation in the western part of Joetsu City. –Studies on the molluscan fossils from the western part of Joetsu district, Niigata Prefecture (Part2)-. Bull Joetsu Univ Educ 6: 157-170 (in Japanese with English abstract)

Amano K, Vermeij GJ, Narita K (1993) Early evolution and distribution of the gastropod genus *Nucella*, with special reference to Miocene species from Japan. Trans Proc Palaeont Soc Japan NS 171: 237-248

Amano, K, Ukita, M, Sato, S (1996) Taxonomy and distribution of the subfamily Ancistrolepidinae (Gastropoda: Buccinidae) from the Plio-Pleistocene of Japan. Trans Proc Palaeont Soc Japan NS 176: 467-477

Amano, K, Lutaenko, KA, Matsubara, T (1999) Taxonomy and distribution of *Macoma* (*Rexithaerus*) (Bivalvia: Tellinidae) in the northwestern Pacific. Paleont Res 3: 95-105

Baba K (1990) Molluscan fossil assemblages of the Kazusa Group, south Kanto, central Japan. Keio Gijuku Yochisha, Tokyo. 445 pp. (in Japanese)

Baird W (1863) Descriptions of some new species of shells, collected at Vancouver Island and in British Columbia by J.K. Lord Esq., naturalist to the British North American Boundary Commission in the years 1858-1862. Proc Zool Soc London for 1863: 66-70

Cecca F (2002) Paleobiogeography of marine fossil invertebrates. Concepts and methods. Taylor & Francis, New York 273 pp

Coan EV, Scott PV, Bernard FR (2001) Bivalve sea shells of western North America. Marine bivalve mollusks from Arctic Alaska to Baja California. Santa Barbara Mus Nat Hist, Santa Barbara, 764 pp

Crosse H (1861) Description de deux *Murex* nouveaux. J de Conchyl 9: 351-354

Dall WH (1907) Descriptions of new species of shells, chiefly Buccinidae, from the dredgings of the U.S.S. "Albatros" during 1906, in the northwestern Pacific, Bering, Okhotsk and Japanese Seas. Smithsonian Misc Col 48: 341-373

Dall WH (1916) Prodrome of a revision of the chrysodomoid whelks of the boreal and Arctic regions. Proc Biol Soc Washington 29:7-8

Dall WH (1918) Notes on *Chrysodomus* and other mollusks from the North Pacific Ocean. Proc US Nat Mus 54: 207-234

Dall WH (1919) Descriptions of new species of Mollusca from the North Pacific Ocean in the collection of the United States of National Museum. Proc U S Nat Mus 56: 293-371

Durham JW, MacNeil FS (1967) Cenozoic migrations of marine invertebrates through the Bering Strait region. In Hopkins DM (ed) The Bering Land Bridge, Stanford Univ, Stanford 326-349

Egorov R (1993) Trophoninae (Muricidae) of Russian and adjacent waters.Ruthenica (Suppl) 1: 1-48. Faustman, WF (1964) Paleontology of the Wildcat Group at Scotia and Centerville Beach, California. Univ Calf Pubs Geol Sci 41: 97-139

Fujii Environmental Geology Institute (1998) The Japan Sea in the Cenozoic and geology and fossils from the Japan Sea borderland. Fujii Environmental Geology Institute, Toyama, 363 pp (in Japanese, title translated)

Gladenkov AYu (1998) Oligocene and lower Miocene diatom zonation in the North Pacific. Stratigr Geol Correlation 6: 150-163

Gladenkov AYu, Gladenkov YuB (2004) Onset of connections between the Pacific and Arctic Oceans through the Bering Strait in the Neogene. Stratigr Geol Correlation 12: 175-187

Gladenkov AYu, White LD, Gladenkov YuB, Blueford JR (2000) Cenozoic biostratigraphy of the Pogranichnyi Region, eastern Sakhalin, Russia. Palaeogeogr, Palaeoclimat, Palaeoecol 158: 45-64

Gladenkov YuB, Sinelnikova VN (1990) Miocene mollusks and Climatic Optimum in Kamchatka. Trans Acad Sci USSR Geol Inst 453: 1-172 (in Russian)

Gladenkov YuB, Gladenkov AYu, Titova LV (2000) Lower Miocene of East Kamchatka region. Geol Pac Ocean 15: 685-714

Golikov AN (1962) New species of gastropods of the genus *Neptunea* Bolten from the Far-Eastern seas of the USSR. Trans Acad Sci USSR Zool Inst 30: 3-10 (in Russian)

Goryachev VN (1987) To the history of the formation of *Neptunea* (Gastropoda, Buccinidae) fauna of the North Pacific. In Kafanov A (ed) Fauna and distribution of molluscs: North Pacific and Polar Basin, Far East Sci Center, Vladivostok 57-64

Grant US, Gale HR (1931) Catalogue of marine Pliocene and Pleistocene Mollusca of California and adjacent regions. Mem San Diego Soc Nat Hist 1: 1-1036

Hatai K, Nisiyama S (1952) Check list of Japanese Tertiary marine Mollusca. Sci Rep Tohoku Univ, Spec Vol 3: 1-464

Hickman CS (1972) Review of the bathyal gastropod genus *Phanerolepida* (Homalopomatinae) and a description of a new species from the Oregon Oligocene. Veliger 15: 107-112

Hickman CS (1974) *Nehalemia hieroglyphica*, a new genus and species of archaeogastropod (Turbinidae: Homalopomatinae) from the Eocene of Oregon. Veliger 17: 89-91

Hickman CS (2003) Evidence for abrupt Eocene-Oligocene molluscan faunal change in the Pacific northwest. In Prothero DR, Ivany LC, Nesbitt EA (eds) From Greenhouse to Icehouse, The marine Eocene-Oligocene transition, Columbia Univ Press, New York 71-87

Higo S, Callomon, P, Goto, Y (1999) Catalogue and Bibliography of the Marine Shell-bearing Mollusca of Japan. Elle Sci Pub, Yao 749 pp

Houart R (2001) A review of the Recent Mediterranean and northeastern Atlantic species of Muricidae. Arti Grafiche La Moderna, Rome 227 pp

Ilyina AP (1954) Molluscs from the Neogene deposits in South Sakhalin. Trans VNIGRI 10: 188-327 (in Russian)

Kafanov AI (1982) Cenozoic history of the malacofaunas of the North Pacific shelf. In Kussakin OG (ed) Marine biogeography. Object, methods and principles of zonation, Publ. House Nauka, Moscow: 134-176 (in Russian)

Kafanov AI (1999) Recent and fossil Clinocardiinae (Bivalvia: Cardiidae) of the World. III. Genus *Keenocardium* Kafanov, 1974 (Part 1). Bull Mizunami Fossil Mus 26: 1-76

Kafanov AI (2000) Recent and fossil Clinocardiinae (Bivalvia: Cardiidae) of the World. IV. Genus *Keenocardium* Kafanov, 1974 (Part 2). Bull Mizunami Fossil Mus 27: 109-123

Kafanov AI (2001) Recent and fossil Clinocardiinae (Bivalvia: Cardiidae) of the World. V. Genus *Ciliatocardium* Kafanov, 1974 (Part 1). Bull Mizunami Fossil Mus 28: 139-171

Kafanov AI (2003) Recent and fossil Clinocardiinae (Bivalvia: Cardiidae) of the World. VII. Tribus Serripedini Kafanov, 1975. Bull Mizunami Fossil Mus 30: 1-23

Kamada Y (1955) *Japelion yabei*, a new Miocene gastropod from the Joban coal-field, Japan. Venus 18: 185-188

Kanehara K (1937) On some Tertiary fossil shells from Hokkaido (Yesso). Japan J Geol Geogr 14: 155-161

Kang S (1995) Sedimentary facies and paleoenvironment of the lower Pleistocene Sogwipo Formation, Cheju Island, Korea. Quat Res 34: 19-38

Kanno S (1971) Tertiary molluscan fauna from the Yakataga district and adjacent areas of southern Alaska. Spec Pap, Palaeont Soc Japan 16: 1-154

Kanno S (1973) Japanese Tertiary cassidids (Gastropoda) and their related molluscs from the west coast of North America. Sci Rep, Tohoku Univ 2nd Ser, Spec Vol 6: 217-233

Kano K, Kato, H, Yanagisawa Y, Yoshida F (1991) Stratigraphy and geologic history of the Cenozoic of Japan. Rep Geol Sur Japan 274: 1-114 (in Japanese with English abstract)

Kase T, Miyauchi T, Amano, K (1996) A new species of the Family Ungulinidae from Wakkanai, Hokkaido, Japan. Venus 55: 97-103

Khomenko I P (1938) Stratigraphy of the Tertiary deposits of Schumidt Peninsula (North Sakhalin). Trans Geol Oil Inst, Ser A 103: 1-78 (in Russian with English summary)

Kitamura A, Kimoto K (2004) A reconstruction of the southern channel of the Japan Sea at 3.9-1.0Ma. Quat Res 43: 417-434 (in Japanese with English abstract)

Kosuge S (1970) Description of a new species of Volutidae from the East China Sea. Venus 29: 111-113

Kurita H, Obuse A, Ogasawara K, Hasegawa S, Amano K, Hisada K (2000) Oligocene-Middle Miocene palynostrsatigraphy (dynoflagellate cysts and pollen) in Sakhalin Island, Far East Russia, and its implications for geochronology and paleoenvironments. J Geogr 109: 187-202 (in Japanese with English abstract)

Kuroda T (1931) Mollusca. In Homma F (ed) Geology of Central Shinano, Kokin-Shoin, Tokyo 4: 1-90 (in Japanese)

MacNeil FS (1960) Tertiary and Quaternary Gastropoda of Okinawa.US Geol Surv Prof Pap 339: 1-148

MacNeil FS (1965) Evolution and distribution of the genus *Mya*, and Tertiary migrations of Mollusca. US Geol Surv Prof Pap 483G: G1-G51

Majima R (1989) Cenozic fossil Naticidae (Mollusca: Gastropoda) in Japan. Bull Amer Paleont 96: 1-159

Majima R, Murata, A (1992) Intraspecific variation and heterochrony of *Phanerolepida pseudotrasenna* Ozaki (Gastropoda: Turbinidae) from the Pliocene Nobori Formation, Pacific side of southwestern Japan. Trans Proc Palaeont Soc Japan NS 165: 1024-1039

Marincovich LJr (1977) Cenozoic Naticidae (Mollusca: Gastropoda) of the northeastern Pacific. Bull Amer Paleont 70: 169-494

Marincovich LJr (1988) The recognition of an earliest middle Miocene warm-water event in the southwestern Alaskan molluscan fauna. Saito Ho-on Kai Spec Pub 2: 1-24

Marincovich LJr (1990) Molluscan evidence for early middle Miocene marine glaciation in southern Alaska. Geol Soc Am Bull 102: 1591-1599

Marincovich LJr (2000) Central American paleogeography controlled Pliocene Arctic Ocean molluscan migrations. Geol 28: 551-554

Marincovich, LJr , Gladenkov, AYu (1999) Evidence for an early opening of the Bering Strait. Nature 397: 149-151

Marincovich LJr, Barinov KB, Oleinik AE (2002) The *Astarte* (Bivalvia: Astartidae) that document the earliest opening of Bering Strait.J Paleont 76: 239-245

Marko PB, Vermeij GJ (1999) Molecular phylogenetics and the evolution of labral spines among eastern Pacific Ocenebrine gastropods. Molec Phyl Evol 13: 275-288

Martin B (1914) Descriptions of new species of fossil Mollusca from the later Neogene of California. Univ Calif Publ Bull Dept Geol 8: 181-202

Masuda K (1986) Notes on origin and migration of the Cenozoic pectinids in the northern Pacific. Palaeont Soc Japan Spec Pub 29: 95-110

Masuda K, Noda H (1976) Check list and bibliography of the Tertiary and Quaternary Mollusca of Japan, 1950-1974. Saito Ho-on Kai Spec Pub 1: 1-494

Matsubara T (1996) Fossil Mollusca of the lower Miocene Yotsuyaku Formation in the Ninohe district, Iwate Prefecture, Northeast Japan. Part 2 (2). Gastropoda. Trans Proc Palaeont Soc Japan NS 181: 361-374

Matsubara T, Amano K (2000) A new species of *Ocinebrellus* (Gastropoda: Muricidae) from the Miocene Muraoka Formation in the Tajima District, Hyogo Prefecture, Southwest Japan. Venus 59: 201-207

McLean JH (1996) 1. The Prosobranchia. Taxonomic atlas of the benthic fauna of the Santa Maria Basin and western Santa Barbara Channel 9: 1-160

Moore EJ (1983) Tertiary marine pelecypods of California and Baja California: Nuculidae through Malleidae. Geol Surv Prof Pap 1228- A: A1-A108

Moore EJ (1984) Tertiary marine pelecypods of California and Baja California: Propeamussiidae through Pectinidae. Geol Surv Prof Pap 1228- B: B1-B112

Nagao T (1928) Paleogene fossils of the Island of Kyushu, Japan. Part 2. Sci Rep Tohoku Imp Univ 2 (12): 11-140

Nelson CM (1978) *Neptunea* (Gastropoda: Buccinacea) in the Neogene of the North Pacific and adjacent Bering Sea. Veliger 21: 203-215

Nemoto N, O'Hara S (1979) Molluscan fossils from the Taga Group in the Futatsu-numa area of Hirono-machi. Taira Chigaku Dokokai Kaiho Spec Vol 52-66 (in Japanese)

Nesbitt EA (2003) Changes in shallow-marine faunas from the northeastern Pacific margin across the Eocene/Oligocene boundary. In Prothero DR, Ivany LC, Nesbitt EA (eds) From Greenhouse to Icehouse, The marine Eocene-Oligocene transition, Columbia Univ Press, New York: 57-70

Noda H (1966) The Cenozoic Arcidae of Japan. Sci Rep Tohoku Univ 2nd Ser 38: 1-161

Noda H (1988) Molluscan fossils from the Ryukyu Islands, Southwest Japan. Part 2. Gastropoda and Palecypoda from the Shinzato Formation in the middle part of Okinawa-jima. Sci Rep Inst Geosci Univ Tsukuba Sec B 9: 29-85

Noda H, Ogasawara K (1976) On the variation of *Phanerolepida rehderi* MacNeil. Contr Dept Geol Miner Niigata Univ 4: 431-436 (in Japanese with English abstract)

Noda H, Watanabe R, Kikuchi Y (1995) Pliocene marine molluscan fauna from the Hitachi Formation in the northwestern part of Ibaraki Prefecture, Japan. Sci Rep Inst Geosci Univ Tsukuba Sec B 16: 39-93

Nomura S (1935) On some Tertiary Mollusca from Northeast Honshu, Japan. Pt.1. Fossil Mollusca from the vicinity of Narusawa Hot Spring, northeastern part Kurikoma Voilcano, Iwate-ken. Saito Ho-on Kai Mus Res Bull 5: 71-100

Nomura S, Hatai K (1940) The marine fauna of Kyuroku-sima and its vicinity, Northeast Honshu, Japan. Saito Ho-on Kai Mus Res Bull 19: 57-115

Ogasawara K (1986) Paleoenvironmental siginificance of the late Cenozoic "*Astarte*" (Bivalvia) in the northern Pacific region. Monogr Mizunami Foss Mus 6: 183-198 (in Japanese with English abstract)

Ogasawara K (1994) Neogene paleogeography and marine climate of the Japanese Islands based on shallow-marine molluscs. Palaeogeogr, Palaeoclimat, Palaeoecol 108: 335-351

Okutani T (1968) Systematics, ecological distribution and paleoecological implication of archibenthal and abyssal Mollusca from Sagami Bay and adjacent areas. J Fac Sci Univ Tokyo Sec 2 17: 1-98

Oleinik AE (1996) Genus *Arctomelon* (Gastropoda, Volutidae) in the Tertiary of the northwestern Pacific: evolution and adaptations. J Paleont 70: 236-246

Oleinik AE, Marincovich L Jr (2003) Biotic response to the Eocene-Oligocene transition: Gastropod assemblages in the high-latitude North Pacific. In Prothero DR, Ivany LC, Nesbitt EA (eds) From Greenhouse to Icehouse, The marine Eocene-Oligocene transition, Columbia Univ Press, New York: 36-56

Otuka Y (1935) The Oti Graben in southern Noto Peninsula, Japan. Part 3. Bull Earthq Res Inst 13: 846-909

Otuka Y (1940) Miocene Mollusca from Tesio Province, Hokkaido. Japan J Geol Geogr 17: 91-99

Ozaki H (1956) Some new and interesting molluscs from the Miocene Nobori Formation in the eastern Koti Prefecture, Japan. Bull Nat Sci Mus 3: 1-6

Ozawa T, Hayashi S, Endo M, Kumagai, T (2001) A molecular phylogenetic perspective on the origin of the Japanese marine molluscan fauna. Biol Sci (Tokyo) 53: 144-147 (in Japanese)

Pilsbry HA (1901) New Japanese marine, land and freshwater Mollusca. Proc Acad Nat Sci Philadelphia 53: 385-408

Popov SV (1983) Late Cenozoic and Recent bivalve molluscs of the Family Carditidae of the USSR. Trans PIN 203: 1-120 (in Russian)

Prothero DR (2001) Chronostratigraphic calibration of the Pacific coast Cenozoic: A summary. In Prothero DR (ed) Magnetic stratigraphy of the Pacific coast Cenozoic, Pacific Section SEPM Spec Pub 91: 377-394

Qi Z, Ma X, Wang Z, Ling G, Xu F, Dong Z, Li, F, Li, R (1989) Mollusca of Huanghai and Bohai. Agr Pub House, Beijing 309 pp (in Chinese)

Reid DG (1996) Systematics and evolution of *Littorina.* The Ray Society, London 463 pp

Roeding PF (1798) Museum Boltenianum sive catalogus cimeloriume tribus regnis naturae quae olim collegerat Joa. Fried Bolten, MD pd per XL, Annos Protophysicus Hamburgensis, pars secunda continens conchylia sive testacea univalvia, bivalvia et multivalvia, Pub J C Trapii, Hamburg 199 pp

Sato T, Kameo K (1996) Pliocene to Quaternary calcareous nannofossil biostratigraphy of theArctic Ocean, with reference to late Pliocene glaciation. Proc ODP 151: 39-59

Sato T, Higuchi T, Ishii T, Yuguchi S, Amano,K, Kameo K (2003) Calcareous nannofossil biostratigraphy of the upper Pliocene to lowermost Pleistocene distributed in the northern part of Akita Prefecture, Honshu, Japan –with special reference to the late Pliocene paleoceanography event. J Geol Soc Japan 109: 280-292 (in Japanese with English abstract)

Sato, T, Saito, T, Yuguchi, S, Nakagawa, H, Kameo, K, Takayama T (2002) Late Pliocene calcareous nannofossil paleobiogeography of the Pacific Ocean: evidence for galaciation at 2.75Ma. Rev Mexicana de Cienc Geol 19: 175-189

Sato T, Yuguchi S, Takayama T, Kameo K (2004) Drastic change in the geographical distribution of the cold-water nannofossil *Coccolithus pelagicus* (Wallich) Schiller at 2.74Ma in the late Pliocene, with special reference to glaciation in the Arctic Ocean. Mar Micropaleont 52: 181-193

Schrenck L (1862) Vorläufige Diagnosen eininger neuer Molluskenarten aus der Meerenge der Tartarei und dem Nordjapanischen Meere. Bull Acad Imp Sci St Petersbourg 4: 408-413

Sirenko BI, Gryachev VN (1990) A new species of the genus Ancistrolepis (Gastropoda, Buccinidae) from the South-China Sea. Zool J 69: 137-140 (in Russian)

Smith JT (1970) Taxonomy, distribution, and phylogeny of the cymatiid gastropods *Argobuccinum*, *Fusitriton*, *Mediargo*, and *Priene*. Bull Amer Paleont 56: 445-573

Squires RL (2003) Turnovers in marine gastropod faunas during the Eocene-Oligocene transition, west coast of the United States. In Prothero, DR, Ivany, LC, Nesbitt, EA (eds) From Greenhouse to Icehouse, The marine Eocene-Oligocene transition, Columbia Univ Press, New York: 14-35

Strauch F (1972) Phylogenese Adaptation und Migration eininger nordischer mariner Mollukangenera (*Neptunea*, *Panomya*, *Cyrtodaria* und *Mya*). Abh Senckenberg Naturf Gesell 531: 1-210

Swainson W (1840) A treatise on malacology, or the natural classification of shells and shell-fish. London 419 pp

Tiba R, Kosuge S (1985) North Pacific shells (15) *Trophon* Montfort (s. l.). Inst Malacology, Tokyo 36 pp (in Japanese and English)

Titova LV (1993) The early history of the North Pacific Ancistrolepidinae (Gastropoda: Buccinidae). Ruthenica 3:1-16

Titova LV (1994) Cenozoic history of Turritelloidea and Buccinoidea (Mollusca: Gastropoda) in the North Pacific. Palaeogeogr, Palaeoclimat, Palaeoecol 108: 319-334

Tuchida T (1975) Mean temperature, salinity and dissolved oxygen of East China Sea water. Mar Sci 63:33-37 (in Japanese, title translated)

Tsuda K (1959) New Miocene molluscs from the Kurosedani Formation in Toyama Prefecture, Japan. J Fac Sci Niigata Univ 2 (3): 67-110

Vermeij GJ (1991) Anatomy of invasion: The trans-Arctic interchange. Paleobiol 17: 281-307

Vermeij GJ (2001) Community assembly in the sea. Geologic History of the living shore biota. In Bertness MD, Gaines SD, Hay ME (eds) Marine community ecology. Sinauer Associates Inc, Massachusetts pp 39-60

Watson RB (1879) Trochidae continued, viz. the genera Basilissa and Trochus, and the Turbinidae, viz. the genus Turbo. J Linn Soc London Zool 14: 692-719

Weaver CS, du Pont JE (1970) Living volutes. A monograph of the Recent Volutidae of the World. Delaware Mus Nat Hist Mongr 1: 395 pp

Yamashita H (1978) Studies on the benthic organisms collected from the East China Sea and the Yellow Sea –V. Distribution of shellfish, Gastropoda, Scaphopoda and Pelecypoda. Bull Seikai Reg Fish Res Lab 51: 45-95 (in Japanese with English abstract)

Yokoyama M (1924) Molluscan remains from the lowest part of the Joban coal-field. J Coll Sci Imp Univ Tokyo 45: 1-22

Yokoyama M (1926) Tertiary Mollusca from southern Totomi. J Fac Sci Imp Univ Tokyo 2 (1): 313-364

Yoon S (1988) The Seoguipo molluscan fauna of Jeju Island, Korea. Saito Ho-on Kai Spec Pub 2: 539-545

Zhao R, Cheng J, Zhao D (1982) Marine Mollusca from Dalian. Marine Pub, Beijing 167pp (in Chinese, title translated)

7 Migration in amphibians and reptiles: An overview of patterns and orientation mechanisms in relation to life history strategies

Anthony P. Russell[1], Aaron M. Bauer[2] and Megan K. Johnson[1]

[1]Department of Biological Sciences, University of Calgary, 2500 University Drive N.W., Calgary, Alberta, Canada, T2N 1N4, [2]Department of Biology, Villanova University, 800 Lancaster Drive, Villanova, Pennsylvania 19085-1699, USA, corresponding author's e-mail: arussell@ucalgary.ca

7.1 Abstract

Differences in integumentary permeability dictate alternative life history strategies in amphibians and reptiles. Limiting resources for amphibians are chiefly associated with availability of water and, as a consequence, amphibian migrations are chiefly associated with movements to and from aquatic breeding habitats. These cyclic migrations from breeding to overwintering sites may be direct, or may be interrupted by periods of residence at foraging sites. In general, migrations take place over relatively short distances and are constrained by the problem of water balance associated with exposure during longer overland journeys. Many amphibians exhibit complex mechanisms of orientation involving multiple sensory modalities and are capable of precise homing abilities.

Among reptiles, migrations are chiefly associated with travel to and from egg laying sites, as displayed by turtles, or communal hibernacula, as is characteristic of some snakes. Marine turtles, in particular, undertake long distance migrations (up to thousands of kilometers) to reach nesting beaches and, sometimes, foraging grounds. In these chelonians, complex patterns of movement vary ontogenetically, as well as by gender and species. A variety of cues are used to locate destinations, including celestial, geomagnetic, olfactory, auditory, thermal, wave, and current pattern signals; however, evidence of a map-compass system of navigation is equivocal.

The migratory patterns of amphibians and reptiles often bring them into conflict with human resource utilization. For amphibians, breeding migrations that cross busy roads or areas turned over to agricultural production lead to mass mortality. Marine turtle migratory pathways often result in conflicts with fishery activities. Knowledge of life history strategies and associated migratory behavior is essential for effective conservation measures.

Key words: Conservation, foraging, hibernation, homing, life history, orientation.

7.2 Introduction

Although amphibians and reptiles do not constitute a monophyletic group, they are often treated together under the rubric of herpetology because they are terrestrial vertebrates that share the plesiomorphic physiological traits of ectothermy and poikilothermy. Although they share some aspects of their general biology because of these traits, they also differ considerably in terms of basic aspects of anatomy, physiology, behavior and reproductive biology. One major contrast is associated with water balance and the range of habitats that can be occupied as a result of osmoregulatory constraints. Amphibians have an integument that is highly permeable to water, and the rate of evaporation from it is similar to that from a free water surface (Katz and Nagel 1994). It is also a route of water absorption and ion uptake, and plays a major role in osmoregulation. Such properties of the amphibian integument render them behavioral osmoregulators, largely controlling water flux across the skin by occupancy of appropriate microhabitats. Conversely, the integument of reptiles shows low permeabililty to water, ions, and gases, and forms an effective permeability barrier (Lillywhite and Maderson 1982) that allows them to be physiological osmoregulators.

The above-stated differences in integumentary function, coupled with the differences in breeding biology of amphibians and reptiles, place differing demands upon their ability and their requirements to migrate although, of course, different species in each taxon will manifest this in slightly different ways. Many species of amphibian have a biphasic lifestyle with typically an aquatic larval stage and a terrestrial adult stage. Although there are many exceptions to this, most amphibian species that undergo metamorphosis undertake breeding migrations (of short to relatively long distances) between foraging and/or overwintering sheltering sites and breeding sites, and back again. These migrations can be physiologically challenging, and are usually triggered by particular combinations of environmental circumstances. Distances traveled are proportional to survivability in what can be osmotically very challenging conditions. Thus, although amphibian migrations may not encompass vast distances, they nonetheless require sophisticated behavioral and navigational mechanisms to undertake these repeated movements.

Reptiles, as amniotes, are able to cover considerably greater distances in migratory movements than amphibians, but their breeding biology, centered around the cleidoic egg or derivatives thereof (Russell and Bauer 2000) does not generally require migrations for reproductive purposes (although many of the longest and most spectacular reptile migrations are for reproductive purposes -- see below). Reptilian migrations are, therefore, much more likely to be based upon trophic resource exploitation, although common breeding sites may be the focus of periodic aggregations.

Dispersal and migration are two important and distinct types of animal movement. When an animal disperses it moves from one home range to another home range, and does not subsequently return to the original location. Furthermore, dispersal involves undirected movements of animals to unknown locations (Zug et al. 2001). Conversely, migration has been described as the movement of an animal from one location to another, followed by its subsequent return to the original location (Alcock 1975; Linzey 2001). Furthermore, migratory movements tend to be directed towards specific locations or goals. However, the migration of animals can be very complicated and highly variable between species, and thus broader definitions of migration as simply "the act of moving from one spatial-unit to another" (Baker 1978) or "a specialized behavior especially evolved for the displacement of the individual in space" (Dingle 1980) have also been adopted. These definitions indicate that migration occurs when the resources required by an animal are spatially or temporally separated, thus forcing movement in order to exploit these resources (Baker 1978; Semlitsch 1985; Pilliod et al. 2002). In this way migration also differs from the foraging movements of animals within their home range, as foraging involves movement within a home range from one feeding site to another, rather than movement between different types of resources (Baker 1978). Generally, migratory movements can be defined by five key features: 1) they are persistent and of greater duration than movements within the animal's home range; 2) movement is directional and takes place along relatively straight line paths (that is, having the properties of directed movement); 3) movements are purposeful and may be related to a scarcity of resources; 4) movements have a defined beginning and end; and 5) animals allocate energy specifically towards migratory movements (Dingle 1996). It is important to note that although these movements may be bidirectional, under this definition this is not necessary for a movement to be considered a migration. Furthermore, although these five characteristics are common components of many migrations, they may not be evident in all examples (Dingle 1996).

Although migrations do not have to be extensive, they do involve the movement of an animal beyond the boundaries of its normal home range, and thus can be very strenuous and energetically expensive (Alcock 2001). Thus, if an animal can acquire all the resources it requires without leaving the boundaries of its home range it will not migrate. Animals will only migrate if the benefits of moving outweigh the costs of the migration (Alcock 2001), but distance alone may not be the only arbiter of such costs as some turtle and tortoise species bypass seemingly good nesting sites on the way to much more distant ones. There are three primary reasons that an animal will migrate. Firstly, animals may undertake annual migrations between feeding areas and breeding, nesting (Luschi 2003), or hibernation sites if the feeding area does not contain suitable breeding or overwintering sites, although an assessment of suitability is sometimes difficult to define. Secondly, animals may migrate in response to habitat changes. For instance, migrations may occur in order to avoid drought or flood conditions. Finally, animals may migrate in response to habitat shifts in their prey, although this is often related to climatic changes as well (Pough et al. 1998; Linzey 2001).

It is evident from the foregoing that migratory movements will have both life history and long-term populational influences as they play out over time and over changing landscapes at several scales. Over the short term, amphibians and reptiles (with the exception of marine turtles -- see below) generally cover only short distances in their migrations (in contrast to many fish, birds and mammals), so that the genetic consequences of these movements may well be swamped out by occasional migrants from adjacent breeding areas. Dispersal results in the establishment of metapopulations and promotes genetic variability (Kozlowski 1999; Brede and Beebee 2004). The potential for dispersal may differ between sexes in some taxa (for example in the frog *Rana temporaria*), in which males are more philopatric than females (Palo et al. 2004). Dispersal in the short term can thus influence the genetic structure of populations and alter the consequences of fundamental migratory patterns. Longer-term patterns are influenced by geologic history and the combined effects of vicariance and waif dispersal (Garcia-París et al. 2003; Vences et al. 2003, 2004; Martinez-Solano et al. 2004; Pauly et al. 2004). In this contribution, we focus on the life history strategies of migration in amphibians and reptiles, but recognize that there are additional short-term and long-term consequences of such behaviors.

7.3 Homing and Orientation

One of the most important aspects of migration is that it generally involves directed, goal-oriented movement. Furthermore, animals frequently return to the same location year after year with great accuracy. It is thus essential to understand how reptiles and amphibians are able to accomplish these feats, as well as why homing and orientation abilities have developed in these animals. It is likely that orientation behavior evolved in parallel with migratory behavior, as animals that are able to follow a direct course towards the destination expend fewer resources searching, and are less at risk of predation (Sinsch 1990a). In addition to this, the ability to home allows animals to return to a familiar, reliable area, and may even serve to stabilize and equalize the distribution of a population, thereby decreasing competition between individuals (Stebbins and Cohen 1995). There are, however, some disadvantages to homing behavior. For instance, animals are vulnerable to habitat destruction, and may decline in reproductive success if their natal breeding site is destroyed. An example of this is evident in a population of toads that was found to return to the same location for years after their breeding pond had been destroyed (Stebbins and Cohen 1995). In general, homing and orientation abilities are advantageous to migrating animals, although the mechanism by which they accomplish this is poorly understood.

One of the difficulties in studying the sensory basis of orientation in amphibians and reptiles is that new sensory capabilities are constantly being identified. For example, Twitty's (1966) work on olfaction in blinded newts and their homing ability did not consider polarized light effects because these had not yet been revealed. The effect of magnetic fields on orientation is another relatively new area of research. The precise system used to detect magnetic information is still being investigated.

In general, orientation can be broken down into three levels (Fig. 1) which show increasing complexity, and which probably function using different mechanisms. Depending on the distance, and the complexity of the migration an animal undertakes, it may employ only the simplest level of orientation, or any combination of the three levels. The first level of orientation is piloting. An animal that pilots uses a fixed reference, such as visual, auditory, or chemical cues to guide itself towards the destination.

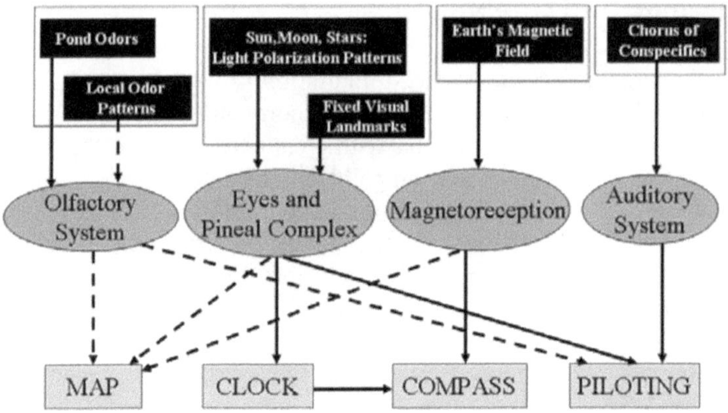

Fig. 1. Relationships between orientation cues (upper row, black boxes), perception systems (middle row, shaded ovals) and possible functional contributions to the map-compass concept (lower row, shaded boxes) for anuran amphibians. Solid arrows indicate demonstrated relationships; dashed arrows represent assumed relationships. After Sinsch (1990a, Fig. 4)

The second level of orientation involves compass orientation, which allows an animal to be aware of the direction in which it is moving. Finally the most complex level is map-compass orientation. This level of orientation involves a map-step, whereby the animal is able to determine its location relative to the goal, and a compass-step, whereby the animal determines the direction that will lead to the goal (Sinsch 1991; Phillips 1998). Map-compass orientation generally requires the use of more than one type of sensory input (Fig. 1), and an animal employing this orientation is capable of true navigation, meaning that it is able to orient towards a specific location even in the absence of familiar landmarks (Philips 1987).

7.4 Patterns of Amphibian Migration

Short-distance, or microgeographic migrations, are fairly typical of amphibians including many salamanders and anurans (Glandt 1986; Linzey 2001), though nothing is known about the migratory habits of caecilians (Stebbins and Cohen 1995). Many salamanders and frogs require foraging areas, aestivation or hibernation sites, and are dependent upon water for reproduction. If these resources are spatially separated, then an amphibian must migrate (Sinsch 1990a) (Fig. 2). However, the migratory behavior of amphibians is more constrained than that of other vertebrates, and they generally have much smaller migratory ranges (Sinsch 1990a). For instance, the migratory distance covered by amphibians may be as small as a few meters in some salamanders, and in some frog species like the arboreal *Hyla arborea* (Stumpel 1987). Migrations are generally not longer than 500m in salamanders, and not more than 1500m in anurans (Sinsch 1990a; Johnson 2003) (Fig. 2), and do not exceed speeds of 400m/night in salamanders (*Taricha rivularis*: Twitty et al. 1967) or 500m/night in anurans (*Bufo bufo*: Gelder et al. 1986). Even so, several anuran species, such as *Rana areolata aesopus* and *Bufo bufo*, have been observed to migrate up to 2-3km (Heusser 1968, Glandt 1986; Franz et al. 1988; Sinsch 1990b). The longest migration ever recorded for an amphibian is 15km for two species of European water frogs, *Rana lessonae,* and *R. esculenta* (Tunner 1992). However, most individuals of these species undertake much shorter migrations, and generally migrate only a few hundred meters (Tunner 1992). Even the longest amphibian migrations are very short compared to the much more extensive migrations of other vertebrates, including humpback whales (7000km), arctic terns (16000km), and sea turtles (2200km) (Sinsch 1990a).

The migrations of amphibians are limited for several reasons. Firstly, their skin is permeable to water, and thus they are highly susceptible to water loss. For this reason the migration of amphibians is limited to very humid periods, or they must stop frequently to rehydrate (Sinsch 1990a). Furthermore, amphibians are ectothermic, which further limits their migrations to a narrow range of environmental conditions (Sinsch 1990a).

The most common type of migrations among amphibians are seasonal breeding migrations from terrestrial habitats or hibernacula to aquatic breeding sites, and then back again (Martof 1953; Heusser 1968; Twitty et al. 1967; Orr 1982; Semlitsch 1985; Verrell 1987; Stebbins and Cohen 1995; Sinsch 1990a; Pough et al. 1998) (Fig. 2), although the occurrence of migration varies between species and populations. For instance, there is a great deal of variation in the migratory patterns observed among different groups of salamanders (Glandt 1986).

Breeding migrations are well documented for ambystomatids, salamandrids, and newts of the genera *Taricha* and *Notophthalmus* (Shoop 1965; Twitty et al. 1967; Hurlburt 1969; Gill 1978; Douglas and Munro 1981; Semlitsch 1985; Feldmann 1987; Petranka 1998; Johnson 2003), whereas most plethodontids do not migrate (Petranka 1998) and are not tied to water for reproduction.

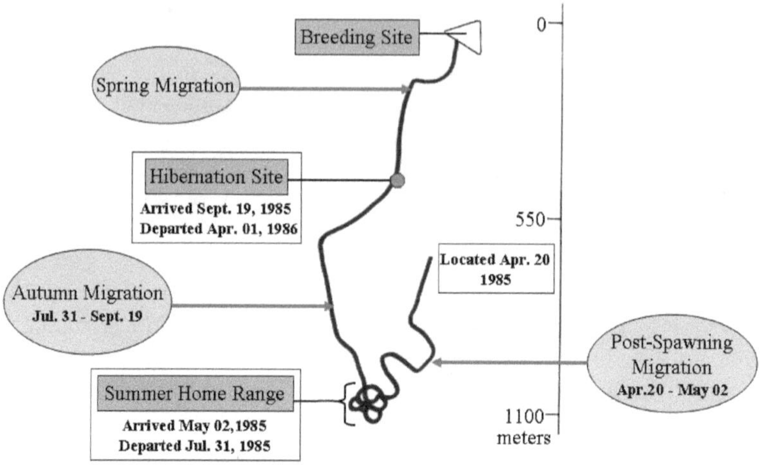

Fig. 2. Annual migrations of a male common toad (*Bufo bufo*) in Bavaria, Germany. Modified after Sinsch (1990a, Fig. 3)

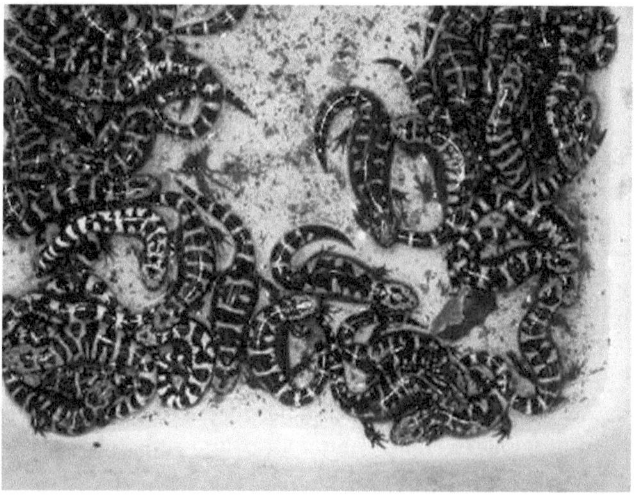

Fig. 3. Marbled salamanders (*Ambystoma opacum*) caught in a pitfall trap during their fall breeding migration at Jug Bay Wetlands Sanctuary, Maryland, USA. Picture reproduced with permission from Chris Swarth, Director and Karyn Molines, Sanctuary Naturalist, Jug Bay Wetlands Sanctuary

In amphibians that do migrate, breeding migrations are often highly synchronized, nocturnal, mass migrations (Fig. 3) that seem to be triggered by certain environmental conditions (Sinsch 1990a; Pough et al. 1998), but some salamanders and newts may migrate during the day (Hurlburt 1969; Semlitsch and Pechman 1985). Precipitation is the primary environmental factor triggering the migration of many amphibians, although falling barometric pressure in the absence of precipitation may be a trigger for some. Temperature is of secondary importance (Hurlburt 1969; Healy 1975; Semlitsch 1985; Sinsch 1990a; Palis 1997), although the specific cues that trigger the onset of migration vary between species and populations. For example, although precipitation is the most important factor for most amphibian migrations, for the salamanders *Eurycea quadridigitata* and *Ambystoma maculatum*, a drop in air temperature plays a major role in triggering the onset of migration (Semlitsch and McMillan 1980; Sexton et al. 1990). Within North America, there is also some variation in the migratory cues of northern and southern salamander populations (Fig. 3). Southern species, like the Mole salamander (*Ambystoma talpoideum*) tend to migrate during the fall and winter, and the onset of migration is generally associated with decreased temperatures and sometimes with falling barometric pressure. Conversely, northern salamander species migrate during the spring, when temperatures increase (Semlitsch 1985). However, the migrations of both northern and southern species are associated with increased precipitation, and the breeding migrations of numerous other salamander species, including *Ambystoma talpoideum* (Semlitsch 1985), *Ambystoma macrodactylum* (Beneski et al. 1986), and *Ambystoma tigrinum,* have also been found to be limited to periods of increased rainfall (Semlitsch 1985; Beneski et al. 1986; Loredo et al. 1996). Precipitation has also been found to be an important factor in the migration of anurans. For example, the migration of the African ranid frog *Hoplobatrachus occipitalis* from permanent bodies of water to ephemeral breeding ponds corresponds to the arrival of the first rains of the wet season (Spieler and Linsenmair 1998).

Although many amphibians undertake breeding migrations, the pattern of these is not always consistent within populations, and thus migration in these animals may be a conditional tactic (Glandt 1986; Alcock 2001). For instance, among Columbian Spotted Frogs (*Rana luteiventris*) some individuals migrate while others do not (Pilliod et al. 2002). Whether or not an individual migrates seems to be related to the resources available to it. If a certain site has all of the resources that a frog needs to survive, such as food and breeding sites, then there is no necessity to expend energy migrating (Pilliod et al. 2002). Variability among migratory patterns has also been observed among Spotted salamanders (*Ambystoma maculatum).* This species exhibits facultative emigration, and thus the timing of the emigration of salamanders from the breeding ponds back to terrestrial home ranges varies from year to year, and may occur in either the fall or spring, or may be split between them, depending on the environmental conditions (Madison 1997).

In some amphibian species, both adults and juveniles undertake migrations that may have different patterns. This is especially true for newts, which have both aquatic and terrestrial juvenile stages, and a terrestrial adult stage (Hurlburt 1969; Johnson 2003). Terrestrial juveniles, or efts, and adults of the species *Notophthalmus viridescens* both undertake migrations in the spring, but efts will also migrate in the fall (Hurlburt 1969). Furthermore, adult females generally migrate later than female efts (Hurlburt 1969). Conversely, in the anuran *Bufo hemiophrys* juveniles migrate later than adults (Kelleher and Tester 1969).

Among amphibians, variation between the migratory patterns of males and females of a species is also common. In the salamanders *Ambystoma jeffersonianum* and *A. maculatum* males tend to migrate earlier than females (Douglas 1979; Sexton et al. 1990). Males seem to have a lower threshold for environmental cues, and will migrate as soon as the minimal conditions for migration are met. Females, on the other hand, will migrate only when conditions have improved and stabilized, such that there is reduced environmental danger during their migration (Douglas 1979). This difference in the migratory patterns of male and female salamanders relates to the differing reproductive strategies of the sexes. Females are a limiting resource to male reproductive success, and thus males arrive at the breeding site early in order to enhance their reproductive success by ensuring that they are present when the females arrive. Conversely, females do not need to compete for access to mates, and thus are able to arrive later, allowing them to travel when conditions are safer, and also allowing them to have a greater choice of mates when they arrive at the breeding pond (Douglas 1979). In the Chinhai salamander, *Echinotriton chinhaiensis,* only females migrate to spawning ponds, as fertilization appears to take place when the animals are dispersed (Xie et al. 2000).

Male and female migrations also differ in several species of anurans, including *Rana luteiventris*, *Rana clamitans, Bufo hemiophrys* and *Bufo microscaphus californicus* (Martof 1953, Sweet 1993; Pilliod et al. 2002). For instance, male *Rana clamitans* have been observed to visit more than one breeding area in order to meet reproductive demands, while females remain in only one breeding area (Martof 1953), and male *Bufo hemiophrys* and *B. bufo* have been found to migrate earlier than females (Kelleher and Tester 1969; Gittens 1983). Another example is evinced by Columbian Spotted frogs and Wood frogs, in which females are found to undertake longer migrations than males (Pilliod et al. 2002). This difference may be related to a physical limitation, as males of these species tend to be smaller than the females, and thus may not be capable of long migrations; or it may be related to the differing reproductive strategies of males and females, with males remaining near breeding sites so that they can more easily compete for females in the spring. Another possible explanation is that males require less energy than females and thus they do not have to migrate to better foraging sites (Pilliod et al. 2002).

In addition to breeding migrations, some amphibians undertake migrations from summer feeding ranges or breeding areas, to overwintering or hibernation sites (Fig. 2). This behavior has been observed in numerous anuran species, including *Rana clamitans* (Martof 1953; Lamoreux and Maddison 1999), *R. sylvatica* (Bellis 1965), *Scaphiopus holbrookii* (Pearson 1955 1957) and *Bufo hemiophrys* (Kelleher and Tester 1969). *Rana clamitans* overwinters in streams, which have high oxygen levels, and which rarely freeze (Lamoreux and Maddison 1999). Other species, like *Bufo bufo,* which undertake relatively long breeding migrations, will overwinter in locations along the migratory route between their summer ranges and breeding ponds (Sinsch 1988) (Fig. 2). Generally the hibernacula of *Bufo bufo* are relatively close to their breeding sites (Sinsch 1990a) (Fig. 2). The newt *Notophthalmus viridescens* also overwinters at sites near the breeding pond (Gill 1978). Italian crested newts (*Triturus carnifex*) have been observed to migrate from their breeding site to hibernation sites in an old growth spruce forest, with individuals seeking refuge in small mammal burrows (Schabetsberger et al. 2004). Females migrated significantly further than males (median distance 168 m versus 53 m).

Other migrations may occur seasonally or sporadically in association with changing environmental conditions. Tailed frogs, *Ascaphus montanus*, may migrate along stream courses seasonally to avoid high water temperatures (Adams and Frissell 2001), and amphiumas may escape loss or degradation of their aquatic habitats by short overland migrations (Aresco 2002). Mass overland migrations of relatively long distance by aquatic frogs of the genus *Xenopus* have been observed in various parts of Africa and have been associated with the abandonment of drying ponds and dams (Hewitt and Power 1913; Loveridge 1953; Brass 1983). As these frogs are entirely dependent upon suitable aquatic habitats for feeding, breeding, development, and all other life functions, this represents perhaps the ultimate in resource-driven migration.

Despite the variation in migratory patterns that exist among amphibians, most species demonstrate a high level of fidelity to certain sites (Johnson 2003), often their natal pond, and are able to find these sites year after year. Some may even continue to return after the site has been destroyed (Twitty et al. 1967). This behavior is common in most groups of amphibians. For example, marking experiments have shown that the newt *Taricha rivularis* returned to the same stream sections to breed for 11 years, even when numerous other, apparently suitable alternative streams exist in the area (Twitty 1966; Twitty et al. 1967). *Notophthalmus viridescens* shows fidelity to breeding sites, and adults will return to their natal pond to breed after spending 2-4 years on land (Hurlburt 1969; Gill 1978). A metapopulation of the crested newt (*Triturus cristatus*) in Germany was shown to display a high level of breeding pond fidelity, with a very low rate of inter-pond migration (1.3 - 9.0%). Exchange of first-time breeders provided the main source of interaction between breeding ponds within the metapopulation (Sinsch et al. 2003). Fidelity to breeding sites is also common among anurans. For example, *Rana sylvatica* has been found to be 100% faithful to its breeding ponds, even when other apparently suitable ponds occur in the vicinity (Berven and Grudzien 1990).

In addition to breeding site fidelity, some amphibians demonstrate hibernation site fidelity. This has been documented for *Salamandra salamandra,* which has been observed to return to the same hibernation site for 20 years (Feldman 1987), and for the toad, *Bufo hemiophrys* (Kelleher and Tester 1969). Many migrating amphibians also tend to enter and exit their breeding ponds in the same location (Shoop 1965; Douglas and Munro 1981; Phillips and Sexton 1989), although some species may show considerable annual variation that is dependent upon sex and life stage (Dodd and Cade 1998). Some species have been observed to follow relatively straight migratory courses (Twitty 1966; Douglas and Munro 1981; Sinsch 1988), even when displaced over great distances (Twitty 1966). For *Taricha rivulina* the return journey to the home site is very slow, and after displacements of several kilometers may take two or more years (Twitty 1966). These characteristics, and site fidelity exhibited by amphibians, indicate that these animals are capable of at least some degree of homing and orientation (Young 1981; Semlitcsh 1985; Phillips 1987; Sinsch 1990a; Pasanen and Sorjonen 1995; Stebbins and Cohen 1995; Susuma et al. 1995; Pough et al. 1998).

7.5 Homing and Orientation in Amphibians

Experimental evidence of homing and orientation to breeding ponds or terrestrial sites has been documented for at least 13 species of salamanders, including ambystomatids, plethodontids, and salamandrids, and at least 16 species of anurans, including hylids, bufonids, pelobatids and the genus

Ascaphus. Since homing is common to both salamanders and anurans, it is possible that the common ancestor of lissamphibians was capable of homing, a postulate supported by the commonality of this behavior with fishes (Sinsch 1992). Even though many amphibians are capable of homing to and from their breeding ponds, most are not, however, able to home from outside their natural migratory range. This implies that most amphibians are not capable of true navigation, and that they must be familiar with an area in order to orient and home properly (Sinsch 1990a; Zug et al. 2001). One exception to this is *Taricha rivularis,* which has the ability to home from unfamiliar locations in excess of 30 km from their natural home range (Philips 1987). Experiments have shown that amphibians use a variety of cues for homeward orientation, and that the cues used may vary considerably between species (Sinsch 1987 1990b 1992; Papi 1992; Wilson 2001) (Fig. 1).

Olfaction is an important cue for most migrating amphibians (Fig. 1). Many species of amphibian, including newts (*Taricha rivularis*: Grant et al. 1968; *Notophthalmus viridescens*: Hershey and Forester 1980; *Triturus alpestris*: Joly and Miaud 1993), and anurans (*Bufo japonicus:* Susuma et al. 1995), are unable to orient, and show random movements, when their olfactory nerves are severed. Furthermore, when given a choice between foreign and home pond substrate, the salamanders *Ambystoma maculatum* and *Triturus alpestris,* and several anuran species, including *Bufo nebulifer* (as *B. valliceps*), *Pseudacris clarkii, Pseudacris streckeri,* and *Rana sphenocephala utriculara,* preferred their home substrate, suggesting olfaction to be an important aspect of homing in these species (Martof 1962; Grubb 1973a 1975 1976; McGregor and Teska 1989; Joly and Miaud 1993). It is possible that these odor preferences develop in the embryonic stages and are retained into adulthood (Hepper and Waldman 1992). In the toad *Bufo nebulifer*, reproductive hormones may be responsible for triggering odor preferences (Grubb 1973b). It has also been suggested that amphibians establish an olfactory map of their migratory range (Fig. 1) based on a gradient of odors, which allows them to orient in familiar areas (Sinsch 1990a; Susuma et al. 1995). Therefore, an amphibians' ability to locate and perceive a pond depends on the intensity of the olfactory gradient (Ambrogio and Gillis 1998). For species such as *Ambystoma opacum,* however, which construct nests in dry vernal ponds that later fill with water, olfactory cues may be less important (Shoop and Doty 1972).

In general, olfactory cues seem to be more important in species that undertake long distance migrations, like those of the genus *Bufo*, than for those that live close to the breeding site (Semlitsch 1987; Sinsch 1987

1990b), and may play a role in the initial orientation towards the breeding pond (Grubb 1973a,b 1975; Sinsch 1987 1990a,b).

In addition to olfactory cues, visual cues also play an important role in amphibian orientation and migration. There are two main types of visual cues: celestial cues and fixed visual landmarks (Sinsch 1990a) (Fig. 1). Fixed visual landmarks are probably important for short-distance piloting (Sinsch 1987; 1990b). For instance, Great Crested Newts (*Triturus cristatus*) use cues in the vicinity of their breeding pond to orient (Malmgren 2002), and *Bufo bufo* uses visual landmarks to help it maintain a straight course (Heusser 1969). On the other hand, celestial cues may be used by amphibians for compass or Y-axis orientation (Sinsch 1990a). The home shore is an important landmark for amphibians, and many tend to orient 90 degrees to it. This is Y-axis orientation, and it allows amphibians to orient along the most direct route to and from the shore, which is important for breeding migrations (Taylor and Auburn 1978; Stebbins and Cohen 1995; Zug et al. 2001). This type of orientation has been observed in numerous amphibian species, including the newt, *Notophthalmus viridescens* (Philips 1987). The establishment of an X-Y compass for Y-axis orientation requires light information from the sky, a familiarity with the home shore, and an internal clock phased to local time (Newcomer et al. 1974; Stebbins and Cohen 1995). However, the type of celestial cues used varies between species; light information may be received from the sun or stars using the eyes, or polarized light may be perceived by extraocular photoreceptors in the upper part of the brain, such as the pineal or frontal organ (Taylor and Ferguson 1970; Taylor 1972; Adler 1976; Demian and Taylor 1977) (Fig. 1). Some amphibians require a clear sky to orient, whereas others do not, and amphibians may or may not (depending upon species) be able to orient when blinded (Twitty et al. 1967; Stebbins and Cohen 1995). Many species, including *Notophthalmus viridescens* and *Ambystoma tigrinum,* are able to orient when blinded, but become disoriented when their pineal organ is covered (Taylor and Adler 1978; Hershey and Forester 1980; Hairston 1994; Stebbins and Cohen 1995), suggesting that these species use polarized light perceived by extraocular photoreceptors for orientation. This is also the case in eyeless red-spotted newts (Demian and Taylor 1977).

Furthermore, experiments have suggested that some species of amphibians use visual cues in conjunction with olfactory cues to orient. For example, with its pineal gland covered, *Notophthalmus viridescens* is unable to orient, but still exhibits directional movement. However, when its olfactory nerves are severed, this species exhibits random movements, suggesting that both senses in combination are important for orientation (Hershey and Forrester 1980). Furthermore, *Bufo valiceps* is able to orient if either its vision or olfactory sense is impaired, but not if both are damaged, indicating that it can use either sense for orientation (Grubb 1970). It is possible that visual cues are less important in amphibian species that migrate over the course of a single rainy or overcast night (Duellman and Trueb 1986).

For anurans, short-distance orientation to the breeding pond may be accomplished using acoustic cues (Bogert 1947; Brattstrom 1962; Sinsch 1990a). For example frogs may use the calls of conspecifics to guide them to the pond (Fig. 1). These cues are generally only useful for short distance movements of 10-100m, although the calls of chorus frogs (*Pseudacris triseriata)* and natterjack toads (*Bufo calamita)* can be heard from up to 1km away (Sinsch 1990a). Studies of natterjack toads have shown that displaced females will orient towards the nearest pond that has calling males, and are unaffected by a deprivation of olfactory cues. Conversely, males were found to return faithfully to the original breeding pond, and were disoriented when deprived of olfactory or visual cues (Sinsch 1992). Thus, in this species, auditory cues seem to play an important role in the orientation and piloting of females, whereas males use a different set of cues, including magnetic, olfactory and visual cues, to orient (Sinsch 1992). However, since calling is energetically expensive, and since these cues cannot be used for orientation during migration away from the breeding pond, acoustic cues are probably of secondary importance in the orientation of most anurans.

Some amphibians may use magnetic cues for orientation (Phillips and Adler 1978; Phillips 1986a,b) (Fig. 1), this ability being found in widely disparate taxa of amphibians, and vertebrates in general (Phillips 1977; Sinsch 1990a). The ability to detect magnetic fields was first identified in the salamander, *Eurycea lucifuga* (Phillips 1977), and has since been found in other salamanders, newts and in some frogs. For instance, male natterjack toads were disoriented when bar magnets were glued to their heads, disrupting the magnetic field (Sinsch 1992). Magnetic cues are likely used for simple compass orientation in such species, and are probably used in conjunction with other sensory information, including visual and olfactory cues (Phillips 1986a; Diego-Rasilla 2003). However, in *Notophthalmus viridescens* this ability may be related to their capability of using true navigation (Phillips 1986a,b 1987; Pough et al. 1998; Zug et al. 2001). Newts of this species are capable of orienting towards their home pond after being displaced 30km away in the absence of all types of cues during the displacement (Phillips 1987; Phillips et al. 1995). They are able to detect the earth's magnetic field, and may form a magnetic map based on the spatial variation in the magnetic field, which, combined with a compass sense, allows them to navigate (Phillips and Borland 1992a,b

1994; Phillips et al. 2001; Zug et al. 2001). This species may detect the magnetic field either via visual centres in the brain, a trigeminal nerve system or a magnetite-based receptor in the head (Zug et al. 2001). Recent work has suggested that these newts have two separate magnetic receptors, with one being sensitive to changes in the wavelength of ambient light and related to the visual system, and the second a non-light dependent detector sensitive to the polarity of the magnetic field (Phillips and Borland 1994). It is possible that both types of receptors are necessary for true navigation (Deutschlander et al. 1999).

The newt *Taricha torosa* is the only species so far that has been shown to use kinesthetic senses in migration. Newts that were spun around showed erratic migratory courses, indicating body position is important for maintaining a straight migratory course (Endler 1970). Geotactic and hygrotactic responses may also be important, as some amphibians, including juvenile *Notophthalmus viridescens,* have been shown to migrate along depressions, following humidity gradients (Hurlburt 1969; Duellman and Trueb 1986). However, these senses probably play only a secondary role in the homing and orientation of most amphibians.

Overall, amphibians employ a multisensory system of orientation that involves the complex integration of many different cues (Sinsch 1991) (Fig. 1). They most likely use olfactory or visual cues, and in the case of newts, magnetic cues, to determine their location, and then use celestial cues to establish a compass direction (Sinsch 1990a 1991). However, it is important to note that amphibians are likely capable of using alternate cues if the preferred cue is unavailable (Sinsch 1987 1990b 1992; Papi 1992), and that cues are used in combination, but with one assuming dominance over others in a pattern that varies by species.

7.6 Patterns of Reptilian Migration

The scale of reptile migrations varies dramatically, from a few hundred metres to thousands of kilometers. Unlike amphibians, few reptiles migrate to common areas to breed, and most do not migrate at all (Cogger and Zweifel 1998; Pough et al. 1998). This is because reptiles do not depend upon water for reproduction, and thus most terrestrial reptiles do not need to travel long distances and instead generally lay their eggs at nesting sites within their normal home range (Pough et al. 1998). However, there are many other reasons for which reptiles may migrate. For instance, they may migrate in response to seasonal habitat changes, or to follow their prey. Migration may also be a consequence of laying their eggs on land, with journeys being undertaken to reach suitable, terrestrial nesting sites (Pough et al. 1998), as is the case for many aquatic reptiles, including sea turtles, freshwater turtles, and some crocodilians. The most extensively studied of these nesting migrations are those of sea turtles.

Sea turtles often undertake resource-driven migrations of several thousand kilometers between nesting and feeding grounds (Meylan 1982) (Fig. 4). Marine turtles undertake these long-distance, or macro-geographic, migrations, because their nesting sites are often distantly separated from their feeding grounds (Bowen et al. 1989; Linzey 2001). Different species of sea turtles exhibit different migratory patterns (Luschi et al. 2003) (Fig. 4). For example, Green Turtles (*Chelonia mydas)* from various colonies in the Indian and Pacific oceans regularly migrate over 1000 km between their feeding grounds and nesting beaches; Olive Ridley turtles (*Lepidochelys olviacea)* tagged in Surinam have been found up to 1900 km away; and Loggerhead turtles (*Caretta caretta)* tagged in South Africa have been observed in Zanzibar, almost 3000 km away (Hughes 1974; Pough et al. 2002). Highly nomadic Leatherback sea turtles (*Dermochelys coriacea)* tagged in French Guiana have been recaptured in a variety of locations up to 5000 km away, including Texas, Mexico, New Jersey, New England, Nova Scotia and Africa (Luschi et al. 1996; Pough et al. 2002). The migratory patterns of these species may be fixed or not, depending on whether they exploit predictable or unpredictable resources. However, one common characteristic of marine turtle migrations is that despite the enormous spans of open ocean they traverse, these turtles are able to return with great precision to the same nesting beaches year after year (Meylan 1982) (Fig. 5).

The most well studied of these migrations is that of Green Sea Turtles (*Chelonia mydas*) (Fig. 4A). These turtles inhabit tropical and subtropical waters, where they forage in the neritic zone of inshore waters (Carr 1965; Luschi et al. 1998; Godley, Lima et al. 2003). Every 2-4 years these turtles traverse the open ocean, using fast, directed movements along migration corridors to reach nesting sites thousands of kilometers away (Miller 1998; Luschi et al. 1997; Luschi et al. 2001). In the Caribbean Sea and North Atlantic Ocean, Green Turtles utilize four major nesting sites located on Tortuguero, Aves Island, the Surinam coast, and Ascension Island. Post-nesting females migrate hundreds or thousands of kilometers to return to foraging areas, although some may remain near the nesting beach for a time (Luschi et al. 1998; Garduno et al. 2000; Papi et al. 2000; Hays et al. 2001). Females from distantly separated nesting beaches may use the same feeding grounds (LeGall and Hughes 1987).

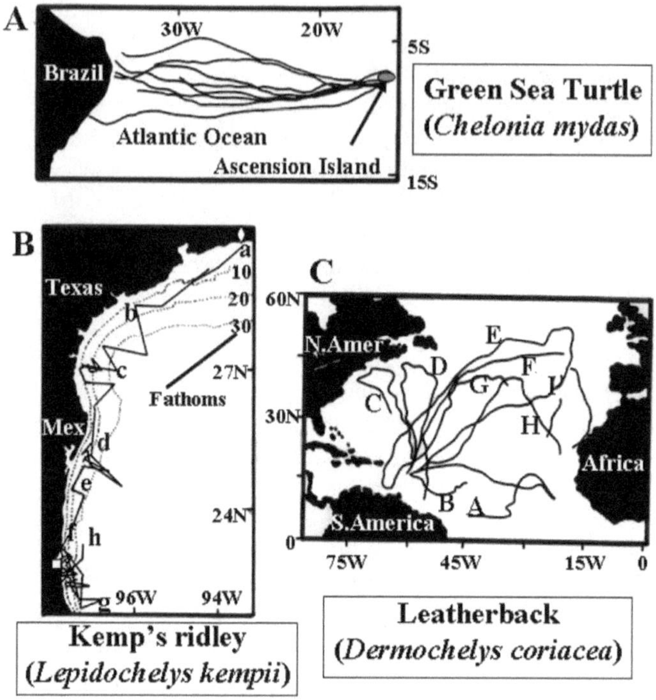

Fig. 4. Representative migration patterns of marine turtles. **A.** Routes of eight Ascension turtles (*Chelonia mydas*) undertaking a transoceanic shuttling migration from their breeding grounds on Ascension Island to their feeding grounds along the Brazilian coast. Modified from Luschi, Hays and Papi (2003, Fig. 1). Elapsed time for the journey between 33 and 74 days. **B.** Movement of a satellite-tracked Kemp's ridley (*Lepidochelys kempii*) from Cameron, Louisiana (a, white diamond) to a nesting site at Rancho Nuevo, Mexico (white square). The tracking period lasted from 13 August 1994 (Cameron, Louisiana -- a) to May 16 1995 (point h). The time course of the migration is indicated by the letters a-h: a - August 13 1994; b - December 10 1994; c - December 18 1994; d - January 17 1995; e - March 8 1995; f - March 10 1995; g - March 18 1995; h - May 16 1995. Modified from Renaud et al. (1996, Fig. 1). **C.** Pan-Atlantic foraging movements of nine leatherback turtles (*Dermochelys coriacea*) tracked after nesting in the Caribbean. Individuals A and B were tracked for 12 months (July 2002 - July 2003). Individuals C to I were tracked for 6-8 months (May-July 2003 to January 2004). Modified after Hays, Houghton and Myers (2004, Fig. 1a)

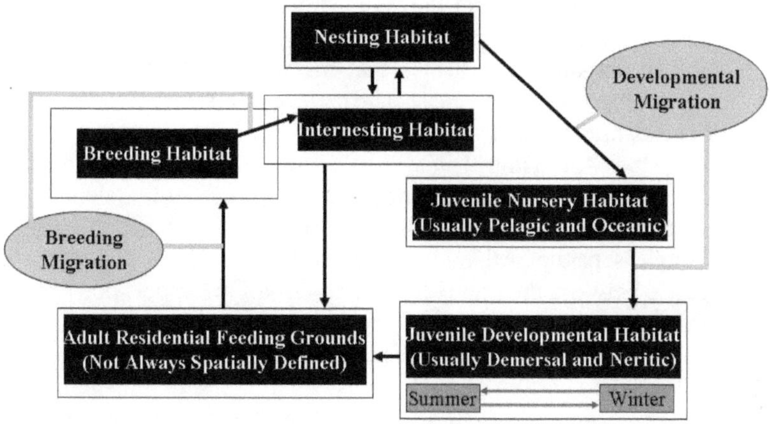

Fig. 5. Conceptual model of a generalized sea turtle life cycle, after Luschi (2003, Fig. 6.1) and Musick and Limpus (1997, Fig. 4.1)

As many as 50,000 *Chelonia mydas* come ashore to nest on the beaches Tortuguero, Costa Rica each year. Tens of thousands of these turtles have been tagged, and have been recovered at feeding grounds as far away as Panama, Colombia, Venezuela and the Yucatan Peninsula (Carr et al. 1978). The reproductive circuits identified for this Tortugeura – Western Caribbean population of Green Turtles (Carr and Giovannoli 1957) are very similar to other populations located near Yemen in the Western Indian Ocean (Hirth and Carr 1970), the central South Pacific (Craig et al. 2004), the Mediterranean (Godley et al. 2002), and southern Chinese waters (Cheng 2000).

The Ascension Island population of *Chelonia mydas* has also been extensively studied (Fig. 4A). These turtles migrate 2200 km from their feeding grounds off the coast of Brazil, where their population mixes with turtles that nest on the coast of Surinam and French Guiana, to their nesting beaches on Ascension Island (Carr 1962; Mortimer and Carr 1987). Ascension Island is a tiny island of only 20 km diameter located along the mid-Atlantic ridge (Pough et al. 2002), and does not possess adequate resources for feeding (Rebel 1974). Turtles return to it with incredible precision every 2-3 years in order to breed and lay their eggs, a journey which takes six weeks to complete (Bowen et al. 1989; Luschi et al. 1998; Pough et al. 2002). Females demonstrate nest site fidelity, and return to the same beach, which is often their natal beach. This has been confirmed using mitochondrial DNA evidence (Meylan et al. 1990).

It has been suggested that the isolated breeding population of turtles on Ascension Island is the result of sea-floor spreading (Carr and Coleman 1974). This hypothesis indicates that historically, Ascension Island was located much closer to the coast of Brazil, and that sea floor spreading at the mid-Atlantic ridge has slowly moved the island farther from Brazil, requiring the Ascension Island turtles to migrate across increasingly greater distances to their nesting beach (Carr and Coleman 1974). However, examination of the mitochondrial DNA of female Green turtles from three separate Atlantic rookeries has revealed that the genetic divergence between the populations is too small to support this hypothesis, and instead suggests a more recent origin of the Ascension Island population of turtles, probably in the past few thousand years (Bowen et al. 1989). Thus the question of how turtles began to nest on this remote island remains unanswered. However, although the females from the three rookeries are closely related, the rookeries are genetically distinct from each other (Bowen et al. 1989; Avise and Bowen 1994). This suggests that not only do females show nest site fidelity, but that they return to their natal beach year after year (Bowen et al. 1989; Avise and Bowen 1994).

The range of Eastern Pacific Green Turtles (*Chelonia mydas agassizi*) is restricted to the west coast of the Americas, from Baja California to southern Peru. Females migrate from foraging grounds to nesting sites along the Mexican coast every 3-4 years (Alvarado et al. 2003). The primary nesting site of these turtles is Michoacan. After breeding the turtles undertake a return migration to their foraging grounds in the Gulf of California and coastal Central and South America (Byles et al. 1995).

Hawksbill turtles (*Eretmochelys imbricata*) show a similar migratory pattern to that of Green Turtles. These turtles are widely distributed, and generally inhabit coral reefs. This species was originally thought to be non-migratory, but radio-telemetry studies have revealed that hawksbill turtles may migrate long distances (Parmenter 1993; Meylan 1999; Horrocks et al. 2001; Buitrago and Guada 2002). There is evidence that males of the species are also highly migratory (Nietschmann 1981). The nesting migrations of this species are not well understood, but it appears that females return to their natal beaches to nest every 2-3 years (Witzell 1983; Bass 1999), and that females inhabiting the same foraging areas do not migrate to the same nesting beaches. After laying their eggs Hawksbill turtles immediately migrate back to their foraging grounds (Horrocks et al. 2001). There is some variation in the length of migration that the females undertake; females nesting on the same beaches exploit different foraging areas that may be either very close to, or a long distance from the nesting beach, and thus migrations may take place across deep oceans or through shallow water (Ellis et al. 2000).

Kemp's ridley turtles (*Lepidochelys kempii*) exhibit a different migratory pattern from that of Green Turtles or Hawksbill turtles (Fig. 4B). These turtles have a restricted range in the Gulf of Mexico and the western Atlantic, and migrate in shallow waters along the United States and Mexican coast along a well-defined migration corridor (Byles and Plotkin 1994; Marquez 1994). Mature females migrate to, and aggregate at the nesting beaches prior to the breeding season, whereas males appear to be non-migratory, and remain at the nesting beach year round (Plotkin 2003). The migration of females takes place annually, and the majority of the population of Kemp's ridley turtles migrates to the same beach at Rancho Nuevo, Mexico (Fig. 4B), although some turtles nest at a few adjacent beaches in Vera Cruz, as well as in Texas and other areas of the southern United States (Renaud et al. 1996). After breeding, females migrate back to their restricted feeding ranges (Marquez 1994; Byles and Plotkin 1994; Renaud et al. 1996).

Adult Loggerhead turtles (*Caretta caretta*) inhabit subtropical and temperate waters (Polovina et al. 2004) around continental shelves and estuarine areas, and spend the majority of their time in near shore and inshore waters (Dodd and Byles 2003; Godley, Broderick et al. 2003). Unlike some other marine turtle species, female loggerhead turtles are not restricted to any one, fixed foraging area, but instead move continuously between a series of foraging areas (Dodd 1988; Papi et al. 1997; Plotkin and Spotila 2002). Females migrate hundreds to thousands of kilometers (Limpus and Limpus 2001; Bentivegna 2002) to breeding sites approximately once every three years, and generally migrate in inshore waters, with only brief offshore movements (Papi et al. 1997; Plotkin and Spotila 2002), although some populations may undergo trans-Pacific migrations along specific migratory corridors with particular thermal characteristics (Resendiz et al. 1988; Nichols et al. 2000). Male and female Loggerhead turtles migrate asynchronously, with males arriving at the breeding site weeks in advance of the females (Limpus 1985; Dodd 1988). However, it is possible that some males are non-migratory, and remain near the nesting beaches year round (Henwood 1987).

Another species of turtle that exhibits primarily inshore migrations is the Flatback turtle (*Natator depressus*). These turtles have the most restricted migratory range of all sea turtles. They have feeding grounds around Indonesia and New Guinea, and migrate with a high degree of fidelity to the nesting beaches on the northern coast of Australia every 1-3 years (Limpus et al. 1981 1983 1984; Parmenter 1994). Postnesting migrations may be hundreds to thousands of kilometers in length, and are generally restricted to inshore waters around northern Australia, and as far as Papua (western New Guinea).

Leatherback turtles (*Dermochelys coriacea*) demonstrate a completely different migratory pattern to those of the species of marine turtles discussed above. This species is highly nomadic, undertakes transoceanic migrations, and is capable of diving to great depths (Goff et al. 1994; Morreale et al. 1996; Eckert 1998; Hays, Houghton et al. 2004). Leatherback turtles do not exploit fixed feeding grounds (Hays, Houghton and Myers 2004, but some areas of the North Atlantic may be regularly visited as feeding grounds); instead they migrate widely between sites (Keinath and Musick 1993; Steyermark et al. 1996; Hays, Houghton et al. 2004) (Fig. 4C), and are often found in cold northern waters far from their nesting sites (Pritchard 1976; Hays, Houghton and Myers 2004). The pre-nesting migrations of leatherback turtles are not well understood; however, females are known to migrate to nearshore waters adjacent to nesting beaches a few weeks prior to the nesting season every 2-3 years. Unlike other marine turtles, female leatherbacks do not demonstrate strong nest site fidelity, although they do return to the same general nesting area. After nesting, females migrate extremely long distances, sometimes across oceanic basins (Morreale et al. 1996; Eckert 1998; Hughes et al. 1998). It is not known, however, whether or not these long distance movements follow generalized tracks. Indeed, major oceanographic processes, such as main currents and eddies, have been shown to have a major influence on Leatherback movements, placing into question whether these journeys are really migrations or just prolonged stays in vast feeding areas (Luschi et al. 2003b).

Like Leatherback turtles, Olive ridley turtles (*Lepidochelys olivacea*) are highly migratory (Polovina et al. 2004), and spend most of their non-reproductive life in the open ocean (Beavers and Cassano 1996; Plotkin 2003). During the nesting season reproductively active males and females migrate towards the coast, and aggregate in breeding grounds near the nesting beaches. Females then emerge onto the beaches to lay their eggs, although they do not exhibit strong nest site fidelity (Kalb 1999). Some males of this species appear to be non-migratory, and instead attempt to intercept females along their migratory route to the nesting beaches, and thus different males utilize different reproductive strategies (Plotkin et al. 1995; Kopitsky et al. 2000). After mating, females remain in the breeding area for several weeks or even months. The post-nesting migration of Olive ridley turtles is unique and complex. The routes taken by individual females and by groups of females vary annually, and do not seem to follow any sort of migratory corridor (Plotkin 2003). Olive ridley's are highly nomadic, and traverse vast stretches of open ocean without exploiting any specific, fixed feeding grounds.

Overall, sea turtles demonstrate three distinct migratory patterns (Fig. 4). Some species, including Leatherbacks and Olive Ridley turtles, do not have a fixed foraging area, and roam widely and unpredictably before returning to their breeding areas (Fig. 4C). Conversely, Kemp's ridley turtles, Loggerheads, and Flatback turtles migrate between breeding areas and highly productive neritic foraging areas on continental shelves (Fig. 4B). Finally, Green turtles and Hawksbill turtles exploit well established, fixed foraging areas, with very little variation (Fig. 4A). Superimposed upon this are smaller migratory movements made by marine turtles on a seasonal basis as they travel to warmer foraging grounds during winter periods (Gitschlag 1996).

Many aspects of marine turtle migrations remain uncertain. For instance, studies have focused primarily on postnesting migrations, as these are easier to study than prenesting migrations. Furthermore, the movements of male sea turtles are not well understood, because they do not come ashore, and are thus difficult to tag. Limited radiotracking studies of males indicate that they return faithfully to the same breeding areas each year (FitzSimmons et al. 1997). However, other studies have revealed that the males of some species may be non-migratory (Henwood 1987; Plotkin et al. 1994; Kopitsky et al. 2000). Finally, the movement patterns of young turtles after hatching are not very well understood (Fig. 5). It is thought that hatchlings of some species drift around gyres, or circular current systems, that serve as moving, open ocean nursery grounds (Luschi, Hughes et al. 2003). Hatchlings thus gradually migrate around the Atlantic Ocean, a round trip that takes up to 12 years, and eventually return to their natal beaches to breed (Hairston 1994; Lohmann et al. 2001; Pough et al. 2002). Juveniles and subadults of many populations reside in coastal feeding areas located hundreds or thousands of kilometers from their natal beaches (Fig. 5). For example, juvenile Loggerhead turtles from nesting beaches in Japan and Australia traverse the entire Pacific Ocean (Bowen et al. 1995), and young Loggerheads from nesting beaches in the western Atlantic will traverse the entire Atlantic Ocean in the course of their developmental migrations (Bolten et al. 1998).

Some other types of chelonians also undertake nesting migrations, although they are not as extensive as those of sea turtles. For instance, European Tortoises (*Testudo hermanni)* (Swingland et al. 1986), and some freshwater turtles, such as *Glyptemys insculpta* (Quinn and Tate 1991), *Kinosternon* (Gibbons et al. 1990), *Chelydra serpentina* (Gibbons et al. 1990), and *Chelodina longicollis* (Graham et al. 1996) have been found to undertake seasonal migrations of 50m to 5km from ponds or forested areas to nesting sites. Females of some species exhibit strong fidelity to general nesting areas, and will return to the same aquatic habitats year after year

(Lindeman 1992; Dodd 2001). However, turtles have also been observed to migrate for reasons other than finding suitable nesting sites. For example, Snapping Turtles (*Chelydra serpentina*) may migrate up to 4 km beyond their normal home range to find suitable overwintering sites in flowing water (Cagle 1944; Brown and Brooks 1994). Aldabra Giant Tortoises (*Geochelone gigantea)* migrate from inland areas to coastal regions during the rainy season to gain access to high quality food resources available on the coast (Swingland and Lessells 1974; Swingland et al. 1989). Galapagos Tortoises also migrate between different habitats (Rodhouse et al. 1975). Many freshwater turtles have been observed to migrate to new ponds as a result of fluctuating environmental conditions such as droughts (Bennett et al. 1970; Yeomans 1995; Graham et al. 1996). However, as in amphibians, there is some degree of variation in the migratory patterns within populations and between species of turtles. For instance, if suitable nesting sites are available nearby, European Tortoises will not migrate (Swingland et al. 1986). Furthermore, as migration can be dangerous, energetically expensive, or both, only about 20% of the Aldabra Giant Tortoise population studied migrated to the coast to exploit rich food resources (Swingland et al. 1989).

In addition to turtles, some crocodilians also undertake regular, seasonal migrations. Some species, including Nile crocodiles (*Crocodylus niloticus)* may swim great distances to find suitable nesting sites (Chelazzi 1992). Females of this species have been observed to swim to breeding beaches on Central Island, Lake Rudolf to lay their eggs (Modha 1967 1968). However, this type of behavior is rare among crocodilians, and most species, including the Australian *Crocodylus johnstoni,* remain within their home range to nest (Webb et al. 1983a,b). Some species of alligator and caiman may increase their home range during the breeding season, but like crocodiles, their nesting sites are generally located within the home range, and the animals do not undertake nesting migrations (Pough et al. 1998). Furthermore, although some crocodilians travel very long distances, for example the island hopping behavior exhibited by saltwater crocodiles, these movements are not considered migrations (Rodda 1985). Among some crocodilians, migrations in response to environmental changes are more common than are nesting migrations. Some species of crocodiles (*Crocodylus johnstoni)*, caimans (*Caiman crocodilus),* and alligators migrate seasonally from drying swamps to more permanent bodies of water (Neill 1971; Pough et al. 1998).

Fig. 6. Part of a massive accumulation of red-sided garter snakes (*Thamnophis sirtalis parietalis*) emerging from a communal den in the spring in the Interlakes region of Manitoba, Canada. Photograph courtesy of Dr. Herb Rosenberg

Most oviparous squamates lay their eggs within their normal home ranges, and thus do not undertake breeding migrations, although a few species of snakes have been reported to make directional movements to specialized oviposition or breeding sites (Pough et al. 2002) For example, *Vipera berus* in Finland moves from basking sites to a common breeding area several hundred metres away (Pough et al. 2002). Many species of snakes also exhibit dramatic migrations between summer feeding ranges and winter denning sites. This is common in northern temperate zone snakes, including garter snakes (*Thamnophis)* (Fig. 6) and some rattlesnakes (e.g. *Crotalus atrox)* (Landreth 1973; Klauber 1982; Lawson 1989; Lawson and Secoy 1991; Bauer and Russell 2001; LeMaster et al. 2001; Linzey 2001). These snakes generally migrate about 1-10km to their dens in the fall for winter hibernation. Snakes congregate at the denning sites, and mating occurs at the dens before the snakes migrate back to their summer feeding areas in the spring (Klauber 1982; Lawson 1989; LeMaster et al. 2001) (Fig. 6), although the patterns of seasonal movement may differ between males and females (King and Duvall 1990). Like other migrating reptiles and amphibians, garter snakes and rattlesnakes tend to show fidelity to both their den sites and their feeding areas, returning to the same locations year after year (Lawson 1989). The sea snake, *Pelamis platurus,* also undertakes long distance breeding migrations in the Indian and Pacific oceans (Graham et al. 1971).

Other types of snakes demonstrate different migratory patterns. For example, water pythons (*Liasis fuscus)* in Australia migrate considerable distances each year to follow their most important source of prey, the dusky rat (*Rattus colletti)*. Although many large mammals migrate to follow their prey, this type of migration is difficult for terrestrial ectotherms because they are generally small, have a limited capacity for sustained activity, and are often inactive for much of the year due to thermal constraints (Madsen and Shine 1996). However, water pythons are large animals that live in a tropical environment, and are thus able to undertake long migrations to track their prey. These snakes have been observed to follow their primary prey up to 12km from the backswamp area to the floodplain during the wet season (Madsen and Shine 1996). Other types of aquatic snake, like Arafura filesnakes (*Acrochordus arafurae*), also undergo seasonal migrations, moving from restricted ponds to flooded grasslands at the onset of the wet season. They thus show a strong seasonal shift in habitat utilization which may be due to the snakes following their prey (fish) out of the ponds, or may simply be a result of changing environmental conditions (Shine and Lambeck 1985).

Among reptiles, migration is least common among lizards. These terrestrial animals often show strong, territorial attachments to specific sites, and their small size and ectothermic nature tends to preclude migration (Madsen and Shine 1996). However, a few species of lizards have been observed to migrate. Large iguanids that live in environments where nesting sites are scarce may migrate up to 15km to reach suitable nesting sites (Pough et al. 1998). For example, Green Iguanas on Barro Colorado Island swim to small offshore islands to nest, although most move less than 1km beyond their normal home range (Rand 1968; Bock et al. 1985). A similar pattern is seen among female Ground Iguanas (*Cyclura spp.*), which migrate up to 6.5 km to reach suitable nesting sites (Wiewandt 1982). Galapagos Land Iguanas (*Conolophus subcristatus*) also migrate to reach nesting sites, and some females have been observed to climb 1400m to the rim of a volcanic crater, and then descend 900m into the crater to nest (Werner 1983).

Somewhat unusual migratory behavior has also been observed in *Ctenophorus* lizards in Western Australia. Juveniles emigrate away from rock outcrops when they are 2-3 months old to avoid harassment by adult males. Once the juveniles reach sexual maturity they migrate back to the main outcrop (Bradshaw 1971). Most other lizards do not migrate, but many lacertids and iguanids exhibit homing abilities (Mayhew 1963; Weintraub 1970).

7.7 Homing and Orientation in Reptiles

Homing and orientation play very important roles in the migration of reptiles, and particularly in the very long-distance migrations of sea turtles. The lives of sea turtles consist of a series of migrations beginning with hatchlings swimming from their natal beaches to the open sea, and then later the migrations of adult turtles between foraging areas and their nesting beaches (Fig. 5). These animals migrate long distances in the open ocean where landmarks are largely unavailable, yet they demonstrate very precise orientation to specific locations, including some very small islands (Zug et al. 2001). The lifestyle of sea turtles is thus inextricably linked to the ability to orient accurately across spans of seemingly featureless ocean, indicating that these animals possess a sophisticated orientation system.

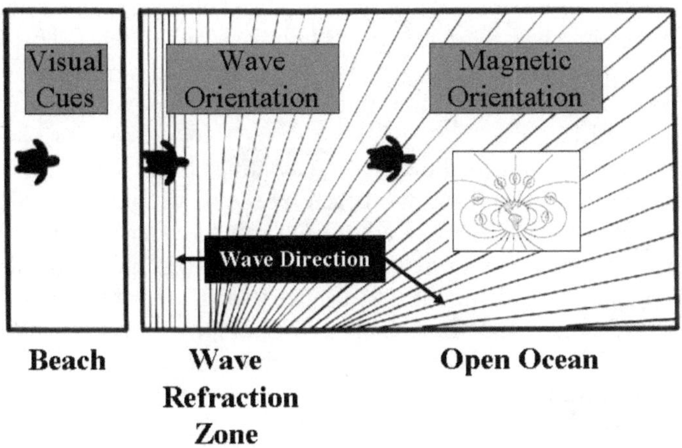

Fig. 7. Sequential orientation cues hypothesized to guide hatchling Loggerhead turtles from the nesting beach to the open ocean. Modified after Lohmann et al. (1997, Fig. 5.8)

The migratory movements of hatchling sea turtles are no less impressive than those of adults (Fig. 7). These small animals crawl to the sea, and immediately establish an offshore heading which is maintained long after the sight of land has been lost. After hatching, young sea turtles use three different types of cues to orient (Pough et al. 2002) (Fig. 7). First, upon emerging from the nests, hatchlings use visual cues to orient towards the sea. Studies have shown that hatchling Loggerhead and Green Turtles exhibit a strong phototropotactic response, and will orient towards the brightest light in the vicinity, which, in the absence of artificial lights, is the horizon over the ocean, highlighted by the reflection of light off the ocean (Carr 1962; Kingsmill and Mrosovsky 1982; Mrosovsky and Kingsmill 1985; Salmon and Wyneken 1990; Peters and Verhoeven 1994) (Fig. 7). The color or wavelength of light is also used by hatchling turtles to help them find the sea. For this reason, artificial lighting near beaches can be a problem for hatchlings, causing them to become disoriented and disrupting their sea-finding ability (Salmon 2003). Green turtles exhibit a preference for blue or purple wavelengths over red, orange or yellow light (Witherington and Bjorndal 1991; Lohmann et al. 1997). This preference causes them to avoid orienting towards the rising or setting sun. Polarized light may also be used by hatchlings to orient towards the sea. Finally, hatchlings tend to avoid dark, elevated silhouettes, which may be caused by dunes or vegetation, objects typically located in a direction opposite to that of the shoreline (Godfrey 1995; Lohmann et al. 1997). This behavior also helps to guide hatchlings towards the sea, and similar cues may be used by nesting adults to return to the sea following oviposition (Caldwell and Caldwell 1962).

Once they have reached the ocean, hatchlings begin to swim vigorously, and are carried 5-10m from the beach by the wave undertow. When they surface, hatchlings establish a course towards the open ocean, using wave propagation direction to orient (Fig. 7). They swim perpendicular to the waves, which leads them out into the ocean and toward the global current systems (Salmon and Lohmann 1989; Lohmann et al. 1990 1995; Lohmann 1992; Pough et al. 2002; Nagelkirken et al. 2003).

However, further from shore, wave direction no longer provides a reliable indication of the offshore direction, yet hatchlings continue to follow the same seaward course. Furthermore, hatchlings carried thousands of kilometers from their natal beaches find their way back to breed, sometimes 30-50 years later. This indicates that they use other, more complex sources of directional information. Studies have shown that hatchling Loggerhead and Leatherback sea turtles are sensitive to the earth's magnetic field, and it has been suggested that they develop a magnetic compass as they leave their natal beaches (Lohmann 1992; Lohmann et al. 1997; Irwin et al. 2004) (Fig. 7). For example, hatchlings trained to swim in a particular direction, towards a light will maintain the same compass direction even in the dark. Furthermore, if the magnetic field is reversed, the turtles will reverse their swimming direction (Lohmann et al. 1997).

Young turtles at sea also show abilities beyond a simple sensitivity to the earth's magnetic field. There is evidence that they are capable of detecting changes in the inclination angle of the earth's magnetic field relative to the earth's gravitational field (Light et al. 1993), as well as changes in the local intensity of the magnetic field. For example, hatchling turtles with no migratory experience have been shown to orient relative to a magnetic field in a direction that would keep them within the warmer North Atlantic Gyre current system, and on a safe migratory route (Lohmann and Lohmann 1998; Lohmann et al. 2001). The ability to detect both the inclination angle and the intensity of the earth's magnetic field would allow turtles to approximate the latitude and longitude of their global position, providing them with a bi-coordinate geomagnetic map sense (Lohmann et al. 1997).

These studies on hatchlings may provide insight into the navigational cues employed by adult sea turtles, although it is possible that juveniles and adults respond to different types of cues. The migration of adult sea turtles has been extensively studied over the past 10 years, but the mechanism of orientation and navigation employed by these animals remains uncertain (Papi and Luschi 1996; Papi et al. 2000), although it likely involves a variety of sensory cues, as is the case for amphibians (Hays et al. 2002; Pough et al. 2002). There is even some debate over whether or not these animals are capable of true navigation. It has been suggested that since they follow fairly straight routes to their goal, simple compass orientation is adequate to guide their migrations (Luschi, Hughes et al. 2003). This may be true for migrations towards large targets, such as the return migration of Green Sea Turtles to the Brazilian coast (Fig. 4A), but probably not for the nesting migration, which is directed towards a very small target. Displaced adult Loggerhead turtles in the Indian Ocean were found to be unable to compensate for the post-nesting displacement and appeared to rely on simple orientation mechanisms, such as the coastline, as a guide (Luschi, Hays and Papi 2003). The nesting migration is much more complex, and probably requires the animals to have a map-sense, and true navigational abilities (Papi et al. 2000; Hays et al. 2002). Furthermore, the ability of displaced turtles to return to or orient towards their original location (Papi and Luschi 1996; Luschi et al. 1998; Avens and Lohmann 2004), and the ability of turtles to correct deviations in their migratory course caused by ocean currents (Luschi et al. 1998; Papi et al. 2000), suggest that simple compass orientation is not adequate, and that a more sophisticated mechanism is being used.

If the ability to detect magnetic fields that has been identified in hatchlings is retained in adults, then it is likely that adult sea turtles also use a magnetic map for orientation and navigation (Lutcavage 1996; Lohmann et al. 1997; Papi et al. 2000; Lohmann et al. 2004). However, if such an ability exists, it is not yet known if it provides very precise or only approximate information about the natal beach, with other types of cues being used closer to the natal beach. Furthermore, a study tracking the migration routes of Green Sea Turtles found that the routes taken by magnetically disturbed turtles did not differ significantly from those of the control turtles, suggesting that magnetic cues are not essential for the migration of sea turtles (Papi et al. 2000). On the other hand, this study focused on the return migration of the sea turtles, and thus it is possible that turtles rely more heavily on magnetic cues while undertaking the much more complex pre-nesting migration (Papi et al. 2000).

In addition to magnetic cues, many other mechanisms have been proposed to explain the orientation abilities of sea turtles. For instance, as in amphibians, olfaction may play an important role in the migration of these animals (Carr 1967; Manton et al. 1972; Bowen et al. 1989; Chelazzi 1992; Papi et al. 2000). Carr (1967) found evidence that sea turtles have an odor memory of at least one year, which led others to suggest that juvenile sea turtles imprint on chemical cues associated with natal beaches, and then use a chemical plume along ocean currents, combined with compass orientation, to guide them to their beach when they return as adults (Lohmann et al. 1997; Papi et al. 2000). However, the role of chemical cues in long-distance navigation is almost unknown, and there is no convincing experimental evidence to support this hypothesis. Furthermore, the straight migration routes of adults are not consistent with tracking a chemical plume, and oceanic currents resulting from trade winds may disrupt chemical cues, thus preventing the formation of chemical trails that turtles could follow (Brown 1990). It is also debatable whether the time of exposure to natal beach odors is long enough for their imprinting on hatchlings to occur. Åkesson et al. (2003) reviewed the evidence relating to the methods Green Sea Turtles (*Chelonia mydas*) may use to locate Ascension Island in their migration from the Brazilian coast. Investigating the behavior of displaced females, they concluded that chemical information transported by ocean currents did not account for navigation, nor did bi-coordinate geomagnetic methods, other magnetic gradients or celestial cues. Instead, females found the island by searching and beaconing, and then used wind-borne cues for final location.

Another possibility is that turtles orient using wave direction or ocean currents, as observed in hatchling Loggerhead sea turtles, and that they calibrate this system using celestial cues such as the stars or the sun (Morreale et al. 1996; Papi et al. 2000). Other potential mechanisms of navigation include information from the wind above the ocean surface (Luschi et al. 2001; Åkesson et al. 2003; Hays et al. 2003), bathymetric features of the ocean floor (Morreale et al. 1994), or water temperature (Plotkin 2003). However, evidence is sparse to support these mechanisms, and some have been shown to be unimportant in the migrations of some species of sea turtles. For instance, Green Sea Turtles do not rely on ocean surface temperatures while migrating (Hays et al. 2001). Overall, much more research is necessary before any definite conclusions can be drawn regarding the orientation and navigation system of sea turtles.

In addition to sea turtles, orientation and homing mechanisms are also used by other types of reptiles including freshwater and terrestrial turtles, snakes, crocodilians and some lizards. Although few species of freshwater and terrestrial turtles undertake extensive migrations, many have been found to possess impressive homing abilities. For example, map turtles (*Graptemys pulchra*) returned home after being displaced up to 24km along a river, which puts them far beyond the boundaries of their natural home range (Shealy 1976). However, the total number of turtles that returned successfully was low. Many other freshwater turtle species, including *Clemmys guttata, Chrysemys picta, Glyptemys insculpta* and *Apalone mutica,* have been shown to demonstrate good homing performance following displacement (Cagle 1944; Ernst 1968 1970; Plummer and Shirer 1975; Carroll and Ehrenfeld 1978). The mechanism by which freshwater turtles orient is not well understood. Experiments have shown that the orientation abilities of turtles of the genera *Terrapene* and *Chrysemys* are reduced on overcast days, although some are still able to home for short distances under cloudy skies (Emlen 1969; Yeomans 1995; Graham et al. 1996). These results indicate that these turtles orient using a sun compass calibrated with an internal clock (Gould 1957; DeRosa and Taylor 1980 1982; Yeomans 1995; Graham et al. 1996). However, olfactory cues may also play an important role in homing and orientation of freshwater turtles. Olfaction has been suggested as an important factor in the homing ability of *Glyptemys insculpta* (Carroll and Ehrenfeld 1978), and olfaction and geotaxis could not be ruled out as factors in the homing of *Chrysemys picta* (Ernst 1970). Furthermore, the Australian Snake-necked Turtle, *Chelodina longicollis,* has been found to orient towards its home substrate in laboratory experiments (Graham et al. 1996), suggesting that olfaction also plays a key role in the orientation of these animals. It has furthermore been suggested that some freshwater turtles have the ability to detect and orient towards water, indicating that they may be using surface-reflected polarized light to orient, as polarization tends to be higher over moist surfaces (Yeomans 1995; Zug et al. 2001). Finally, the species *Terrapene carolina* has been shown to use both local visual cues (Lemkau 1970) and magnetic cues to aid in orientation (Mathis and Moore 1988).

The mechanism of homing and orientation of tortoises is even less well understood. The tortoise *Testudo hermanni* is capable of orienting at distances up to 1560m outside its home range (Calzolai and Chelazzi 1981; Chelazzi and Calzolai 1986). It has been suggested that tortoises use olfactory cues to orient towards their home ranges, and to locate bodies of water following displacement (Chelazzi and Delfino 1986; Pough et al. 2002). Tortoises may also utilize visual cues based on topographic relief (Auffenberg and Weaver 1969; McCoy et al. 1993), but for the most part no definitive conclusions about tortoise orientation have been reached.

Homing behavior has also been demonstrated among crocodilians. The most impressive of these abilities have been observed in juvenile alligators. In general, alligators are fairly sedentary animals, but will migrate in order to avoid unfavorable environmental conditions. Studies have shown that alligators are able to home from over 5km away from their capture site, and are able to orient towards home from up to 34km away (Rodda 1985). Like sea turtles, alligators are able to detect magnetic fields, and thus they use magnetic cues for orientation (Rodda 1984). Furthermore, they are able to home from unfamiliar areas, suggesting that they are capable of true navigation (Rodda 1984 1985). Other species of crocodilians, including *Caiman crocodilus* and the saltwater crocodile (*Crocodylus porosus*) have also been found to return home following displacements of 2km and 30km respectively (Gorzula 1978; Ouboter and Nanhoe 1988; Pough et al. 2002).

Homing abilities also appear to be common in some snakes, but results regarding homing in North American snakes have been contradictory. Homing behavior has been observed in several species of garter snake (*Thamnophis sirtalis*, *T. radix)*, water snakes (*Nerodia sipedon, Regina septemvittata)* and a rattlesnake (*Crotalus atrox)* when they are displaced from their den sites or from rich food sources (Landreth 1973; Newcomer et al. 1974; Brown and Parker 1976; Lawson 1989 1994; Lawson and Secoy 1991). However, garter snakes did not demonstrate the same evidence of homeward orientation when displaced from their normal home ranges (Lawson 1994). Other species of snake, including *Sisturus catenatus* and *Carphophis amoenus*, also do not exhibit homing behavior, and will establish new home ranges when displaced (Barbour et al. 1969; Reinert and Kodrich 1982). Based on these results it is possible that some snakes may be capable of homing, but they may not always be motivated to use these abilities.

Some snakes that do exhibit homing behavior are able to orient along a specific compass direction, even when displaced great distances from their home range or den sites (Landreth 1973; Newcomer et al. 1974; Brown and Parker 1976; Lawson 1989 1994; Lawson and Secoy 1991). Furthermore, studies have shown that the orientation direction taken by snakes shifts if they are transported in the dark, suggesting that these animals use celestial cues for orientation (Newcomer et al. 1974; Gregory et al. 1987; Lawson 1989; Lawson and Secoy 1991). The type of celestial cue used is likely solar, as there is no evidence that nocturnal snakes utilize lunar or stellar cues for orientation (Chelazzi 1992). Finally, there is some evidence to suggest that some snake species use olfactory cues for short-distance orientation (*Diadophis punctatus*: Dundee and Miller 1968), and that juvenile *Thamnophis* and *Crotalus* follow the pheromone trails of conspecifics to guide them towards their goals (Ford 1986; Graves et al. 1986; Lawson and Secoy 1991; Ford and Burghardt 1993; Pough et al. 1998).

Among lizards, migratory behavior is very rare and very little is known about their homing and orientation. Homing abilities in lizards are found almost exclusively in lacertids and iguanids (Mayhew 1963; Weintraub 1970). Several species of *Sceloporus, Phrynosoma* and *Uta are* capable of homing following displacements of less than 300m from their normal home ranges (Spoekker 1967; Guyer 1978; Bissinger 1983; Ellis-Quinn and Simon 1989; Chelazzi 1992). Some other lizard species have demonstrated the ability to return home following short displacements (*Dipsosaurus dorsalis*: Krekorian 1977; *Takydromus takydromoides*: Ishihara 1969; *Podarcis sicula*: Foà et al. 1990; *Tilqua rugosa*: Freake 1998). However, many other lizards are not capable of homing from similar distances (*Anolis lineatopus*: Rand 1967; *Sceloporus occidentalis*: Fitch 1940; *Uta stansburiana*: Tinkle 1967). It has been suggested that the strength of attachment to a home range affects the homing ability of lizards. For example, male and female *Sceloporus jarrovii* are both territorial, and demonstrate similar homing abilities. Conversely, non-territorial female *Sceloporus orcutti* exhibit poorer homing abilities than territorial conspecific males (Ellis-Quinn and Simon 1989; Weintraub 1970). Sex-based differences in homing behavior have also been observed in *Tilqua rugosa* (Freake 1998). Similarly, the territorial species *Sceloporus graciosus* has better homing abilities than the non-territorial species *Phrynosoma douglasii*, even though they occupy the same habitat (Guyer 1991).

The mechanism of orientation employed by lizards is not well understood. Experiments using directionally trained lizards have suggested the use of a celestial compass in *Lacerta viridis* (Fischer 1961), *Uma notata* (Adler and Phillips 1985) and *Tiliqua rugosa* (Freake 1999). It is possible that the pineal organ of lizards acts as a polarized light detector, and thus some lizards may be able to orient using polarized light, as is the case for some species of salamanders (Ellis-Quinn and Simon 1991; Freake 1999; Zug et al. 2001). However, although many lizard species are capable of homing and orientation, most do not undertake migrations, and these abilities may be useful for some other function in these animals.

7.8 Conclusion

Migration plays a very important role in the lives of many amphibians and reptiles, as it allows them to acquire the resources they need to survive, or to avoid unfavorable environmental conditions. In order to undertake these migrations amphibians and reptiles employ a wide variety of orientation mechanisms, which likely involve the use of olfactory, visual, celestial and magnetic cues. However, the exact mechanisms used by many of these animals are still poorly understood. Furthermore, the migratory routes and patterns of many amphibians and reptiles are not completely known. Such patterns are strongly dependent on a diversity of factors, including phylogeny, body size, physiological and energetic constraints, and especially life history traits. As numerous amphibian and reptile species are highly endangered, it is essential that we develop an understanding of the migratory patterns of these species. This is especially crucial for some taxa, such as sea turtles, whose extensive marine migrations increase their likelihood of interactions with longline fishing operations (Plotkin and Spotila 2002; Luschi 2003; Hays, Houghton and Myers 2004) or fishing weirs (Godley, Lima et al. 2003), or many amphibians, whose migrations cross busy roadways (Langton 1989; Laufer 1997; Scoccianti 2000; Clavenger et al. 2001), or areas subject to habitat fragmentation as a result of human activities (de Maynadier and Hunter 1999). Without this knowledge, and that of the consequences of and potential for dispersal (Farrier et al. 2000; Lehtinen and Galatowitsch 2001), it is impossible to adequately protect and conserve these animals.

References

Adams SB, Frissell CA (2001) Thermal habitat use and evidence of seasonal migration by Rocky Mountain tailed frogs, *Ascaphus montanus*, in Montana. Canadian Field-Naturalist 115: 251-256

Adler K (1976) Extraocular photoreception in amphibians. Photochemistry and Photobiology 23: 275-298

Adler K, Phillips JB (1985) Orientation in a desert lizard (*Uma notata*): time-compensated compass movement and polarotaxis. Journal of Comparative Biology 156: 547-552

Åkesson S, Broderick AC, Glen F, Godley BJ, Luschi P, Papi F, Hays GC (2003) Navigation by green turtles: which strategy do displaced adults use to find Ascension Island? Oikos 103: 363-372
Alcock J (1975) Animal Behaviour: An Evolutionary Approach. Sinauer Associates, Sunderland, Massachusetts
Alcock J (2001) Animal Behaviour 7th Ed. Sinauer Associates, Sunderland, Massachusetts
Alvarado Diaz J, Arias Covotl E, Delgado Trejo C (2003) Clutch frequency of the Michoacan Green Seaturtle. Journal of Herpetology 37: 183-185
Ambrogio A, Gillis, L (1998) Il Tritone Alpestre. Anfibi Italiani I. Edizioni Planorbis, Reggio Emilia, Italy
Aresco MJ (2002) *Amphiuma means* overland migration. Herpetological Review 33: 296-297
Aufenberg W, Weaver WG (1969) *Gopherus berlandieri* in southeastern Texas. Bulletin of the Florida State Museum, Biological Sciences 13: 141-203
Avens L, Lohmann KJ (2004) Navigation and seasonal migratory orientation in juvenile sea turtles. Journal of Experimental Biology 207: 1771-1778
Avise JC, Bowen BW (1994) Investigating sea turtle migration using DNA markers. Current Opinions in Genetics and Development 4: 882-886
Baker RR (1978) The Evolutionary Ecology of Animal Migration. Hodder and Stoughton, London
Barbour RW, Harvey MJ, Hardin JW (1969) Home range, movement, and activity of the eastern worm snake *Carphophis amoenus amoenus*. Ecology 50: 470-476
Bass AL (1999) Genetic analysis to elucidate the natural history and behaviour of hawksbill turtles (*Eretmochelys imbricata*) in the wider Caribbean: a review and re-analysis. Chelonian Conservation Biology 3: 195-199
Bauer AM, Russell AP (2001) The first record of reptiles in Alberta: Aemilius Simpson's journal of 1826. Herpetological Review 32: 174-176
Beavers SC, Cassano ER (1996) Movements and dive behavior of a male sea turtle (*Lepidochelys olivacea*) in the eastern tropical Pacific. Journal of Herpetology 30: 97-104
Bellis ED (1965) Home range and movement of the wood frog in a northern bog. Ecology 46: 90-98
Beneski JTJ, Edward JZ, Larsen JHJ (1986) Demography and migratory patterns of the eastern long-toed salamander *Ambystoma macrodactylum columbianum*. Copeia 1986: 398-408
Bennett DH, Gibbons JW, Franson JC (1970) Terrestrial activity of aquatic turtles. Ecology 51: 738-740
Bentivegna F (2002) Intra-Mediterranean migrations of loggerhead sea turtles (*Caretta caretta*) monitored by satellite telemetry. Marine Biology 141: 795-800
Berven KA, Grudzien TA (1990) Dispersal in the wood frog (*Rana sylvatica*): implications for genetic population structure. Evolution 44: 2047-2056

Bissinger B (1983) Homing behavior, suncompass orientation and thermoregulation in the lizard *Sceloporus jarrovii*: the role of the pineal eye. PhD thesis, The City University of New York

Bock BC, Rand AS, Burghardt GM (1985) Seasonal migration and nesting site fidelity in the green iguana. In Rankin MA (ed) Migration Mechanism in Adaptive Significance. Supplement to Contributions to Marine Science, University of Texas Marine Science Institute, Texas, USA 27: 435-443

Bogert CM (1947) Results of the Archbold Expeditions. No. 57. A field study of homing in the Carolina toad. American Museum Novitates 1355: 1-24

Bowen BW, Abreu-Grobois FA, Balazs GH, Kamezaki N, Limpus CJ, Ferl RJ (1995)

Trans-Pacific migrations of the loggerhead turtle (*Caretta caretta*) demonstrated with mitochondrial DNA markers. Proceedings of the National Academy of Science USA 92: 3731-3734

Bowen BW, Meylan AB, Avise, JC (1989) An odyssey of the green sea turtle: Ascension Island revisited. Proceedings of the National Academy of Sciences, USA 86: 573-576

Bradshaw SD (1971) Growth and mortality in a field population of *Amphibolurus* lizards exposed to seasonal cold and aridity. Journal of Zoology, London 165: 1-25

Brass D (1983) Day of the frogs. Bulletin of the East African Natural History Society 1983: 99-89

Brattstrom BH (1962) Homing in the giant toad, *Bufo marinus*. Herpetologica 18: 176-180

Brede EG, Beebee TJC (2004) Contrasting population structures in two sympatric anurans: implications for species conservation. Heredity 92: 110-117

Brown CW (1990) The significance of the South Atlantic Equatorial Countercurrent to the ecology of the green turtle breeding population of Ascension Island. Journal of Herpetology 24: 81-84

Brown WS, Parker MJ (1976) Movement ecology of *Coluber constrictor* near communal hibernacula. Copeia 1976: 225-242

Brown WS, Brooks RJ (1994) Characteristics of and fidelity to hibernacula in a northern population of snapping turtles, *Chelydra serpentina*. Copeia 1994: 222-226

Buitrago J, Guada HJ (2002) La tortuga carey (*Eretmochelys imbricata*) en Venezuela. Interciencia 27: 393-399

Byles RA, Plotkin PT (1994) Comparison of the migratory behavior of congeneric sea turtles *Lepidochelys olivacea* and *L. kempii*. In Schroeder BA, Witherington BE (compilers) Proceedings of the 13th Annual Symposium on Sea Turtle Biology and Conservation. NOAA Technical Memorandum NMFS-SEFSC-341 39

Byles RA, Alvarado J, Rostal DC (1995) Preliminary analysis of post-nesting movements of the black turtle (*Chelonia agassizi*) from Michoacan Mexico. In Richardson JI, Richardson TH (compilers) Proceedings of the 12th Annual Workshop on Sea Turtle Biology and Conservation. NOAA Technical Memorandum NMFS-SEFCS-361 12-13

Cagle FR (1944) Home range, homing behavior, and migration in turtles. Miscellaneous Publications of the University of Michigan Museum of Zoology 61: 1-33

Caldwell MC, Caldwell DK (1962) Factors in the ability of the northeastern Pacific green turtle to orient toward the sea from the land, a possible coordinate in long-range navigation. Los Angeles County Museum Contributions to Science 60: 1-27

Calzolai R, Chelazzi G (1991) Habitat use in a central Italy population of *Testudo hermanni* Gmellin (Reptilia, Testudinidae). Ethology, Ecology and Evolution 3: 53-166

Carr A (1962) Orientation problems in the high seas travel and terrestrial movements of marine turtles. American Scientist 50: 360-374

Carr A (1965) The voyages of the green sea turtle. Nature and Science 3: 2-6

Carr A (1967) So Excellente a Fische. Natural History Press, New York

Carr A, Coleman P (1974) Seafloor spreading theory and the odyssey of the green turtle. Nature 249: 128-130

Carr A, Giovannoli L (1957) The ecology and migrations of sea turtles, 2. Results of field work in Costa Rica 1955. American Museum Novitates 1835: 1-32

Carr A, Carr MH, Meylan AB (1978) The ecology and migrations of sea turtles 7. The western Caribbean green turtle colony. Bulletin of the American Museum of Natural History 162: 1-46

Carroll TE, Ehrenfeld DW (1978) Intermediate-range homing in the wood turtle, *Clemmys insculpta*. Copeia 1978: 117-126

Chelazzi G (1992) Reptiles. In Papi F (ed) Animal Homing. Chapman and Hall, New York 235-261

Chelazzi G, Calzolai R (1986) Thermal benefits from familiarity with the environment in a reptile. Oecologia 68: 557-558

Chelazzi G, Delfino G (1986) A field test on the use of olfaction in homing by *Testudo hermanni* (Reptilia, Testudinidae). Journal of Herpetology 20: 451-455

Cheng IJ (2000) Post-nesting migrations of green turtles (*Chelonia mydas*) at Wan-An Island, Peng Hu Archipelago, Taiwan. Marine Biology, Berlin 137: 747-754

Clevenger AP, Melvor M, Chruszcz B, Gunson K (2001) Tiger salamander, *Ambystoma tigrinum*, movements and mortality on the Trans-Canada highway in southwestern Alberta. Canadian Field-Naturalist 115: 251-256

Cogger HG, Zweifel RG (1998) Encyclopedia of Reptiles and Amphibians 2nd Ed. Weldon Owen Pty Ltd, Singapore

Craig P, Parker D, Brainard R, Rice M, Balazs GH (2004) Migrations of green turtles in the central South Pacific. Biological Conservation 116: 433-438

de Maynadier PG, Hunter ML Jr (1999) Forest canopy closure and juvenile emigration by pool-breeding amphibians in Maine. Journal of Wildlife Management 63: 441-450

Demian JJ, Taylor DH (1977) Photoreception and locomotor rhythm entrainment by the pineal body of the newt, *Notophthalmus viridescens* (Amphibia, Urodela, Salamandridae). Journal of Herpetology 11: 131-139

DeRosa CT, Taylor DH (1980) Homeward orientation mechanisms in three species of turtles (*Trionyx spinifer, Chrysemys picta*, and *Terrapene carolina*). Behavior, Ecology and Sociobiology 7: 15-23

DeRosa CT, Taylor DH (1982) A comparison of compass orientation mechanisms in three turtles (*Trionyx spinifer, Chrysemys picta* and *Terrapene carolina*). Copeia 1982: 394-399

Deullman WE, Trueb L (1986) Biology of Amphibians. McGraw-Hill Book Company, New York

Deutschlander ME (1999) The case for light-dependant magnetic orientation in animals. Journal of Experimental Biology 202: 891-908

Diego-Rasilla FJ (2003) Homing ability and sensitivity to the geomagnetic field in the alpine newt (*Triturus alpestris*). Ethology, Ecology, Evolution 15: 251-259

Dingle H (1980) Ecology and Evolution of Migration. In Gauthreaux SA Jr (ed) Animal Migration, Orientation and Navigation. Academic Press, London 2-103

Dingle H (1996) Migration. Oxford University Press, Oxford

Dodd CK Jr (1988) Synopsis of the biological data on the loggerhead sea turtle *Caretta caretta* (Linnaeus 1758). United States Fish and Wildlife Service, Biological Report 88: 1-110

Dodd CK Jr (2001) North American Box Turtles. A Natural History. University of Oklahoma Press, Norman

Dodd CK Jr, Byles RA (2003) Post-nesting movements and behavior of loggerhead turtles (*Caretta caretta*) departing from east-central Florida nesting beaches. Chelonian Conservation Biology 4: 530-536

Douglas ME (19790 Migration and selection in *Ambystoma jeffersonianum*. Canadian Journal of Zoology 57: 2303-2310

Douglas ME, Monroe BLJ (1981) A comparative study of topographical orientation in *Ambystoma* (Amphibia: Caudata). Copeia 1981: 460-463

Dundee HA, Miller MC (1968) Aggregative behavior and habitat conditioning by the prairie ring-necked snake (*Diadophis punctatus arnyi*). Tulane Studies in Zoology and Botany 15: 41-58

Eckert SA (1998) Perspectives on the use of satellite telemetry and other electronic technologies for the study of sea turtles. In Epperly SP, Braun J (compilers) Proceedings of the 17th Annual Sea Turtle Symposium. NOAA Technical Memorandum NMFS-SEFSC-415 44-46

Ellis DM, Balazs GH, Gilmartin SKK, Katahira LK (2000) Short-range reproductive migrations of hawksbill turtles in the Hawaiian Islands as determined by satellite telemetry. In Abreu-Grabois FA, Briseño-Dueñas R, Márquez-Millán R, Sarti-Martínez R (compilers) Proceedings of the 18th International Sea Turtle Symposium, NOAA Technical Memorandum NMFS-SEFSC-436 252-253

Ellis-Quinn BA, Simon CA (1989) Homing behavior of the lizard *Sceloporus jarrovii*. Journal of Herpetology 23: 146-152

Ellis-Quinn BA, Simon CA (1991) Lizard homing behaviour: the role of the parietal eye during displacement and radiotracking, and time-compensated

celestial orientation in the lizard *Sceloporus jarrovii*. Behavior, Ecology, Sociobiology 28: 397-407
Emlen ST (1969) Homing ability and orientation in the painted turtle, *Chrysemys picta marginata*. Behaviour 33: 58-76
Endler J (1970) Kinesthetic orientation in the California newt (*Taricha torosa*). Behaviour 37: 15-22
Ernst CH (1968) Homing ability of the spotted turtle *Clemmys guttata* (Schneider). Herpetologica 24: 77-78
Ernst CH (1970) Homing ability of the painted turtle *Chrysemys picta* (Schneider). Herpetologica 26: 399-403
Feldman R (1897) Überwinterung, Ortstreue und Lebensalter des Feuersalamanders, *Salamandra salamandra terrestris*. Schlussbericht einer Langzeituntersuchung. Jahrbuch für Feldherpetologie 1: 33-44
Ferriere R, Belthoff JR, Olivieri I, Krackow S (2000) Evolving dispersal: where to go next? Trends in Ecology and Evolution 15: 5-7
Fischer K (1961) Untersuchung zur Sonnenkompassorientierung und Laufaktivität van Smaragdeidechsen (*Lacerta viridis* Laur.). Zietschrift für Tierpsychologie 18: 450-470
Fitch HS (1940) A field study of the growth and behavior of the fence lizard. University of California Publications in Zoology 44: 151-172
FitzSimmons NN, Limpus CJ, Norman JA, Goldizen AR, Miller JD, Moritz C (1997) Philopatry of male marine turtles inherited from mitochondrial DNA markers. Proceedings of the National Academy of Sciences USA 94: 8912-8917
Foà A, Bearzi M, Balbaccini NE (1990) A preliminary report on he size of the home range and in the orientational capabilities in the lacertid lizard *Podarcis sicula*. Ethology, Ecology, Evolution 2: 310
Ford NB (1986) The role of pheromone trails in the sociobiology of snakes. Duvall D, Müller-Schwarze D, Silverstein RM (eds) Chemical Signals in Vertebrates 4: Ecology, Evolution, and Comparative Biology. Plenum Press, New York 261-278
Ford NB, Burghardt GM (1993) Perceptual mechanisms and the behavioral ecology of snakes. In Seigel R, Collins JT (eds) Snakes. McGraw-Hill, New York 117-164
Franz R, Dodd CK Jr, Jones C (1988) *Rana oreolata aesopus* (Florida gopher frog) Movement. Herpetological Review 19: 33
Freake MJ (1998) Variation in homeward orientation performance in the sleepy lizard (*Tiliqua rugosa*): effects of sex and reproductive period. Behavior, Ecology and Sociobiology 43: 339-344
Freake MJ (1999) Evidence for orientation using the e-vector direction of polarized light in the sleepy lizard (*Tiliqua rugosa*). Journal of Experimental Biology 202: 1159-1166
García-París M, Alcobendas M, Buckley D, Wake DB (2003) Dispersal of viviparity across contact zones in Iberian populations of fire salamanders (*Salamandra*) inferred from discordance of genetic and morphological traits. Evolution 57: 129-143

Garduño M, Maldonado A, Márquez R, Schroeder BA, Balazs GH (2000) Satellite tracking of an adult male and female turtle from the Yucatan in the Gulf of Mexico. In Kalb H, Wibbles T (compilers) Proceedings of the (19th Annual Symposium on Sea Turtle Biology and Conservation. NOAA Technical Memorandum NMFS-SEFSC-443 158-159

Gelder JJ van, Aarts HMJ, Staal WM (1986) Routes and speed of migratory toads (*Bufo bufo* L.): a telemetric study. Herpetology Journal 1: 111-114

Gibbons JW, Greene JL, Congdon JD (1990) Temporal and spatial movement patterns of sliders and other turtles. In Gibbons JW (ed) Life History and Ecology of the Slider Turtle. Smithsonian Institution Press, Washington DC 201-215

Gill DE (1978) Effective population size and interdemic migration rates in a metapopulation of the red-spotted newt, *Notophthalmus viridescens* (Rafinesque). Evolution 32: 839-849

Gitschlag (1996) Migration and diving behaviors of Kemp's ridley (Garman) sea turtles along the U.S. southeastern Atlantic coast. Journal of Experimental Marine Biology and Ecology 205: 115-135

Gittens SP (1983) The breeding migration of the common toad (*Bufo bufo*) to a pond in mid-Wales. Journal of Zoology, London 199: 555-562

Glandt D (1986) Die saisonalen Wanderungen der mitteleuropaischen Amphibien. Bonner Zoologische Beitrage 37: 211-238

Godfrey MH, Barreto R (1995) Beach vegetation and seafinding orientation of turtle hatchlings. Biological Conservation 74: 29-32

Godley BJ, Richardson S, Broderick AC, Coyne MS, Glen F, Hays GC (2002) Long-term satellite telemetry of the movements and habitat utilization by green turtles in the Mediterranean. Ecography 25: 352-362

Godley BJ, Broderick AC, Glen F, Hays GC (2003a) Post-nesting movements and submergence patterns of loggerhead marine turtles in the Mediterranean assessed by satellite tracking. Journal of Experimental Marine Biology and Ecology 287: 119-134

Godley BJ, Lima EHSM, Åkesson S, Broderick AC, Glen F, Godfrey MH, Luschi P, Hays GC (2003b) Movement patterns of green turtles in Brazilian coastal waters described by satellite tracking and flipper tagging. Marine Ecology Progress Series 253: 279-288

Goff GP, Lien J, Stenson GB, Fretey J (1994) The migration of a tagged leatherback turtle, *Dermochelys coriacea*, from French Guiana, South America, to Newfoundland, Canada, in 128 days. Canadian Field-Naturalist 108: 72-73

Gorzula SJ (1978) An ecological study of *Caiman crocodilus crocodilus* inhabiting savannah lagoons in the Venezuelan Guayana. Oecologia 35: 21-34

Gould E (1957) Orientation in box turtles, *Terrapene c. carolina* (Linnaeus). Biological Bulletin 112: 335-348

Graham JB, Rubinoff I, Hecht MK (1971) Temperature physiology of the sea snake, *Pelamis platurus*: and index of its colonization potential in the Atlantic Ocean. Proceedings of the National Academy of Science USA 68: 1360-1363

Graham T, Georges A, McElhinney N (1996) Terrestrial orientation by the eastern long-necked turtle *Chelodina longicollis* from Australia. Journal of Herpetology 30: 467-477

Grant D, Anderson O, Twitty VC (1968) Homing and orientation by olfaction in newts (*Taricha rivularis*). Science 160: 1354-1355

Graves BM, Duvall D, King MB, Linstedt SL, Gern WA (1986) Initial den location by neonatal prairie rattlesnakes: functions, causes and natural history in chemical ecology. In Duvall D, Muller-Schwarze D, Silverstein RM (eds) Chemical Signals in Vertebrates IV: Ecology, Behaviour and Comparative Biology. Plenum, New York 285-304

Gregory PT, McCartney JM, Larsen KW (1987) Spatial patterns and movements. In Seigel RA, Collins JT, Novak SS (eds) Snakes: Ecology and Evolutionary Biology. MacMillan, New York 366-395

Grubb JC (1970) Orientation in post-reproductive Mexican toads, *Bufo valliceps.* Copeia 1970: 674-680

Grubb JC (1973a) Orientation in newly metamorphosed Mexican toads, *Bufo valliceps.* Herpetologica 29: 95-100

Grubb JC (1973b) Olfactory orientation in breeding Mexican toads, *Bufo valliceps.* Copeia 1973: 490-497

Grubb JC (1975) Olfactory orientation in southern leopard frogs, *Rana utricularia.* Herpetologica 31: 219-221

Grubb JC (1976) Maze orientation of Mexican toads, *Bufo valliceps* (Amphibia, Anura, Bufonidae) using olfactory and configurational cues. Journal of Herpetology 10: 97-104

Guyer C (1978) Comparative ecology of the short-horned lizard (*Phrynosoma douglasi*) and the sagebrush lizard (*Sceloporus graciosus*) in south eastern Idaho. MSc Thesis, Idaho State University

Guyer C (1991) Orientation and homing behaviour as a measure of affinity for a home range in two species of Iguanid lizards. Amphibia-Reptilia 12: 373-384

Hairston NGS (1994) Vertebrate Zoology: An Experimental Field Approach. Cambridge University Press, Cambridge

Hays GC, Dray M, Quaife T, Smyth TJ, Mironnet NC, Luschi P, Papi F, Barnsley MS (2001) Movements of migrating green turtles in relation to AVHRR derived sea surface temperature. International Journal of Remote Sensing 22: 1403-1411

Hays GC, Broderick AC, Godley BJ, Lovell P, Martin C, McConnell BJ, Richardson S (2002) Biphasal long-distance migration in green turtles. Animal Behavior 64: 895-898

Hays GC, Åkesson S, Broderick AC, Glen F, Godley BJ, Papi F, Luschi P (2003) Island-finding ability of marine turtles. Proceedings of the Royal Society of London, Series B (Supplement) 270: S5-S7

Hays GC, Houghton JDR, Myers AE (2004) Pan-Atlantic leatherback turtle movements. Nature 429: 522

Hays GC, Houghton JDR, Isaacs C, King RS, Lloyd C, Lovell P (2004) First records of oceanic dive profiles for leatherback turtles, *Dermochelys coriacea,* indicate behavioral plasticity associated with long-distance migration. Animal Behavior 67: 733-743

Healy WR (1975) Breeding and postlarval migrations of the red-spotted newt, *Notophthalmus viridescens*, in Massachusetts. Ecology 56: 673-680

Henwood TA (1987) Movements and seasonal changes in loggerhead turtle (*Caretta caretta*) aggregations in the vicinity of Cape Canaveral, Florida (1978-1984). Biological Conservation 40: 191-202

Hepper PG, Waldman B (1992) Embryonic olfactory learning in frogs. Quarterly Journal of Experimental Psychology 44: 179-197

Hershey JL, Forrester DC (1980) Sensory orientation in *Notophthalmus v viridescens* (Amphibia: Salamandridae). Canadian Journal of Zoology 58: 266-276

Heusser, H (1968) Die Liebensweise der Erdkröte (*Bufo bufo* L) Wanderungen und Sommerquartiere. Revue Suisse de Zoologie 75: 928-982

Heusser, H (1969) Die Liebenweise der Erdkröte (*Bufo bufo* L): Orientierungsproblem. Revue Suisse de Zoologie 76: 444-517

Hewitt J, Power JH (1913) A list of South African Lacertilia, Ophidia and Batrachia in the McGregor Museum, Kimberley, with field notes on various species. Transactions of the Royal Society of South Africa 3: 147-176

Hirth HF, Carr A (1970) The green turtle in the Gulf of Aden and Seychelles Islands. Verhandlingen Kononklijke Nederlandse Akademie van Wetenschappen, Afdeling Natuurkunde 58: 1-41

Horrocks JA, Vermeer LA, Krueger B, Coyne MS, Schroeder BA, Balazs GH (2001) Migration routes and destination characteristics of post-nesting hawksbill turtles satellite-tracked from Barbados, West Indies. Chelonian Conservation Biology 4: 107-114

Hughes GR (1974) The sea turtles of South-East Africa 2. The biology of the Tongaland loggerhead turtle *Caretta caretta* L. with comments on the leatherback turtle *Dermochelys coriacea* L. and the green turtle *Chelonia mydas* L. in the study region. Investigative Reports of the Ocean Research Institute, Durban, South Africa 36: 1-96

Hughes GR, Luschi P, Mencacci R, Papi F (1998) The 7000 km journey of a leatherback turtle tracked by satellite. Journal of Experimental Marine Biology and Ecology 229: 207-217

Hurlburt SH (1969) The breeding migrations and interhabitat wandering of the vermillion-spotted newt *Notophthalmus viridescens* (Rafinesque). Ecological Monographs 39: 465-488

Irwin JP, Horner AJ, Lohmann KJ (2004) Magnetic field distortions produced by protective cages around sea turtle nests: unintended consequences for orientation and navigation? Biological Conservation 118: 117-120

Ishihara S (1969 Homing behaviour of the lizard, *Takydromus tachydromoides* (Schlegal). Bulletin of the Kyoto University of Education, Series B 36: 11-23

Johnson SA (2003) Orientation and migration distances of a pond-breeding salamander (*Notophthalmus perstriatus,* Salamandridae). Alytes 21: 3-22

Joly P, Miaud C (1993) How does a newt find its pond? The role of chemical cues in migrating newts (*Triturus alpestris*). Ethology, Ecology, Evolution 5: 447-455

Kalb HJ (1999) Behavior and physiology of solitary and arribada olive ridley sea turtles (*Lepidochelys olivacea*) during the internesting period. PhD thesis, Texas A&M University, USA

Katz U, Nagel W (1994) Biophysics of ion transport across amphibian skin. In Heatwole H, Barthalmus GT (eds) Amphibian Biology. Volume 1. The Integument Surrey Beatty & Sons Pty Limited, Chipping Norton, New South Wales, Australia 98-119

Keinath JA, Musick JA (1993) Internesting movements and behavior of a leatherback turtle, *Dermochelys coriacea.* Copeia 1993: 1010-1017

Kelleher KE, Tester JR (1969) Homing and survival in the Manitoba toad, *Bufo hemiophrys*, in Minnesota. Ecology 50: 1040-1048

King MB, Duvall D (1990) Prairie rattlesnake seasonal migrations: episodes ofmovement, vernal foraging and sex differences. Animal Behaviour 39: 924-935

Kingsmill SF, Mrosovsky N (1982) Sea-finding behaviour of loggerhead hatchlings: the time course of transient circling following unilateral and asynchronous bilateral blindfolding. Brain, Behaviour and Evolution 20: 29-42

Klauber LM (1982) Rattlesnakes: Their Habits, Life Histories and Influence on Mankind. University of California Press, Berkeley

Kopitsky K, Pitman RL, Plotkin PT (2000) Investigations on mating and reproductive status of olive ridleys, *Lepidochelys olivacea*, captured in the eastern tropical Pacific. In Kalb H, Wibbles T (compilers) Proceedings of the 19th Annual Symposium on Sea Turtle Biology and Conservation NOAA, Technical Memorandum NMFS-SEFSC-443 160-162

Kozlowski J (1999) Adaptation: a life history perspective. Oikos 86: 185-194

Krekorian CO (1977) Homing in the desert Iguana, *Dipsosaurus dorsalis.* Herpetologica 23: 123-127

Lamoreux VS, Maddison DM (1999) Overwintering habitats of radio-implanted green frogs, *Rana clamitans.* Journal of Herpetology 33: 430-435

Landreth HF (1973) Orientation and behavior of the rattlesnake, *Crotalus atrox.* Copeia 1: 26-31

Langton TES (ed) (1989) Amphibians and Roads Proceedings of the Toad Tunnel Conference, Rendsburg, Federal Republic of Germany, 7-8 January 1989. ACO Polymer Products Ltd, Shefford, UK

Laufer H (1997) Methodik beu Untersuchungen von Amphibienwanderungen zum Laichgewasser im Rahmen von Strassenausbauverfahren - am Beispeil der Erdkröte (*Bufo bufo*). Mertensiella 7: 35-60

Lawson PA (1989) Orientation abilities and mechanisms in a northern migratory population of the common garter snake (*Thamnophis sirtalis*). Musk-Ox 37: 110-115

Lawson PA (1994) Orientation abilities and mechanisms in non-migratory populations of garter snakes (*Thamnophis sirtalis* and *T. ordinoides*). Copeia 1994: 263-274

Lawson PA, Secoy DM (1991) The use of solar cues as migratory orientation guides by the plains garter snake, *Thamnophis radix*. Canadian Journal of Zoology 69: 2700-2702

Le Gall JY, Hughes GR (1987) Migrations de al tortue verte *Chelonia mydas* dans l'Océan Indien Sud-Ouest observées a partir des marquages sur les sites de ponte Europa et Tromelin (1970-1985). Amphibia-Reptilia 8 : 277-282

Lehtinen RM, Galatowitsch SM (2001) Colonization of restored wetlands by amphibians in Minnesota. American Midland Naturalist 145 : 388-396

LeMaster MP, Moore IT, Mason RT (2001) Conspecific trailing behaviour of red-sided garter snakes, *Thamnophis sirtalis parietalis*, in the natural environment. Animal Behavior 61: 827-833

Lemkau PJ (1970) Movements of the box turtle, *Terrapene c. carolina* (Linnaeus) in unfamiliar territory. Copeia 1970: 781-783

Light P, Salmon M, Lohmann KJ (1993) Geomagnetic orientation of loggerhead sea turtles: evidence for an inclination compass. Journal of Experimental Biology 182: 1-10

Lillywhite HB, Maderson PFA (1982) Skin structure and permeability. In Gans C, Pough HF (eds) Biology of the Reptilia. Vol. 12. Academic Press, London 397-442

Limpus CJ (1985) A study of the loggerhead sea turtle, *Caretta caretta*, in eastern Australia. PhD thesis. University of Queensland, St Lucia, Queensland

Limpus CJ, Limpus DJ (2001) The loggerhead turtle, *Caretta caretta*, in Queensland: breeding migrations and fidelity to a warm temperate feeding area. Chelonian Conservation Biology 4: 142-153

Limpus CJ, Parmenter CJ, Parker R, Ford NB (1981) The flatback turtle, *Chelonia depressa*, in Queensland: The Peak Island rookery. Herpetofauna 13: 14-16

Limpus CJ, Parmenter CJ, Baker U, Fleay A (1983) The flatback turtle, *Chelonia depressa*, in Queensland: post-nesting migration and feeding ground distribution. Australian Wildlife Research 10: 557-561

Limpus CJ, Fleay A, Guinea M (1984) Sea turtles of the Capricornia Section of the Great Barrier Reef Marine Park. In Ward WT, Saenger P (eds) The Capricornia Section of the Great Barrier Reef Past, Present and Future. The Royal Society of Queensland, Brisbane, Australia 61-78

Lindeman PV (1992) Nest-site fixity among painted turtles (*Chrysemys picta*) in northern Idaho. Northwestern Naturalist 73: 27-30

Linzey D (2001) Vertebrate Biology. McGraw Hill, New York

Lohmann KJ (1992) How sea turtles navigate. Scientific American (Jan): 100-106

Lohmann KJ, Cain SD, Dodge SA, Lohmann CMF (2001) Regional magnetic fields as navigational markers for sea turtles. Science 294: 364-367

Lohmann KJ, Lohmann CMF (1998) Migratory mechanisms in marine turtles. Journal of Avian Biology 29: 585-596

Lohmann KJ, Lohmann CMF, Ehrhart LM, Bagley DA, Swing T (2004) Geomagnetic map used in sea-turtle navigation. Nature 428: 909-910

Lohmann KJ, Salmon M, Wyneken J (1990) Functional autonomy of land and sea orientation systems in sea turtle hatchlings. Biological Bulletin 179: 214-218

Lohmann KJ, Swartz AW, Lohmann CMF (1995) Perception of ocean wave direction by sea turtles. Journal of Experimental Biology 198: 1079-1085

Lohmann KJ, Witherington BE, Lohmann CMF, Salmon M (1997) Orientation, navigation and natal beach homing in sea turtles. In Lutz PL, Musick JA (eds) The Biology of Sea Turtles. CRC Press, Boca Raton, Florida 107-155

Loredo I, VanVuren D, Morrison ML (1996) Habitat use and migration behavior of the California tiger salamander. Journal of Herpetology 30: 282-285

Loveridge A (1953) Zoological results of a fifth expedition to East Africa IV Amphibians from Nyasaland and Tete. Bulletin of the Museum of Comparative Zoology, Harvard 110: 325-406

Luschi P (2003) Migration and conservation: the case of sea turtles. In Festa-Bianchet M, Apollonio M (eds) Animal Behavior and Wildlife Conservation Island Press, Washington, Corelo and London 49-61

Luschi P, Papi F, Liew HC, Chan EH, Bonadonna F (1996) Long-distance migration and homing after displacement in the green turtle (*Chelonia mydas*): a satellite tracking study. Journal of Comparative Physiology A 178: 447-452

Luschi P, Hays GC, Del-Seppia C, Marsh R, Papi F (1998) The navigational feats of green sea turtles migrating from Ascension Island investigated by satellite telemetry. Proceedings of the Royal Society of London, Series B 265: 2279-2284

Luschi P, Åkesson S, Broderick AC, Glen F, Godley BJ, Papi F, Hays GC (2001) Testing the navigational abilities of ocean migrants: displacement experiments on green sea turtles (*Chelonia mydas*). Behavior, Ecology, Sociobiology 50: 528-534

Luschi P, Hays GC, Papi F (2003) A review of long-distance movements by marine turtles, and the possible role of ocean currents. Oikos 103: 293-302

Luschi P, Hughes GR, Mencacci R, DeBernardi E, Sale A, Broker R, Bouwer M, Papi F (2003) Satellite tracking of migrating loggerhead sea turtles (*Caretta caretta*) displaced in the open sea. Marine Biology 143: 793-801

Lutcavage, M (1996) Planning your next meal: leatherback travel routes and ocean fronts In Keinath JA, Barnard DE, Musick JA, Bell BA (compilers) Proceedings of the 15th Annual Symposium on Sea Turtle Biology and Conservation. NOAA Technical Memorandum NMFS-SEFSC-387 174-178

Madison DM (1997) The emigration of radio-implanted spotted salamanders, *Ambystoma maculatum*. Journal of Herpetology 31: 542-551

Madsen T, Shine R (1996) Seasonal migrations of predators and prey: A study of pythons and rats in tropical Australia. Ecology 77: 149-156

Malmgren JC (2002) How does a newt find its way from a pond? Migration patterns after breeding and metamorphosis in Great Crested Newts (*Triturus cristatus*) and smooth newts (*T. vulgaris*). Herpetological Journal 12: 29-35

Manton M, Karr A, Ehrenfeld DW (1972) Chemoreception in the migratory sea turtle, *Chelonia mydas*. Biological Bulletin 143: 184-195

Marquèz R (1994) Synopsis of the biological data on the Kemp's Ridley Turtle *Lepidochelys kempii* (Garman, 1880). NOAA Technical Memorandum NMFS-SEFSC-343 91pp

Martinez-Solano I, Goncalves HA, Arntzen JW, García-París M (2004) Phylogenetic relationships and biogeography of midwife toads (Discoglossidae: *Alytes*). Journal of Biogeography 31: 603-618

Martof B (1953) Home range and movements of the green frog, *Rana clamitans.* Ecology 34: 529-543

Martof B (1962) Some observations on the role of olfaction among salientian amphibia. Physiological Zoology 35: 270-272

Mathis A, Moore FR (1988) Geomagnetism and the homeward orientation of the box turtle, *Terrapene Carolina.* Ethology 78: 265-274

Mayhew WW (1963) Biology of the granite spiny lizard, *Sceloporus orcutti.* American Midland Naturalist 69: 310-327

McCoy ED, Mushinsky HR, Wilson DS (1993) Pattern in the compass orientation of gopher tortoise burrows at different spatial scales. Global Ecology and Biogeography Letters 3: 33-40

McGregor JH, Teska WR (1989) Olfaction as an orientation mechanism in migrating *Ambystoma maculatum.* Copeia 3: 779-781

Meylan AB (1982) Sea turtle migration: evidence from tag returns. In Bjorndal KA (ed) Biology and Conservation of Sea Turtles Smithsonian Institution Press, Washington DC 91-100

Meylan AB (1999) Status of the hawksbill turtle (*Eretmochelys imbricata*) in the Caribbean region. Chelonian Conservation Biology 3: 177-184

Meylan AB, Bowen BW, Avise JC (1990) A genetic test of the natal homing versus social facilitation models for Green Turtle migration. Science 248: 724-727

Miller JD (1997) Reproduction in sea turtles. Lutz PL, Musick JA (eds) The Biology of Sea Turtles. CRC Press, Boca-Raton, Florida

Modha ML (1967) The ecology of the Nile crocodile (*Crocodilus niloticus* Laurenti) on Central Island, Lake Rudolf. East African Wildlife Journal 5: 74-95

Modha ML (1968) Crocodile research project, Central Island, Lake Rudolf: 1967 breeding season. East African Wildlife Journal 6: 148-150

Morreale SJ, Standora EA, Paladino FV (1994) Leatherback migrations along deep water bathymetric contours. Schroeder BA, Witherington BE (compilers) Proceedings of the 13th Annual Symposium on Sea Turtle biology and Conservation. NOAA Technical Memorandum NMFS-SEFSC-341 109-110

Morreale SJ, Standora EA, Spotila JR, Paladino FV (1996) Migration corridor for sea turtles. Nature 384: 319-320

Mortimer JA, Carr A (1987) Reproduction and migration of the Ascension Island green turtles (*Chelonia mydas*). Copeia 1987: 103-113

Mrosovsky N, Kingsmill SF (1985) How turtles find the sea. Zeitschrift für Tierpsychologie 67: 237-256

Musick JA, Limpus CJ (1997) Habitat utilization and migration in juvenile sea turtles. Lutz PL, Musick JA (eds) The Biology of Sea Turtles. CRC Press, Boca Raton, Florida

Nagelkirken I, Pors LPJJ, Hoetjes P (2003) Swimming behaviour and dispersal patterns of headstarted loggerhead turtles, *Caretta caretta.* Aquatic Ecology 37: 183-190

Neill WT (1971) The Last of the Ruling Reptiles: Alligators, Crocodiles and their Kin. Columbia University Press, London

Newcomer RT, Taylor DH, Guttman SI (1974) Celestial orientation in two species of water snakes (*Natrix sipedon* and *Regina septemvittata*). Herpetologica 30: 194-200

Nichols WJ, Resendiz A, Seminoff JA, Resendiz B (2000) Transpacific migration of a loggerhead turtle monitored by satellite telemetry. Bulletin of Marine Science 67: 937-947

Nietschmann B (1981) Following the underwater trail of a vanishing species, the hawksbill turtle. National Geographi Society Research Reports 13: 459-480

Orr RF (1982) Vertebrate Biology 5th Ed. Saunders College Publishing, Philadelphia

Ouboter PE, Nanhoe LM (1988) Habitat selection and migration of *Caiman crocodilus crocodilus* in a swamp and swamp-forest habitat in northern Suriname. Journal of Herpetology 22: 283-294

Palis JG (1997) Breeding migration of *Ambystoma cingulatum* in Florida. Journal of Herpetology 31: 71-78

Palo JU, Lesbarreres D, Schmeller DS, Primmer CR, Merila J (2004) Microsatellite marker data suggest sex-biased dispersal in the common frog *Rana temporaria.* Molecular Ecology 13: 2865-2869

Papi F (1992) Animal Homing. Chapman and Hall, London

Papi F, Luschi P (1996) Pinpointing "Isla Meta": The case of sea turtles and albatrosses. Journal of Experimental Biology 199: 65-71

Papi F, Luschi P, Crosio E, Hughes GR (1997) Satellite tracking experiments on the navigational ability and migratory behaviour of the loggerhead turtle, *Caretta caretta.* Marine Biology 129: 215-220

Papi F, Luschi P, Åkesson S, Capogrossi S, Hays GC (2000) Open-sea migration of magnetically disturbed sea turtles. Journal of Experimental Biology 203: 3435-3443

Parmenter CJ (1993) Australian sea turtle research and conservation and management: a 1993 status review. In Lunney D, Ayers D (eds) Herpetology in Australia: A Diverse Discipline. Transactions of the Royal Zoological Society of New South Wales, Mosmon, New South Wales, Australia 321-325

Parmenter CJ (1994) Species review: the flatback turtle, *Natator depressa.* James R (ed) Proceedings of the Australian Marine Turtle Conservation Workshop (1990). Australian Nature Conservation Agency, Canberra 60-62

Pasenen S, Sorjonen A (1995) An orientation experiment conducted during autumn migration of the common frog (*Rana temporaria*). Memoranda Societas Fauna Flora Fennica 71: 113-117

Pauly GB, Hillis DA, Cannatella DC (2004) The history of a Nearctic colonization: molecular phylogenetics and biogeography of the Nearctic toads (*Bufo*). Evolution 58: 2517-2535

Pearson PG (1955) Population ecology of the spade-foot toad, *Scaphiopus h. holbrooki* (Harlan). Ecological Monographs 25: 233-267

Pearson PG (1957) Further notes on the population ecology of the spadefoot. Ecology 38: 580-586

Peters A, Verhoeven KJF (1994) Impact of artificial lighting on the seaward orientation of hatchling loggerhead turtles. Journal of Herpetology 28: 112-114

Petranka JW (1998) Salamanders of the United States and Canada. Smithsonian Institution Press, Washington

Phillips CA, Sexton OJ (1989) Orientation and sexual differences during the breeding migration of the spotted salamander, *Ambystoma maculatum.* Copeia 1989: 17-22

Phillips JB (1977) Use of the earth's magnetic field by orienting cave salamanders (*Erycea lucifuga*). Journal of Comparative Physiology A 121: 273-288

Phillips JB (1986a) Magnetic compass orientation in the Eastern red-spotted newt (*Notophthalmus viridescens*). Journal of Comparative Physiology A 158: 103-109

Phillips JB (1987) Laboratory studies of homing orientation in the eastern red-spotted newt, *Notophthalmus viridescens.* Journal of Experimental Biology 131: 215-229

Phillips JB (1998) Magnetoreception. In Heatwole H, Dawley EM (eds) Amphibian Biology Vol. 3 Sensory Perception. Surrey Beatty and Sons, Chipping Norton, New South Wales, Australia 954-964

Phillips JB (1986b) Two magnetoreception pathways in a migratory salamander. Science 233: 765-767

Phillips JB, Adler K (1978) Directional and discriminatory responses of salamanders to weak magnetic fields. In Schmidt-Koenig K, Keeton W (eds) Animal Migration, Navigation and Homing Springer-Verlag, Berlin 325-333

Phillips JB, Borland SC (1992a) Behavioral evidence for use of a light dependent magnetoreception mechanism by a vertebrate. Nature 359: 142-144

Phillips JB, Borland SC (1992b) Wavelength specific effects of light on magnetic compass orientation in the eastern red-spotted newt *Notophthalmus viridiscens.* Ethology, Ecology, Evolution 4: 33-42

Phillips JB, Borland SC (1994) Use of a specialized magnetoreception system for homing by the eastern red-spotted newt *Notophthalmus viridiscens.* Journal of Experimental Biology 188: 275-291

Phillips JB, Adler K, Borland SC (1995) True navigation of an amphibian. Animal Behavior 50: 855-858

Phillips, JB, Deutschlander ME, Freake MJ, Borland SC (2001) The role of extraocular photoreceptors in newt magnetic compass orientation: parallels between light-dependent magnetoreception and polarized light detection in vertebrates. Journal of Experimental Biology 188: 275-291

Pilliod DS, Peterson CR, Ritson PI (2002) Seasonal migration of Columbia spotted frogs (*Rana luteiventris*) among complementary resources in a high mountain basin. Canadian Journal of Zoology 80: 1849-1862

Plotkin PT (2003) Adult migration and habitat use. In Lutz PL, Musick JA, Wyneken J (eds) The Biology of Sea Turtles Volume II. CRC Press, Boca Raton, Florida

Plotkin PT, Spotila JR (2002) Post-nesting migrations of loggerhead turtles *Caretta caretta* from Georgia, USA: conservation implications for a genetically distinct subpopulation. Oryx 36: 396-399

Plotkin PT, Byles RA, Rostal DC, Owens DW (1995) Independent versus socially facilitated oceanic migrations of the olive ridley, *Lepidochelys olivacea.* Marine Biology 122: 137-143

Plummer MV, Shirer HW (1975) Movement patterns in a river population of the softshell turtle *Trionyx muticus.* Occasional Papers of the Museum of Natural History, University of Kansas 43: 1-26

Polovina JJ, Balazs GH, Howell EA, Parker DM, Seki MP, Dutton PH (2004) Forage and migration habitat of loggerhead (*Caretta caretta*) and olive ridley (*Lepidochelys olivacea*) sea turtles in the central North Pacific Ocean. Fisheries Oceanography 13: 36-51

Pough HF, Andrews RM, Cadle JE, Crump ML, Savitzky AH, Wells KD (1998) Herpetology Prentice Hall Inc., New Jersey

Pough HF, Janis CM, Heiser JB (2002) Vertebrate Life 6th Ed. Prentice Hall Inc, New Jersey

Pritchard PCH (1976) Post-nesting movements of marine turtles (Cheloniidae and Dermochelidae) tagged in the Guyanas. Copeia 1976: 749-752

Quinn NWS, Tate DP (1991) Seasonal movements and habitat of wood turtles (*Clemmys insculpta*) in Algonquin Park, Canada. Journal of Herpetology 25: 217-220

Rand AS (1967) Ecology and social organization in the iguanid lizard *Anolis lineatopus.* Proceedings of the United States National Museum 122: 1-79

Rand AS (1968) A nesting aggregation of Iguanas. Copeia 3: 552-561

Rebel TP (1974) Sea Turtles and the Turtle Industry of the West Indies, Florida, and the Gulf of Mexico. Revised Edition University of Miami Press, Coral Gables, Florida

Reinert HK, Kodrich WR (1982) Movements and habitat utilization by the massassauga, *Sistrurus catenatus catenatus.* Journal of Herpetology 16: 152-171

Renaud ML, Carpenter JA, Williams JA, Landry AMJ (1996) Kemp's ridley sea turtle (*Lepidochelys kempii*) tracked by satellite telemetry from Louisiana to nesting beach at Rancho Nuevo, Tamaulipas, Mexico. Chelonian Conservation Biology 2: 108-109

Resendiz A, Resendiz B, Nichols WJ, Seminoff JA, Kamezaki N (1998) First confirmed east-west transpacific movement of a loggerhead sea turtle, *Caretta caretta*, released in Baja California, Mexico. Pacific Science 52: 151-153

Rodda GH (1984) The orientation and navigation of juvenile alligators: evidence of magnetic sensitivity. Journal of Comparative Biology 154: 648-658

Rodda GH (1985) Navigation in juvenile alligators. Zeitschrift für Tierpsychologie 68: 65-77

Rodhouse P, Barling RWA, Clark WIC (1975) The feeding and ranging behaviour of the Galapagos giant tortoises (*Geochelone elephantopus*): The Cambridge and London University Galapagos Expeditions 1972 and 1973. Journal of Zoology, London 176: 297-310

Russell AP, Bauer AM (2000) The Amphibians and Reptiles of Alberta: A Field Guide and Primer of Boreal Herpetology 2nd Ed. University of Calgary Press, Calgary, Alberta, Canada

Salmon M (2003) Artificial night lighting and sea turtles. Biologist 50: 163-168

Salmon M, Lohmann KJ (1989) Orientation cues used by hatchling loggerhead sea turtles (*Caretta caretta* L.) during their offshore migration. Ethology 83: 215-228

Salmon M, Wyneken J (1990) Do swimming loggerhead sea turtles (*Caretta caretta* L.) use light cues for offshore orientation? Marine Behaviour and Physiology 17: 233-246

Schabetsberger R, Jehle R, Maletzky A, Pesta J, Sztatecsny M (2004) Delineation of terrestrial reserves for amphibians: post-nesting migrations of Italian crested newts (*Triturus c carnifex*) at high altitude. Biological Conservation 117: 95-104

Scoccianti C (2000) Study on road stretches at high risk for the migration of amphibians (focal crossing points) in the Province of Florence: proposals and measures to minimize impact. Rivista di Idrobiologia 38: 323-332

Semlitsch RD (1985) Analysis of climatic factors influencing migrations of the salamander *Ambystoma talpoideum.* Copeia 2: 477-489

Semlitsch RD, McMillan MA (1980) Breeding migrations, population size structure and reproduction of the dwarf salamander, *Eurycea quadridigitata,* in South Carolina. Brimleyana 3: 97-105

Semlitsch RD, Pechmann JHK (1985) Diel patterns of migratory activity for several species of pond-breeding salamanders. Copeia 1985: 86-91

Sexton OJ, Phillips C, Bramble JE (1990) The effects of temperature in the breeding migration of the spotted salamander (*Ambystoma maculatum*). Copeia 1990: 781-787

Shealy RM (1976) The natural history of the Alabama map turtle, *Graptemys pulchra* Baur, in Alabama. Bulletin of the Florida State Museum, Biological Sciences 21: 47-111

Shine R, Lambeck R (1985) A radiotelemetric study of movements, thermoregulation, and habitat utilization of Arafura Filesnakes (Serpentes: Acrochordidae). Herpetologica 41: 351-361

Shoop CR (1965) Orientation of *Ambystoma maculatum*: movements to and from breeding ponds. Science 149: 558-559

Shoop CR, Doty TL (1972) Migratory orientation by marbled salamanders *(Ambystoma opacum*) near a breeding area. Behavioral Biology 7: 131-136

Sinsch U (1987) Orientation behaviour of toads (*Bufo bufo*) displaced from breeding sites. Journal of Comparative Physiology A 161: 715-727

Sinsch U (1988) Seasonal changes in the migratory behavior of the toad *Bufo bufo*: direction and magnitude of movements. Oecologia 76: 390-398

Sinsch U (1990a) Migration and orientation in anuran amphibians. Ethology, Ecology, Evolution 2: 65-79

Sinsch U (1990b) The orientation behaviour of three toad species (genus *Bufo*) displaced from the breeding site. Fortschritte der Zoologie 38: 75-83

Sinsch U (1991) Mini review: The orientation behaviour of amphibians. Herpetological Journal 1: 541-544

Sinsch U (1992a) Sex-biased site fidelity and orientation behaviour in reproductive natterjack toads (*Bufo calamita*). Ethology, Ecology, Evolution 4: 15-32

Sinsch U (1992b) Amphibians. In Papi F (ed) Animal Homing. Chapman and Hall, London

Sinsch U, Lang V, Wiemer R, Wirtz S (2003) Dynamik einer Kammolch-Metapopulation (*Triturus cristatus*) auf militarischem Uebungsgelaende (Schmittenhohe, Koblenz): 1 Phanologie, Wettereinfluss und Orstreue. Zeitschrift für Feldherpetologie 10: 193-210

Spieler M, Linsenmair KE (1998) Migration patterns and diurnal use of shelter in a ranid frog of a west African savannah: a telemetric study. Amphibia-Reptilia 19: 43-64

Spoekker PD (1967) Movements and seasonal activity cycles in the lizard *Uta stansburiana stejnegeri.* American Midland Naturalist 77: 484-494

Stebbins RC, Cohen NW (1995) A Natural History of Amphibians. Princeton University Press, Princeton, New Jersey

Steyermark AC, Williams K, Spotila JR, Paladino FV, Rostal DC, Morreale SJ, Koberg MT, Arauz R (1996) Nesting leatherback turtles at Las Baulas National Park, Costa Rica. Chelonian Conservation Biology 2: 173-183

Stumpel AHP (1987) Distribution and present numbers of the tree frog, *Hyla arborea*, in Zealand Flanders, The Netherlands (Amphibia, Hylidae). Bijdragen tot de Dierkunde 57: 151-163

Susuma I, Kikushi M, Nishio H (1995) Orientation of the toad, *Bufo japonicus* toward the breeding pond. Zoological Science 12: 475-484

Sweet SS (1993) Second report on the biology of the arroyo toad (*Bufo microscaphus californicus*) in the Los Pedres National Forest of southern California. Contract Report to the USDA, Forest Service, Los Padres National Forest, Goleta, California: 73pp

Swingland IR, Lessels CM (1979) The natural regulation of the giant tortoise populations on Aldabra Atoll: movement polymorphism, reproductive success and mortality. Journal of Animal Ecology 48: 639-654

Swingland IR, Stubbs D, Newdick M, Worton B (1986) Movement patterns in *Testudo hermanni* and implications for management. In Rocek Z (ed) Studies in Herpetology. Charles University and Societas Europaea Herpetologica, Prague 573-578

Swingland IR, North PM, Dennis A, Parker MJ (1989) Movement patterns and morphometrics in giant tortoises. Journal of Animal Ecology 58: 971-985

Taylor DH (1972) Extra-optic photoreception and compass orientation in larval and adult salamanders (*Ambystoma tigrinum*). Animal Behavior 20: 233-236

Taylor DH, Adler K (1978) The pineal body: site of extraocular perception of celestial cues for orientation in the tiger salamander (*Ambystoma tigrinum*). Journal of Comparative Physiology A 124: 357-361

Taylor DH, Auburn J (1978) Orientation of amphibians by linearly polarized light. In Schmidt-Koenig K, Keeton W (eds) Animal Migration, Navigation and Homing Springer-Verlag, Berlin 334-346

Taylor DH, Ferguson DE (1970) Extraoptic celestial orientation in the southern cricket frog *Acris gryllus*. Science 168: 390-392

Tinkle DW (1967) The life and demography of the side-blotched lizard, *Uta stansburiana*. Miscellaneous Publications, Museum of Zoology, University of Michigan 132: 1-182

Tunner HG (1992) Locomotory behaviour in water frogs from Neusiedlersee (Austria, Hungary) 15km migration of *Rana lessonae* and its hybridogenetic associate *Rana esculenta*. In Korsòs Z, Kiss I (eds) Proceedings of the 6th Ordinary General Meeting of the Societas Europaea Herpetologica, Budapest, Hungary 449-452

Twitty VC (1966) Of Scientists and Salamanders. WH Freeman and Co, San Francisco

Twitty VC, Grant D, Anderson O (1967) Initial homeward orientation after long-distance displacements in the newt *Taricha rivularis*. Proceedings of the National Academy of Science, USA 57: 342

Vences M, Vieietes DR, Glaw F, Brinkmann H, Kosuch J, Veith M, Meyer A (2003) Multiple overseas dispersal in amphibians. Proceedings of the Royal Society of London Series B 270: 2435-2442

Vences M, Kosuch J. Rodel MO, Lolters S, Channing A, Glaw F, Böhme W (2004) Phylogeography of *Ptychadena mascareniensis* suggests transoceanic dispersal in a widespread African-Malagasy frog lineage. Journal of Biogeography 31: 593-601

Verell PA (1987) The directionality of migrations of amphibians to and from the pond in southern England, with particular reference to the smooth newt, *Triturus vulgaris*. Amphibia-Reptilia 8: 93-100

Webb GJW, Buckworth R, Manolis SC (1983a) *Crocodilus johnstoni* in the McKinlay River area Northern Territory Australia 4. A demonstration of homing. Australian Wildlife Research 10: 403-406

Webb GJW, Manolis SC, Buckworth R (1983b) *Crocodilus johnstoni* in the McKinlay River area Northern Territory Australia 2. Dry season habitat selection and an estimate of the total population size. Australian Wildlife Research 10: 373-382

Weintraub J (1970) Homing in the lizard *Sceloporus orcutti*. Animal Behavior 18: 132-137

Werner DI (1983) Reproduction in the Iguana *Conolophus subcristatus* on Fernandina Island, Galapagos: Clutch size and migration costs. American Naturalist 121: 757-775

Wiewandt TA (1982) Evolution of nesting patterns in iguanine lizards. In Burghardt GM, Rand AS (eds) Iguanas of the World. Their Behavior, Ecology, and Conservation. Noyes Publications, Park Ridge, New Jersey 119-141

Wilson J (2001) A review of the modes of orientation used by amphibians during breeding migration. Journal of the Pennsylvania Academy of Sciences 74: 61-66

Witherington BE, Bjorndal KA (1991) Influence of artificial lighting on the seaward orientation of hatchling loggerhead turtles *Caretta caretta*. Biological Conservation 55: 139-149

Witzell WN (1983) Synopsis of biological data on the hawksbill turtle, *Eretmochelys imbricata* (Linnaeus, 1766). FAO Fisheries Synopsis 137: 78 pp

Xie F, Fei L, Ye C, Cai C, Wang Z, Sparreboom M (2000) Breeding migration and oviposition of the Chinhai salamander, *Echinotriton chinhaiensis*. Herpetological Journal 10: 111-118

Yeomans SR (1995) Water-finding in adult turtles: random search or oriented behaviour? Animal Behavior 49: 977-987

Young JZ (1981) The Life of Vertebrates 3rd Ed Clarendon Press, London

Zug JR, Vitt LJ, Caldwell JP (2001) Herpetology 2nd Ed. Academic Press, San Diego

8 Glacial Retreat and its Influence on Migration of Mitochondrial Genes in the Long-toed Salamander (*Ambystoma macrodactylum*) in Western North America

Mark D. Thompson[1] and Anthony P. Russell[2]

[1]University of Northern British Columbia, College of Science and Management, 3333 University Way, Prince George, B.C., Canada, V2N 4Z9, thompsma@unbc.ca, [2]Department of Biological Sciences, University of Calgary 2500 University Drive N.W., Calgary, Alberta, Canada, T2N 1N4, arussell@ucalgary.ca

8.1 Abstract

The long-toed salamander (*Ambystoma macrodactylum*) is a widespread inhabitant of the Cordilleran Region of western North America. The Cordilleran ice sheet retreated when climates changed at the end of the Pleistocene. This setting provides a natural experiment for phylogeographic tests of post-glacial migration. As migration occurs, the demographics of populations change; these changes are imprinted into the gene frequencies of descendant populations. Species ranges shifted as migrants inhabited tolerable post-glacial environments, and new genealogical mixtures formed as populations came into secondary contact. Historical climate, ecology, and geography impacted the range dynamics and consequent population genetics of the long-toed salamander. This systematic study of mitochondrial DNA tests biogeographic patterns using phylogenetic trees, nested phylogeographic clade analysis, and mismatch distributions. Phylogenetic congruence is tested first in a partitioned versus an intersected arrangement of two mitochondrial loci, including 95 cytochrome b and 103 intergenic spacer sequences. Nested phylogeographic clade analysis provides an explicit system to correlate lineages and their mismatch distributions. Although mismatch distributions operate ideally in high-migration species, and the long-toed salamander migrates little among contemporary populations, there is reason to suspect that waif dispersal increased with changes in fluvial dynamics following glacial retreat. Clade patterns support a deep vicariance across the central interior and reveal seven Pleistocene refugia. Waves in mismatch distributions indicate that population sizes increased in lineages residing in refugia near the ice margins at this time. The phylogenetic identities that spread away from refugia and their genetic patterns are placed into a historical pre- and post-glaciated context.

Key words: biogeography, glaciation; long-toed salamander, mitochondrial genes, phylogenetics, phylogeography, western North American Cordillera.

Migration is sometimes used synonymously with gene flow, which is correct when migration between established populations is the mechanism of gene flow. (Slatkin 1985: 393)

8.2 Introduction

The contemporary mountainous terrain of Western North America originated with the Laramide uplift of the Rocky Mountains, and attained significant prominence in the landscape during the Eocene, 45-36 million years ago (mya) (Brunsfeld et al. 2001). Such geological activity greatly affected species distributions, which shifted as populations tracked suitable climate regimes (Hewitt 1999). Ranges of populations are determined by the types of environment that were aptively (*sensu* Gould and Vrba 1982) tolerated (MacArthur and Connell 1967; Endler 1977; Avise 2000).

The geological activity of an area greatly affects connectivity among populations in space and time, and should therefore be considered and related to phylogeographic patterns (Da Silva and Patton 1998). *Ambystoma macrodactylum* has its ancestral origins in eastern North America, where species richness of ambystomatids is highest. The Ambystomatidae was isolated to the south-east of the Mid-Continental seaway during the Cretaceous (Milner 1983: Duellman and Sweet 1999). After the Paleocene, with the loss of the mid-continental seaway, progenitor populations of *A. macrodactylum* gained access to the western Cordillera. The Rocky Mountains are implicated in this regard as they had attained significant relief by the Eocene (Fig. 1; Brunsfeld et al. 2001). Other species were migrating towards the coast at this time (Daubenmire 1975), and the rain-shadow effect divided and altered the ranges of tropical forests recorded for that time (Daubenmire 1975). This mountainous environment provided the physiographic features that typify the current distribution of *Ambystoma macrodactylum* (Fig. 2). Other range adjustment factors, such as glacial advances during the Oligocene and Miocene (Zachos et al. 1997), are also implicated in the history of extant species distributions (Dynesius and Jansson 2000).

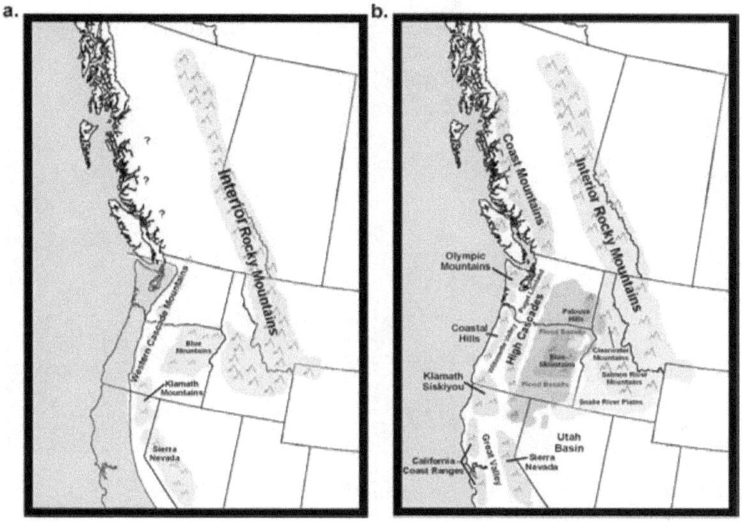

Fig. 1. Contemporary physiographic features of western North America were established during the Paleocene 65.5--55.8 mya (a), and they developed through the Miocene 23--5.3 mya (b). The Cascade and Sierra mountains grew during the Oligocene (Alt and Hyndman 1995), but they did not begin a major uplift until after the Miocene (Brunsfeld et al. 2001). A series of volcanic eruptions released an enormous amount of magma, which covered much of Oregon and Washington with flood basalts during the Miocene (17--15 mya); (see shaded area around the Palouse Hills and Blue Mountains [b]). Lava flows formed the Snake River plains in Southern Idaho (b) 13 mya. The mountain chains running from the Sierra Nevada, in California, to the northern Coast Mountains, in British Columbia, are referred to broadly as the Coastal Mountains, and those running from the Salmon River Mountains to the Interior Rocky Mountains are referred to as the Interior Mountains (b)

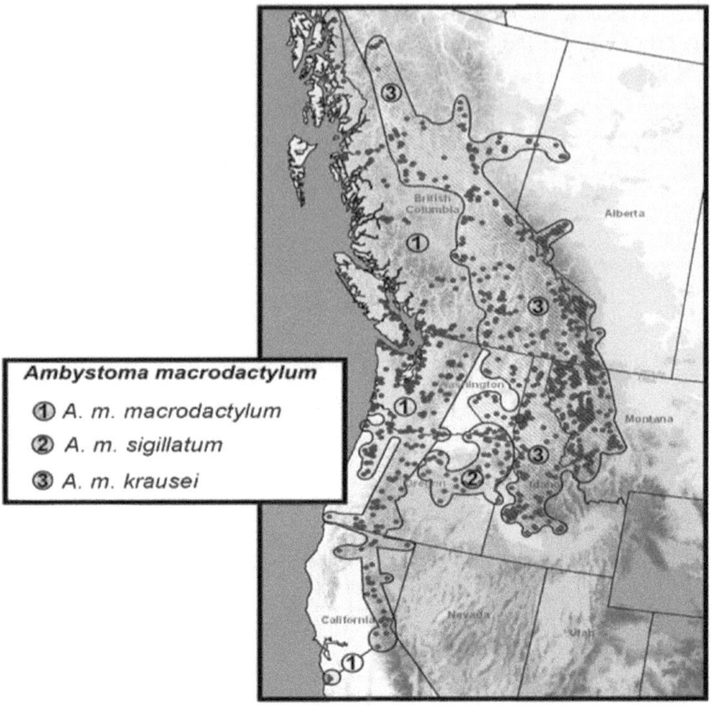

Fig. 2. A range map with the distributions of three subspecies of *Ambystoma macrodactylum*. Registered localities for adults and breeding locations are identified by the circular points. The ranges of the sub-species are based on these places of occupancy, proximal geography, and mitochondrial genealogies (Thompson et al. unpublished)

8.3 Phylogeographical Setting

From the Eocene into the Oligocene a collage of geologic terrains brought about an episode of folding, faulting, and mountain uplifting in the Pacific Northwest (Alt and Hyndman 1995). At this time the Blue Mountains were angled differently, the Cascade Mountains were not high enough to drastically disrupt the flow of Pacific moisture, and the Pacific shoreline was further inland. The non-existent fossil record for *Ambystoma macrodactylum* and the nature of the geological record for this time and place precludes a clear biogeographic interpretation of this deep history and requires reconstruction of this through historically-based patterns of gene flow.

The major uplift of the Cascade/Sierra Mountains during the Pliocene is most commonly associated with the vicariance in this area (Brunsfeld et al. 2001; Nielson et al. 2001; Stone et al. 2002). Another possible cause that has not been considered for this region (see Brunsfeld et al. 2001) are the immense lava flows that spread during the Miocene, and created large flood basalts in Washington, Oregon, and contiguous areas (Alt and Hyndman 1995).

The relief of the Coastal Mountain Ranges, including the Cascade and Sierra chains, produced a rain-shadow that drastically altered the western environment during the Pliocene (5--2 mya). The earth's climate has been cooling since the Paleogene (65.5--23.8 mya), but geologically recent times have experienced increased temperature oscillations. These produced the Pleistocene (1.8--0.01 mya) ice ages, which advanced on a 100 ka cycle, each time being interrupted by a brief (10 ka) interglacial period (Hewitt and Ibrahim 2001). The most profound environmental change to affect range distributions has been the periodic advancement of continental glaciers (Green et al. 1996; Brunsfeld et al. 2001; Hewitt and Ibrahim 2001; Knowles 2001; Tzedakis et al. 2002). The last ice-age retreated about 12 ka, marking the beginning of the Holocene. As glaciers retreated, newly available environments were rapidly occupied by pioneer species (Pielou 1991; Clague and James 2002; Walker and Pellatt 2003).

8.4 Glaciation, Speciation and Range Dynamics

Empirical studies have yielded contradictory results regarding the effects of glaciation on speciation and range dynamics (Knowles 2001). Populations may emigrate from ice-free refugia into (a) a vacant site that was previously glaciated, climatically unsuitable and/or inaccessible; (b) a site already occupied, with a long history of tenure by the same species; or (c) a site that was recently colonized by prior but recent immigrants.

The first scenario would lead to an extension of the ranges of refugial lineages, but founder events would affect the genetic diversity of immigrant populations (Carson and Templeton 1984; Ray et al. 2003). Glacial immigrants expanding along the ecological range margins would consist of populations reduced in size and therefore subject to genetic bottlenecking (Soltis et al. 1997; Hewitt and Ibrahim 2001). Post-glacial founder populations would be reduced in genetic diversity relative to their genealogical origins (Knowles 2001).

Two expectations have been posited for the second post-glacial immigrant scenario. One holds that life persisting south of the glacial margin would experience mixing of lineages, resulting in the loss of localized genetic patterns. Secondary mixing of lineages could therefore inhibit divergence. The second view postulates that divergence would be promoted by the advancement of glaciers by restricting the amount of gene flow. According to this latter hypothesis, the allopatric refugial populations would be buffered against immigrant colonists, and divergence would have increased throughout the Pleistocene (Knowles 2001; Tzedakis et al. 2002).

The third scenario can only be addressed by examining species with modern distributions that transcended the last glacial maximum. Migration into post-glacial environments increases the chance of secondary contact (Cox and Hebert 2001). Throughout the Pleistocene such populations could have either retreated or become extinct in association with periodic glacial advancements. The historical genealogies from either of these events would be essentially erased with each successive glaciation, unless a new refugium was established. However, populations inhabiting post-glacial areas permit investigation of the southern origins of current patterns of distribution (Cox and Hebert 2001; Walter and Epperson 2001).

In this contribution, we explore the effects of glaciation on the range dynamics of the Long-toed salamander (*Ambystoma macrodactylum*), a species widespread in western North America (Fig. 2) that has had a long tenure in this region. We investigate the migration patterns of genes within and between populations to trace connectivity and isolation as they are influenced by climatic and physiographic changes. To do so means studying infraspecific biogeography, which requires an investigation of phylogeography.

8.5 Biogeography and Ecological Demographics of the Long-toed Salamander

Most phylogeographic studies of taxa occupying Western North America have been sub-regionally restricted by the sampling strategy employed or the range limits of the taxon analyzed. The Long-toed salamander is distributed over areas that make it geographically intriguing and particularly suitable for investigating phylogeography in this region (Fig. 2). Not only does its distribution greatly transcend the last glacial maximum, but also much is understood about its ecology. It is found in a variety of vegetational zones, ranging from temperate rainforests, semi-arid sagebrush, and chatgrass plains, to alpine meadows along the rocky shores of mountain lakes (Ferguson 1961; Graham and Powell 1999; Stebbins 2003).

Museum records, consultations with field specialists, field collections, and field guides (Stebbins 1951; Hodge 1976; Nussbaum et al. 1983; Green and Campbell 1984; Powell and Russell 1996; Powell et al. 1996 19997b; Nelson et al. 1995; Oseen et al. 1995a, b; Hamilton et al. 1996) were consulted to produce a comprehensive distribution plot for *A. macrodactylum* (Fig. 2). The Long-toed salamander's eastern range margins are well documented (Powell et al. 1996 1997b; Nelson et al. 1995; Oseen et al. 1995a,b; Hamilton et al. 1996; Marnell 1997; Walsh 1998). The range limit is reached as topographic heterogeneity diminishes in the grassland environments east of the Continental Divide and Rocky Mountain Foothills (Fig. 2) (Graham and Powell 1999). This also coincides with the Cascadian freshwater biogeographic province that meets the Hudson province along Alberta's Rocky Mountain foothills, and the Mississippi in Montana's foothills (Cox and Hebert 2001). While the distribution along the western coast is less precisely understood, there are discrete populations along river basins that drain from more interior areas where the species commonly occurs (Ferguson 1961). This biogeographic pattern has characteristics similar to those of the eastern range margins (Marnell 1997; Walsh 1998; Graham and Powell 1999), and indicates that the periphery of this species results from passive dispersal, or accidental migration, through river systems, limited by as yet poorly understood biotic and/or abiotic factors. California's allopatric populations may also have been, at one time, range marginal, since a historical drainage system is known to have carried mollusks across the Central Valley (Taylor 1985).

8.6 The Biology of Ambystoma macrodactylum

The biology of *Ambystoma macrodactylum* facilitates cohesion among closely related individuals that would maintain phylogeographic structure over time (Cruzan and Templeton 2000). Much of the natural history of *A. macrodactylum* indicates that populations are fairly resilient to extinction and thus should reflect processes that have been operating over evolutionary time (Templeton and Georgiadis 1996). It persists in disturbed habitats (Powell et al. 1997a), thus limiting haplotype loss due to contemporary declines (Templeton and Georgiadis 1996; Fisher and Shaffer 1996; Alford and Richards 1999; Houlahan et al. 2000). Like other salamanders (Petranka 1998), it exhibits philopatric behavior towards breeding ponds (Beneski et al. 1986; Fukumoto and Herrero 1998).

Restricted gene flow has been attributed to semi-prominent landscape features such as glacial basins (Howard and Wallace 1981; Tallmon et al. 2000). Population sizes do fluctuate (Beneski et al. 1986; Powell et al. 1997a), but effective sizes are small (~100 N_e) among restricted basins (Funk et al. 1999). This species has a life span of about six years, on average, in Alberta (Russell et al. 1996), providing for fairly long-lived cohorts to repopulate areas after a temporary disturbance or drought conditions. Males may be more prone to disperse (Sheppard 1977), favoring nuclear over mitochondrial introgression between populations (Jockusch and Wake 2002; García-Parìs et al. 2003).

Ambystoma macrodactylum is phenotypically and behaviorally plastic in response to climatic and ecological gradients (Anderson and Graham 1967; Howard and Wallace 1983 1984; Williams and Larsen 1986). Its range is so topographically and ecologically diverse that much potential exists for contrasting selective mosaics across the landscape (Ohta 1992; Brodie et al. 2002). For example, in lakes in the Cascade Mountains of Oregon (Kezer and Farner 1955), Waterton Lakes National Park (Fukumoto and Herrero 1998), and Montana's Glacier National Park (Marnell 1997) some populations at higher elevations delay metamorphosis until the second season (or later), because the growing season may be shortened in such habitats (Kezer and Farner 1955; Berven et al. 1979).

8.7 Comparative Phylogeography and Distribution of the Long-toed Salamander

Long-toed salamander distribution is so extensive that it covers most of the areas of interest to phylogeographers of Western North America. In addition to the hypothesized vicariance associated with the Central Plateau, there are additional sub-regional patterns of interest. Along the northern British Columbia coast it is found on islands, and its range extends into the Alaskan Archipelago (Fig. 2). An area spanning the northern part of Vancouver Island across the Hecate Strait and into Haida Gwaii may have been a glacial refugium, but interpretations surrounding this issue have been subject to debate (Warner and Mathewes 1982; O'Reilly et al. 1993; Byun et al. 1997; Demboski et al. 1999; Mandryk et al. 2001). The presence of this species in this area makes it a suitable test subject for identifying patterns consistent with a refugium.

The interior of British Columbia and the Rocky Mountains of Alberta remain relatively unexplored phylogeographically. Taylor et al. (1999) identified a phylogeographic break affecting Bull Trout along the Cascade crest and extending into British Columbia's Coast Mountains. An abrupt transition has also been identified in the genetic patterns of several fish species, where the coast wetlands meet the dry interior (Taylor et al. 1999). In this regard, it seems as though coast clades have a relatively restricted dispersal trackway because Rocky Mountain clades have tended to expand over a greater extent in British Columbia (Green et al. 1996; Taylor et al. 1999; Demboski and Cook 2001).

Phylogeographic expectations within Idaho and Montana's Rocky Mountains are not clear. Patterns of endemic diversity have implicated Idaho's Clearwater drainage as a glacial refugium; it is also deep enough to impose unique thermal conditions upon the area (Daubenmire 1975; Brunsfeld et al. 2001; Good and Sullivan 2001). Phylogeography within the remaining parts of the Rocky Mountains is too poorly studied to be able to identify any additional generalized patterns. The Bitteroot Mountains are thought to restrict gene flow in some species, but the biogeography is seen as being more complex, with multiple river valley refugia (Benkman et al. 2001; Brunsfeld et al. 2001; Good and Sullivan 2001). A divergent Palouse clade has been identified in the red-tailed chipmunk, just across Idaho's western border into Washington (Good and Sullivan 2001). The tailed frog similarly exhibits divergence across this region (Nielson et al. 2001). These patterns likely arose from water flow dynamics during the Pleistocene, where Glacial Lake Missoula released immense floods into the Snake and Columbia Rivers. Further possibilities of gene flow restriction in areas exist along Idaho's border south of the Palouse formation, where the Snake River runs through Hell's Canyon (Baker 1983; Alt and Hyndman 1995).

Recent phylogeographic connectivity has been proposed between Cascade Mountain or Coast endemics and interior Rocky Mountain populations of various species (Brunsfeld et al. 2001; Janzen et al. 2002; Richardson et al. 2002). Two inland dispersal hypotheses have been erected, with the first suggesting an eastward dispersal across the Okanogan Highlands, and the second positing a southern connection through central Oregon's highlands (Brunsfeld et al. 2001).

Only recently has phylogeographic data from multiple co-distributed species in the greater extent of western North America been studied to such an extent that a comparative study of phylogeography can be applied to quantify the phylogenetic diversity of multiple taxa. The primary forum for testing phylogeographic pattern is through regional comparative analysis on different groups from the same areas (Gould and Johnston

1972; Avise 1996; Bermingham and Moritz 1998; Schneider et al. 1998; Hewitt 1999; Avise 2000b; Arbogast and Kenagy 2001; Brunsfeld et al. 2001).

The foregoing provides the context that permits us to trace genetic patterns and the migration of populations of *Ambystoma macrodactylum* in a post-glacial landscape. Hence, our overall argument is a natural experiment, wherein the migrations of populations are situated in their natural palaeoenvironment, and the natural history of genetic networking is explained for each population.

8.8 Mitochondria and Phylogeography

Two mechanisms that contribute to the underlying movement of genes are dispersal and sexual introgression. Genes that are inherited through the nucleus flow through populations as somewhat discrete units, because nuclear (autosomal) genes are bi-parentally inherited according to Mendelian principles, and they can "jump" to other chromosomes when these cross over and exchange genetic material. The mitochondrial genome, however, is uni-parentally inherited through matrilineal inheritance and its genes are linked together as a single unit because mitochondria reproduce asexually.

Mitochondria have an ecology and history of their own. They remain confined within the cell cytoplasm and regulate the metabolism of their host organisms. In poikilotherms, however, external temperature is directly experienced by the cell cytoplasm (Ballard and Whitlock 2004). Therefore, it is conceivable that climatic temperature reciprocally governs the geographic mosaic of selection across a species' range and mediates the movement of adept mitochondria into new populations. In this regard, different types of molecular markers are increasingly informative because they have different functions, evolve at different rates, and offer various grades of historical resolution.

In this chapter, we cross-validate our results with two mtDNA markers that include a functional gene, the cytochrome b, and a stable mitochondrial insert, the intergenic spacer region (ISR). We apply multiple analytical methods and explain phylogeographic diversity by testing for: (a) congruent signals between the two molecular markers, (b) mitochondrial lineages segregated by major physiographic breaks, (c) haplotype frequencies in mitochondrial lineages reflecting population growth and range expansion, and (d) physical (or paleoecological) attributes of the landscape that help explain connectivity and dispersal among populations. In so doing, we investigate the genetic structure of populations and reconstruct the history of gene flow by mitochondria and the parallel history of migration in *Ambystoma macrodactylum* across Western North America.

8.9 Methods, Approaches, and Outcomes of Genetic Analyses

8.9.1 Genes, Partitions, and Congruent Molecular Analysis

Historical and geographic patterns of migration can be inferred through genetic data when gene flow is expressed as a migration rate (m). The migration rate is measured as the proportion of haplotypes in a population that are of migrant origin in each generation. Patterns of mutation inheritance are reconstructed according to a phylogenetic gene tree of the haplotypes (one or more uniquely identified gene sequence(s)). Phylogenetics is the branch of science that deals with testing for the relatedness of individuals and heuristically seeks to identify the most parsimonious explanations of inheritance for genetic and morphological attributes; we concentrate on the former. The genetic data are sorted as a parsimonious genealogy and allow exploration of populations that became structured through time by geographic proximity, gene flow, and common ancestry. Our selected markers express the molecular past according to their individual rates and modes of mutation. Cytochrome b codes for a protein that is an integral part of the electron transport chain, and is a "workhorse gene" in phylogeographic research (Avise 2000). In contrast, the ISR is a non-coding segment of DNA that is autapomorphic for ambystomatid salamanders. The ISR has been evolutionarily persistent for approximately 20 ma, and there is a small variance in the length of haplotypes that indicates that this locus does not code for a functionally conserved polypeptide (McKnight and Shaffer 1997; Donovan et al. 2000; Church et al. 2003; Zamudio and Savage 2003; Thompson et al. unpublished). This marker may evolve rapidly in a similar manner to the D-loop region, another non-coding segment of the mitochondrial genome (Clayton 1984).

Given that we sampled mitochondrial genes, our investigation should reveal information about the relatedness of mitochondrial genomes as they drift through the cell cytoplasm of salamander populations. Mitochondrial genes provide only a single estimate of genealogy because they are asexually inherited as a linked group through the female egg cytoplasm; hence, all genes in a single mitochondrial genome share an identical matrilineal history. The effective population size for mitochondria is one quarter the size of sexually inherited genes because they are haploid and only females transmit the genetic information (Avise 2001; Ballard and Whitlock 2004). Simulation studies indicate that these properties of mitochondrial genetics increase the likelihood that spatially restricted populations will exhibit exclusive monophyly (an ancestor and all its descendants) (Avise et al. 1988; Moore 1995; Avise 2000). Such simulations commonly trace descendant relations in reverse through the antecedents until two lineages coalesce as a single ancestor; we discuss the implications of such simulations in more detail below. Although mitochondrial genes trace only a single history for any particular genome, they mutate at different rates. For example, our two markers record different degrees of resolution on the history of migration because the coalescence time for an ISR haplotype is shorter than that for the functionally conserved cytochrome b haplotypes. It is important, however, to check that each marker gives congruent phylogenetic results. Although it is unlikely that different topologies in the gene tree would be due to differences in the lineage sorting process (a naturally occurring phenomenon for sexually recombinant genes), it is possible that different rates of mutation would skew the results of a phylogenetic analysis. If, however, the genetic data are analyzed independently as partitions, and then together as a combined set, we expect the trees to express congruent genealogies. Topographically identical phylogenies allow for the cross-validation of additional biological inferences relating to each lineage, and assessments about populations that are sampled for one locus but not the other (and vice-versa) (Templeton 2004; Thompson et al. unpublished).

When multiple representatives of the population's genetic constitution are sampled, the sampling design frequently becomes unbalanced. The molecular data, however, must be balanced in order to address matters of congruence. Congruence in the phylogenies is examined by reducing the data sets to include only the intersection of genetic sequences. Given two molecular markers, the intersection includes only the set of markers that come from individuals (salamanders in our study) that have been sequenced for both markers. Once the data are balanced as such, the Incongruent Length Difference (ILD) test (Farris et al. 1994) provides a starting point for testing phylogenetic congruence among the partitions (Hipp et al. 2004). The ILD test is executed in PAUP* V4.0b10 (Swofford 2000) and compares two data partitions, x and y, to yield a test statistic (D) that is based on a measure of signal equality. Nucleotide character states (there are four, one for each nucleotide AGCT) are drawn at random from any indiscriminate partition of the genetic sequences. Letting L represent the length of the trees generated from the sample, the test statistic is calculated as follows: Dxy = Lxy − (Lx + Ly). This branching statistic measures the difference between tree lengths for sequences that are combined (Lxy) versus partitioned (Lx + Ly). In our study we used 1000 random partitions and generated a distribution of values that equal W, where W = D_{prior} + D_{rand}; the prior (D_{prior}) refers to the different process partitions, cytochrome b and ISR genes in this study, compared to random (D_{rand}) sample partitions of the combined data. The number of replicates for which D_{rand} is smaller than D_{prior}, designated as S, is used to derive the estimated type I error rate, or P-value, and is equal to 1-(S/W) (Farris et al. 1994; Yoder et al. 2001). This statistic does not test for partition "combinality", but does identify precision in so far as the trees are corroborated. Two possible causes for incongruence are errors in the analytical methods (e.g., long-branch attraction), and differences in lineage sorting (e.g. hybridization) (Felsenstein 1983; Hipp et al. 2004). After performing 1000 sample replicates, we failed to reject the null hypothesis (P = 0.62) that the phylogenetic characters within the cytochrome b and ISR partitions give different phylogenetic from that derivable from a combined single analysis.

Phylogenetic incongruence may also result if an inappropriate phylogenetic model is employed; hence, it is advisable to compare trees that are derived from distinct approaches (Avise 1994). We used MEGA2 (Kumar et al. 2001), MODELTEST (Posada and Crandall 1998), and PAUP* V4.0b10 (Swofford 2000) to study the intersecting of data and produce phylogenetic trees by minimum evolution, maximum likelihood, and parsimony respectively. The relative degree of node support was ranked for each clade by bootstrapping the data with 2000 pseudoreplicates (Felsenstein 1985; Hillis and Bull 1993). Only the phenetically based minimum evolution tree is illustrated (Fig. 3), because the remaining trees exhibit very similar topologies, differing only at the tips where the most closely related haplotypes occur. The tree lengths of minimum evolution trees are proportional to genetic distance.

The probability of selection increases as population sizes become smaller (Ohta 1992). Note that the branch length for haplotype 15 is noticeably longer for the cytochrome b tree (Fig. 3b) than for the combined (Fig. 3a) or ISR (Fig. 3c) trees. This haplotype (H15) is from Santa Cruz California, an area into which *Ambystoma macrodactylum* (and several other salamanders) migrated during the Pleistocene, subsequently becoming isolated by an approximate 300 km gap from the nearest portion of the remainder of the range (Baily 1948; Stebbins 1949; Russell and Anderson 1956; Moritz et al. 1992; Jockusch and Wake 2002; Thompson et al. unpublished). The longer and shorter branch lengths for the cytochrome b tree (Fig. 3b) may indicate regional differences in population size. Moreover, the ISR gene is likely to remain neutral under such conditions; the branch length asymmetry, when compared to cytochrome b (Fig. 3), fits this prediction.

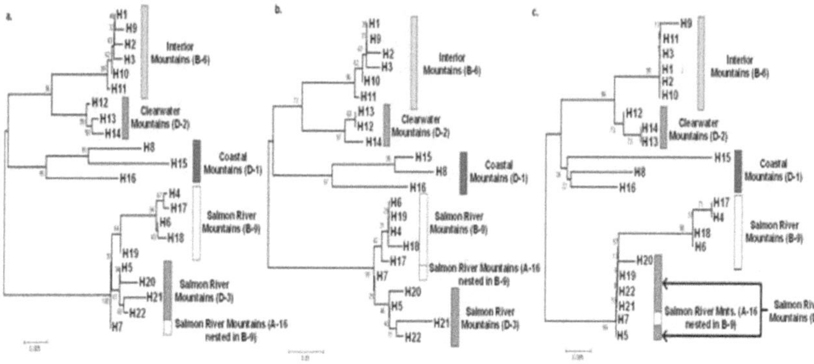

Fig. 3. Three phylogenetic gene trees produced from the intersected data set for (a) cytochrome b and ISR combined, (b) cytochrome b as a partition, and (c) ISR as a partition. The intersected data reduced to 22 haplotypes, seen at the terminal nodes of each branch (H1--H22). Bootstrap values are located at the nodes of each branch, bars to the right of each tree identify regionally based monophyletic clades, and the alpha-numeric labels, in parentheses, refer to clades from the ISR marker. A sub-clade within the Salmon River Mountains (A-16) is incongruent, whereas the Interior, Clearwater, and Coastal Mountain groups are congruent. Note that the haplotype labels in these trees apply only to this figure, because the data were reduced to include the intersected sequence partitions

8.9.2 Nested Cladistic Phylogeographical Analysis

Nested cladistic phylogeographic analysis compares among alternative explanations of the migration and movement of populations. This is accomplished by mapping and contrasting the phylogenetic arrangement of genetic events according to their ancestral and derived spatial distributions (Templeton et al. 1995; Templeton 2004).

The last glacial retreat is a prime phylogeographic calibration point for the study of effects of range expansion on genetics and associated demographic parameters. Comparative phylogeography and molecular clocks provide crude-to-precise time indices for speciation, whereas time indices for geological events, such as glacial retreat, are calibrated and refined from global to local scales through radio isotope dating of material found in stratified sections of pollen, diatoms, chironomids, and ice cores (Walker and Pellat 2003). Paleoecological reconstructions coupled with molecular data enable calibration of molecular rates so that population histories, such as range expansion, can be mapped in accord with prevailing environmental conditions.

Phylogeography includes the analysis of geography and haplotype (or allele) lineages and provides biogeographical explanations for spatial associations. For example, Thompson et al. (unpublished) measured effective geographic distances and established a statistical association between the topographic relief of western North America and genetic distances for the ISR marker. Effective distances account for environmental constraints imposed upon a species by the landscape (Verbeylen et al. 2003) and these were approximated through a measurement of genetic isolation in *Ambystoma macrodactylum* (Thompson et al. unpublished). An effective distance was measured by demonstrating that mitochondrial gene flow, hence the migration of *Ambystoma macrodactylum*, is sufficiently restricted by Cordilleran relief to engender geographical associations among evolutionarily related haplotypes. Closely related individuals are statistically more effectively linked by geographic distances that trace through the valleys of the landscape than those that pass over the mountains and circumvent the topographic features. In other words, genetic patterns are spatially regionalized according to their evolutionary position within the haplotype tree and these are significantly correlated with Cordilleran topography. We narrowed the number of alternate explanations for these statistically supported patterns by applying Templeton et al.'s (1995) Nested Cladistic Phylogeographical Analysis (NCPA), which incorporates principles from both cladistic theory and coalescence simulations. While statistical probabilities are the test criterion for depicting fully resolved networks of haplotype genealogies, a cladistic component is retained in the approach. It is easy to recognize why the haplotype networks are statistical in nature, but it is less clear why the techniques and phylogenetic networks are cladistic in nature.

A genealogical perspective is instructive for comprehending the cladistic basis of Templeton et al.'s (1992) haplotype networks. Simply put, gene trees are the graphical representation of intraspecific genetic relations. Gene trees differ considerably from classical cladogenetic trees used to map interspecific relations (Hennig 1966; Crandall and Templeton 1996; Posada and Crandall 2001), and neutral drift theory becomes relevant for gene trees as this allows genetic simulations to assign probabilities to the branching process (Avise 2000).

Simulations of neutral genetic drift run in reverse to the mutation origin of each haplotype and are used to obtain statistical information about the phylogenetic, or branching, process relative to effective population sizes and mutation rates. This is called coalescence theory. Coalescent simulations and controlled genetic studies reveal that the numerically most abundant haplotype is most likely to display the ancestral state (Castelloe and Templeton 1994; Crandall 1994; Crandall and Templeton 1996). The probability that a polymorphic site from two random haplotypes experienced more than one mutation, the non-parsimonious state (Templeton et al. 1992), can be estimated through the coalescent parameter, $\theta = M\mu$, where M equals N for haploid and 2N for diploid populations (Templeton et al. 1992; Schneider et al. 2000).

While branches in statistical haplotype networks are not ranked according to their relative bootstrap probabilities, it is possible to achieve 95% statistical support with the coalescent parameter. When realistic estimates of θ are obtained, the probabilities for a non-parsimonious relationship between randomly drawn haplotypes cannot be identified frequently enough for general use in phylogenetics (Hudson 1989; Templeton et al. 1992). If attention is focused on non-random haplotypes, however, parsimony increases the effectiveness of statistical estimates of θ. Non-random haplotypes are identified through parsimoniously reconstructed genealogies. In gene tree terminology, a polymorphic site that experienced more than one mutation is a non-parsimonious state, but if two haplotypes differ by only one mutational step, the parsimonious state, they are monophyletic. A statistical parsimony network, with the statistics based on coalescence theory, reduces the frequency of non-monophyletic (i.e. paraphyletic) connections (Fig 4). Given that a pair of haplotypes share m sites, but differ at j sites, the probability of a parsimonious relationship is estimated by

$$\hat{P}_j = \prod_{j}^{i=1}(1 - q_1) \tag{8.1}$$

This equation contains the probability that a mutation occurring between two haplotypes after their point of divergence (q_1), which is estimated by using Hudson's (1989) mathematically derived probability for the non-parsimonious state (H). Hudson's (1989) H statistic serves as the upper bound of a Bayesian prior distribution $(0,H)$. Since non-random haplotypes are compared, q_1 will always remain less than H, and is easily quantified by comparing haplotype sequences. For each non-random haplotype pair the probability of homoplasy among *m* shared sites, plus the probability of change occurring among sites that differ (*j*) is computed. The accumulated probabilities determine the total probability that two haplotypes differ at *j-1* sites in addition to a single site under consideration, while having *m* sites in common. Haplotypes are thus connected within a parsimony network if they satisfy the type *I* error rate (p = 0.05). The procedure starts at *j*=1, runs through the data to yield the set of estimators q_1 ....,q_n, and stops linking haplotypes into the network if the probability exceeds α (Templeton et al. 1992; Crandall 1994).

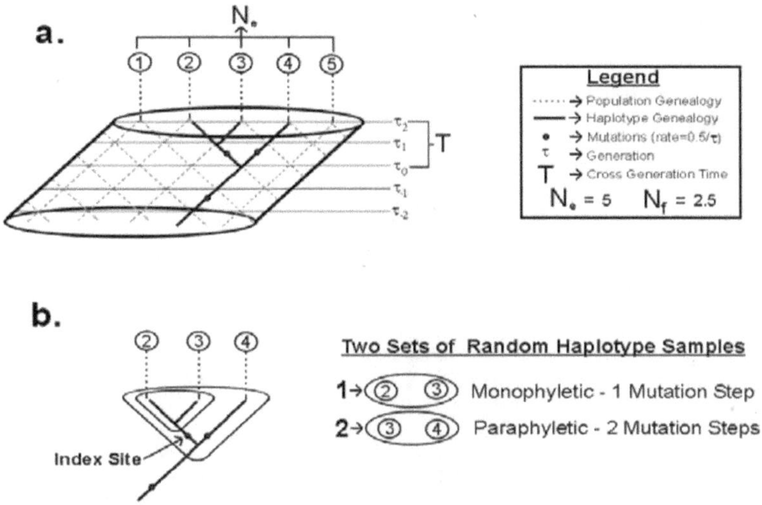

Fig. 4. A graphical illustration of a genealogy evolving within a localized portion of a population tree (a). The illustrated population has an effective breeding number (Ne) of 5 individuals. Five sexual generations (τ) are shown with a haplotype genealogy superimposed. A single line of descent, leading to individuals 2 to 4, is isolated in (b), and these individuals share a haplotype from a single mutation between τ_0 and τ_1 in (a). If individuals 2 and 3 are sampled at random, then a monophyletic clade unites these haplotypes by a single index mutation, the parsimonious state, but if individuals 3 and 4 (or 2 and 4) (b) are sampled, then a non-parsimonious paraphyletic union is constructed and requires two mutational steps, a non-parsimonious state

Three statistical parsimony networks at 95% statistical confidence were resolved for the two mitochondrial markers (Fig. 5a,b). The haplotypes in the network are either interiors or tips. Interiors also include parsimoniously inferred nodes (PINs) (Thompson et al. unpublished) and always have more than one mutation connecting them to the remaining network.

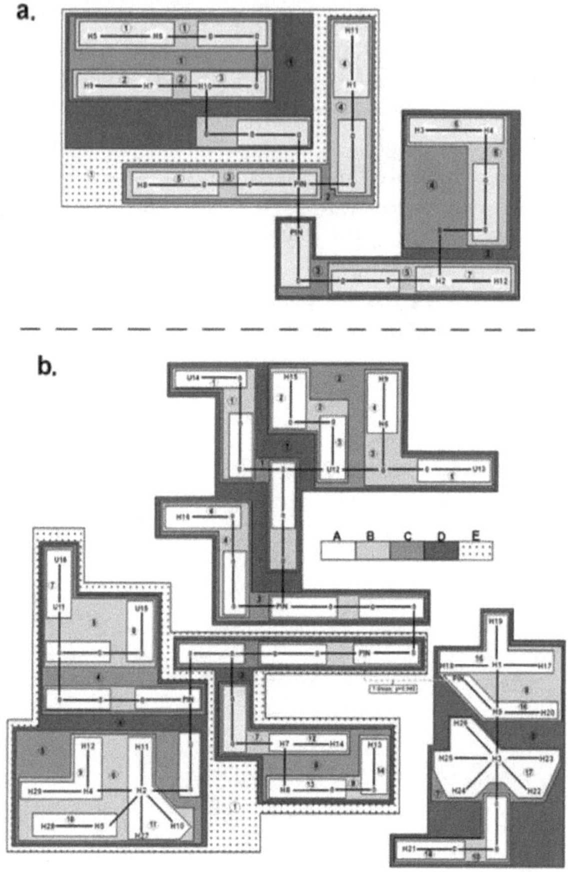

Fig. 5. In (a), the cytochrome b genealogy is resolved as a single TCS network, and the nested design contains 12 haplotypes (H1--H12). In (b), the ISR marker genealogy is statistically resolved, and the relationship among 29 haplotypes (H1--H29) is summarized into two TCS networks with a nested design. The ISR network (b) contains six haplotypes starting with a U, which refer to published GenBank sequences (McKnight and Shaffer 1997). A dashed line connecting two PINs (b) identifies the closest connection in the unresolved relationship between the two ISR networks. A shaded key discerns nesting levels A to E, and numbers in circles reference each nested category. Clades are referenced by the nesting level followed by the nested address (e.g., A-16 contains haplotypes H9 and H20, and one PIN)

The internal haplotypes tend to be ancestral, geographically widespread, and occur more frequently in populations. Tips tend to be the most recently

Table 1. Nested cladistic phylogeographical analysis results for the (a) ISR, and (b) cytochrome b mtDNA markers in *Ambystoma macrodactylum*. Alternative hypotheses (H_a) from the inference key are listed for each clade in which there is more than a single interpretation. Clades with no genetic or geographical variation are not included

			Square Slope	
Loci	Clade	H_a	Inference Key†	Inferred pattern
a.				
	A-9		1-2-11-17-No	Inconclusive
	B-6	1.	1-2-11-12-13-21-No	PFRE -?
		2.	1-2-11-12-13-No	CRE-LDC-PF?
	D-3		1-19-No	AF
	E-1		1-2-3-4-No	RGF-IBD
	Total-1		1-2-?	Inconclusive
	B-9		1-2-11-12-No	CRE
	Total-2		1-2-?	Inconclusive
b.				
	A-6		1-2-11-12-No	CRE
	A-7		1-2-3-4-No	RGF and IBD
	C-1	1.	1-2-11-12-No	CRE
		2.	1-19-20-2-11-12-No	CRE
	E-1		1-19-No	AF
	Total		1-2-?	Inconclusive

* The probability refers to the frequency with which the 1000 permutational chi-square statistics are equal to or greater than the observed

†CRE = contiguous range expansion; RGF = restricted gene flow; IBD = isolation by distance; PFRE = past fragmentation followed by range expansion; LDC = long distance colonization; FR = fragmentation; AF = allopatric fragmentation; ? = indiscriminate; ns = not significant

derived mutations and have only a single connection to the network. Hypothetical unobserved or extinct haplotypes are termed missing intermediates and are represented in the network by zeros (Crandall 1996). A nesting algorithm is applied to the underlying statistical parsimony network that groups haplotypes into a nested set of clades. The nesting design segregates genealogical units that require independent statistical investigation of their spatial patterns, while contrasting older (interior) versus younger (tip) clades (Fig. 5; Templeton et al. 1995). In effect, this allows the investigation of the ancestral migration of a lineage versus the migration of thc descendants of the lineage leading up to the spatial extent of contemporary haplotypes.

Thompson et al. (unpublished) demonstrated, through cross-referencing the combined and cytochrome b nesting designs, that the Clearwater Mountains clade D-2 is more appropriately united with the Rocky Mountain Interior clade D-3 (Fig. 5b). Spatial statistics for the nested clade design test null distributions of clade distance (D_c), nested clade distance (D_n), and interior minus tip clade distance (I-T) against the observed distances for haplotypes (referred to as 0-step clades) and their respective nested clade categories (Templeton et al. 1995; Posada et al. 2000). We used effective geographic distances in our analysis that were weighted by the topography of the landscape to better reflect the habitat grain and effective mobility of *Ambystoma macrodactylum*. Results from the GEODIS analysis were interpreted according to Templeton's inference key (see the preceding Table 1; programs and an up-to-date inference key are available at http://inbio.byu.edu/Faculty/kac/crandallplab/programs.html.

8.10 Relevant Patterns of Population Structure in *Ambystoma macrodactylum* in Space and Time

8.10.1 Regional Patterns of Range Expansion and Isolation

Seven glacial refugia resulting from glacial fragmentation are qualitatively identified from shallow history mtDNA divergence and are accentuated by haplotype distributions emanating from source areas along the coast (1) and interior (2). Refugia were identified (Thompson et al. unpublished) as: 1a) Haida Gwaii; 1b) Coastal/Cascade Mountains; 1c) Santa Cruz and Monterey Counties, California; 2a) Blue Mountains; 2b) Clearwater Mountains; 2c) Montana's Rocky Mountains; 2d) Salmon River Mountains. We use NCPA to describe the genealogy and spatial extent of post-glacial migration from these places for breeding salamander populations (Fig. 6).

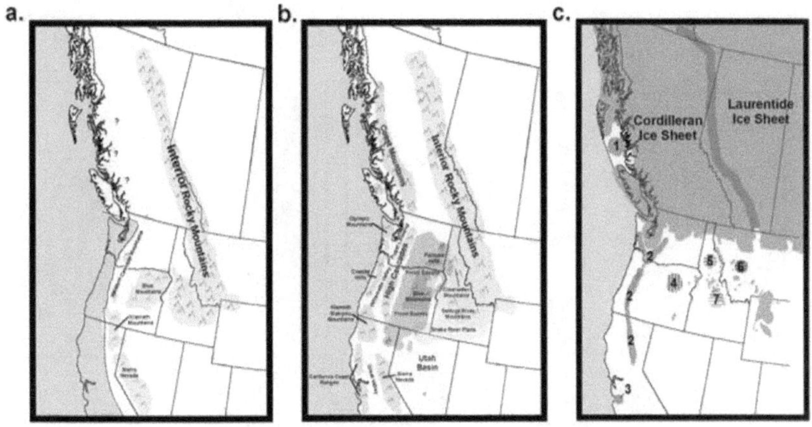

Fig. 6. The Cordilleran Ice sheet (dark grey shading) forced populations into seven isolated refugia during the Pleistocene; these are identified by the stippled regions and associated numbers

During periods of glacial advance sea levels were lower than at present, and isostatic rebounding molded physical geography and shifted hydrological patterns (Pielou 1991; Clague and James 2002). In this dynamic setting multiple Pleistocene refugia occurred in the Pacific Northwest, but the Haida Gwaii refuge is most controversial (Warner and Mathewes 1982; O'Reilly et al. 1993; Byun et al. 1997; Soltis et al. 1997; Demboski et al. 1999; Stone and Cook 2000; Brunsfeld et al. 2001). The Haida Gwaii refugium is an ice-free terrestrial and freshwater environment that existed along Hecate Strait and the continental shelf, and extended to the northern parts of Vancouver Island. A precise size, time and type of ecological system that was supported are all unclear (Warner and Mathewes 1982; Josenhans et al. 1995; Mandryk et al. 2001). Coastal clades A-5 (Fig. 7a) and D-1 (Fig. 7b) suggest that the region harbored salamanders; *Ambystoma gracile* exhibits a similar north-south division along British Columbia's coastal mountains (Titus and Gains 1991). If *A. macrodactylum* or *A. gracile* persisted in this refugium during glaciation, then their descendants emigrated after the glaciers retreated to establish populations that contain the genetic signatures of this history. The possibility of a Haida Gwaii refugium raises an interesting biogeographic question, because it is well established that marine situations effectively isolate amphibians (Darlington 1957; Duellman and Trueb 1986; but see Anderson 1960), yet the Haida Gwaii clade, A-5 (Fig. 7b), spans Georgia

Strait and the continental Interior (clade A-5 overlaps with the ranges of A-4 and C-1, Fig. 7b).

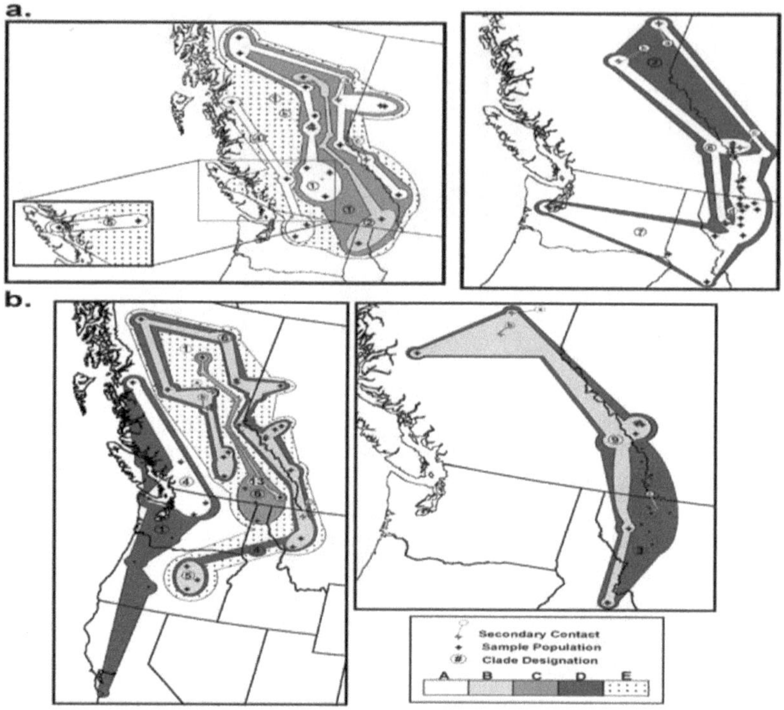

Fig. 7. Nested clade designs from Fig. 5 are spatially overlain onto a map for (a) the cytochrome b genealogy and (b) the ISR genealogy. Numbers in small circles and nesting levels correspond to those in the network design (Fig. 5). An inset of Vancouver Island, in (a), isolates clade A-5 from others for visual clarity. For reasons of simplicity, only the nesting categories that have significant range differences are indicated, and private haplotypes (haplotypes that occur in only one population, *sensu* Hartl and Clark 1989) are also exempt in this regard. Several populations are labeled in the post-glacial regions of Alberta and British Columbia (circles containing letters a--c), which contain a mixture of disparate clades

The occurrence of identical haplotypes throughout these populations (H8 in cytochrome b, clade A-5, Figs. 5a, 7a) resulted from Holocene descendants that migrated into respective areas either via a terrestrial land bridge (Mandryk et al. 2001) or by enduring a marine journey; the latter hypothesis is conditional upon salinity levels that would need to have been sufficiently lowered by the influx of glacial meltwater to permit waif dispersal (Anderson 1960; Walker and Pellatt 2003).Deep genealogical sub-divisions are not expressed in areas further south of the glacial margins, unlike the situation for several other species. For example, there is a close genetic relationship between populations spanning the Klamath-Siskiyou biogeographic boundary (Bury and Pearl 1999), in contrast to the genealogical subdivisions observed for the arionid slug, *Prophysaon coeruleum* (Wilke and Duncan 2004). This suggests that, under certain conditions, salamanders may migrate more readily than certain other species. While California's populations are currently isolated (Baily 1948; Russell and Anderson 1956), the measured divergence between populations that cross this biogeographic boundary reflect recent genealogical connections. Subsequent climate change modified the intervening environments to isolate these populations (Thompson et al. unpublished).

Another salamander species, *Taricha torosa*, migrated southward across the Klamath-Siskiyou boundary and along the Sierra Nevada to finally arrive in California during the middle Miocene (Tan and Wake 1995). The Miocene, however, is too distant in time to account for the small degree of divergence measured between *Ambystoma macrodactylum* populations. Sierra Nevada's populations are geographically nearest, and genetically and phenotypically most similar to those of California (Dr. D. Wake personal communication; Dr. R. Sage unpublished). Hence, the phylogeographic pattern of *A. macrodactylum* matches those of *Batrachoseps attenuatus* and *Ensatina eschscholtzii xanthoptica*, and is hypothesized to have colonized Monterey Bay and Santa Cruz, California during a Pleistocene range expansion from the Sierra Nevada through the Central Valley (Stebbins 1949; Moritz et al. 1992; Jockusch and Wake 2002).

The interior mountain chains have been sampled more intensively and provide greater resolution about migration and endemic history. Oregon's clade, B-5 (Fig. 7b), occurs within a transverse mountain system that is identified as a refugium. This pattern represents an otherwise cryptic contributor to the biodiversity that is endemic to the dry central interior, and is corroborated by the distinctiveness of this clade. To the east of these mountain chains the Snake River is a barrier where it runs through Hell's Canyon along Idaho's western border (Baker 1983; Alt and Hyndman 1995). Several species, in addition to the long-toed salamander, point to Oregon's transverse mountain system as an important genetic refugium. In the eastern section of these mountains, the tailed frog, *Ascaphus truei*, has a distinctive genetic signature, and a spatial distribution that is comparable to that of *Ambystoma macrodactylum*; both species closely match the interior coniferous forest regions (Heusser 1983; Pauken and Metter 1971;

Nielson et al. 2001; Thompson et al. unpublished). The red-tailed chipmunk (*Tamius ruficaudus*) has a divergent genetic clade situated adjacent to Idaho's western border and is isolated by Palouse prairie in Washington (Good and Sullivan 2001). Flycatchers (*Epidonax hammondii*) similarly exhibit a unique pattern of allelic diversity within the Wallawa Mountains (Johnson and Marten 1989). Several factors contribute to the genetic isolation of populations to these mountains. First, Pleistocene environments were colder and drier in Washington (Heusser 1983; Walker and Pellatt 2003), making the conditions even more inhospitable and range marginal, and reducing migration rates. Secondly, meltwater and river flow rates were possibly severe enough to isolate populations during the Pleistocene, as glacial Lake Missoula released floods into the Snake and Columbia Rivers (Alt and Hyndman 1995).

The Clearwater refugium and the crest of the Bitterroot Mountains are two prominent features that have been implicated in the phylogeographic diversity of the interior (Brunsfeld et al. 2001). The Clearwater valley is deep enough to impose unique thermal conditions and has been implicated in explaining the endemism characteristic of this area (Daubenmire 1975; Brunsfeld et al. 2001; Good and Sullivan 2001). Ferguson (1961) suspected secondary contact, or a zone of morphological intergradation between *Ambystoma macrodactylum columbianum* and *A. m. krausei* in central Idaho that implicates the Clearwater populations and a crossing of the Bitteroot Mountains, but secondary contact has not been identified for populations in this area. This has resulted in a revision of the subspecies map (Fig. 2).

Lineages in Montana's Rocky Mountains, however, are mixed with lineages that came from the Salmon River Canyon refugium and migrated through the Bitterroot Mountains. The Salmon River Canyon houses the closest relatives of the coastal lineages, and a large proportion of the genetic diversity evolved in this area. Given the unique nature of this clade and that secondary contact in the Montana's Rocky Mountains has occurred, the mountain topography of the Salmon River Canyon may have only recently changed to permit a connection into Montana, but was otherwise isolated by the Bitteroot Mountains throughout the Pleistocene. This area developed throughout the Pleistocene, with the highest fluvial terraces of the Middle Fork dated at 0.4--1.1 mya (Meyer and Leidecker 1999). Salmon River Mountain's clade A-7 (Fig. 7a) exhibits a significant pattern that is reflective of restricted gene flow and isolation by distance (Table 1a), but this inference is peculiar in the context of qualitative and glacial biogeography. One population of this clade occurs in northern post-glacial British Columbia, while a second is found along the coast in the Puget Trough (Fig. 7a), both of which are distant from the core of this clade. Because Bonneville floods, that ran westward through the Snake River Canyon during the late Pleistocene (15 ka), had incredible discharge through restricted valleys, waif dispersal by way of Pleistocene fluvial dynamics (Baker 1983; Waitt and Thorson 1983; Jarrett and Malde 1987) provides an explanation for a member of this clade being found along the coast. This Bonneville dispersal hypothesis was also proposed by Taylor (1985) in relation to contemporary and fossil mollusk distributions.

The Cordillera was colonized by migrants originating from the Clearwater refugium, Salmon River Mountains, and the Rocky Mountains of Montana (Thompson et al. unpublished) and likely traced along the interior route identified by Conroy and Cook (2000). Populations emigrating from the Clearwater refugium, however, contain a mixture of haplotypes nested below the A-step and may indicate a different type of genetic expansion coming from a larger and more stable population.

8.10.2 Molecular Demographics

Salamander populations that existed proximally south of the ice maximum were likely affected by peri-glacial climates and associated hydrology (Waitt and Thorson 1983; Thompson et al. unpublished). The demographic pattern evolved as a result of the ebb and flow of climate and paleogeography, because these factors influenced the associated migratory connections and carrying capacity of these populations (Knowles 2001). We observed that gene pools originating from peri-glacial environments expanded their ranges northward with the migration routes tracing through the same valleys that the Cordilleran ice sheet flowed through (Thompson et al. unpublished). Although we have an *apriori* expectation of range expansion for these lineages, the NCPA analysis did not necessarily infer this type of explanation a posteriori.

Inferences drawn from Templeton et al.'s (1995) NCPA require a combination of alternate methods for further testing and investigation of alternative causal factors (Masta et al. 2003; Morando et al. 2004). For example, the NCPA spatial statistics for the cytochrome b data set were unable to generate an inference of sudden range-expansion for the lineages emanating from the Salmon River refugium (Thompson et al. unpublished) because two samples of a haplotype in clade A-7 (Fig. 7a) are widely displaced from each other in directions away from the clade center; hence, the effective distance covered by this haplotype, although large, skews the center of the clades' distribution north-westward, and decreases the measured degree of spatial spread of the entire clade. The NCPA approach cannot infer range expansion in this particular instance, because the event is "...older than the coalescence time for the gene region being investigated" (Templeton et al. 1995: 772). In general, we could not test whether range expansion was responsible for genetic patterns for lineages that obviously expanded their ranges and transgressed into the post-glacial Cordillera. Therefore, we compared these results against the inferential capabilities of the mismatch distribution.

A mismatch distribution is the probability (F_i) that two neutral sequences, randomly selected from a population, have accumulated mutations at i nucleotide sites (Rogers and Harpending 1992; Excoffier 2004). The coalescent parameter θ is used to calculate F_i, and is estimated by the observed mean of pairwise genetic differences (Avise et al. 1988; Rogers and Harpending 1992). If a population remains constant in size over successive generations (t), or experiences a temporary reduction in population size (i.e., a bottleneck), then $F_i(t)$ quickly approaches an equilibrium state. The plotted distribution of the number of mutation differences versus $F_i(t)$ has a smooth, L-shaped decline for populations that are in equilibrium. Most empirically observed mismatch distributions, however, do not decline so smoothly, but show waves in the distribution. Single modal waves are produced when populations are simulated to change greatly in effective size from an equilibrium state; this is called the sudden expansion model (Schneider et al. 2000). Waves occur in the distribution because genetic distances tend to increase as population size increases, but after a sudden expansion "...the mean pairwise difference increases much more rapidly than its standard deviation" (Rogers and Harpending 1992: 556). Hence, waves in the mismatch distribution provide a statistical approach for identifying historical instances of increased population size.

Mismatch distributions are commonly employed to identify global migration events that occurred after the Pleistocene, because gene flow increases with migration and the relative degree of connectivity among isolated or restricted populations. The genetics of once restricted populations changing into a state of panmixia is similar to the sudden expansion model because the effective population size increases, but the geographic pattern is affected differently. The geography of genealogical branching depends on the rate of migration (Rogers 1995; Branco et al. 2002; Ray et al. 2003; Morando et al. 2004; Thompson et al. unpublished). Mismatch distributions are ideally limited, however, to species that have high levels of interpopulation gene flow (Avise 2000; Excoffier 2004). Based on autosomal alleles, *Ambystoma macrodactylum* has a small effective size (Ne $\approx$ 100) and exchanges less than one individual per generation among basins in the Bitterroot Mountains of eastern Idaho and western Montana (Funk et al. 1999; Tallmon et al. 2000). These demographic estimates might be even smaller for the mitochondria genome, because it has one-quarter the effective size of autosomal genes, and the females may be less likely to migrate (Sheppard 1977; Moore 1995; Thompson et al. unpublished). However, there are circumstances in vertebrates under which mitochondrially mediated gene flow is more fluid than chromosomal material of the heterogametic (XY : Male) sex (Anderson 2004). During times of glacial retreat, meltwater rendered hydrological connections more dynamic and could have increased the effective migration rate through waif dispersal along river valleys.

Given that the crest of uni-modal waves in the mismatch distribution is centered on the time of expansion τ, and that populations return to an equilibrium after a period of reduced size (Rogers and Harpending 1992; Excoffier 2004), we expect modes to center around 14 ka, the time of abrupt climate change and glacial retreat (Clague and James 2002; Walker and Pellatt 2003). Demographic parameters, Fu's (1997) F_s test, Tajima's (1989) neutral model D-statistic (10,000 simulations), and mismatch distributions (1000 simulations of the sudden expansion model) were calculated in ARLEQUIN ver. 2.000 (Schneider et al. 2000). For these calculations, we selected a subset of the ISR and cytochrome b clades inferred by the NCPA to have experienced a significant period of isolation and/or range expansion. We also investigated lineages that are inferred by the biogeographic interpretation to have expanded into the northern Cordillera after the last glacial retreat.

Multi-modal and uni-modal waves occur in the mismatch distributions (Fig. 8), but multi-modal patterns are more prevalent in the deeper clades. None of the clades deviate significantly from Tajima's (1989) neutral model, but three clades are significant for Fu's (1997) F_s test (Table 2). Both of these tests give significant results for factors other than selection, such as population expansion, bottlenecks, and mutation rate heterogeneity among nucleotides. Tajima's (1989) test for genetic neutrality is overly conservative if effective population size has shifted or if the mutation rate is heterogeneous over the segment of DNA being investigated. This statistic does, however, give negative D values for models of sudden expansion (Tajima 1989; Aris-Brosou and Excoffier 1996; Schneider et al. 2000). It is important to compare these statistics for the two markers under study, because one may reveal information that the other does not. For example, according to the Log Likelihood scores of MODELTEST (Posada and Crandall 1998; a program that identifies the models that statistically fit each data set) the cytochrome b sequences exhibit a uniform mutation pattern, but the mutations of the ISR marker make Tajima's (1989) test unreliable because they vary considerably among sites (gamma shape parameter = 0.0891). Fu's F_s is thought more likely to show significance if range expansion occurred (Ford 2002; Morando et al. 2004).

Only the Salmon River Mountain clade, B-9 (Fig. 8b), is significant for F_s (Table 2b), expanded its range according to the NCPA spatial statistic inferences (Table 1b), and shows a modal wave in the mismatch distribution. Moreover, the mismatch θ_1 value is much larger than that for other clades, and the model of sudden expansion is not rejected by the simulated data (Table 2b). Salmon River Mountain's clade A-6, identified by the cytochrome b marker, is directly comparable because the haplotypes belonging to it are physically linked to the haplotypes in the Salmon River Mountain's clade B-9, identified by the ISR marker (Fig. 8a). For this clade, Tajima's *D* value is positive, but it similarly has a mismatch θ_1 value much larger than that of others, and a raggedness index that is consistent with the demographic-expansion model (Table 2a). However, does clade A-6 of the Salmon River Mountains (Fig. 8a) clearly exhibit a modal wave? This particular clade is unique because it is observed outside of the Salmon River Mountains (Fig. 7a) and it is genealogically distinct within this race.

These observations confirm our suspicions that the Salmon River Mountains are sampled insufficiently. Clade D-2 from the Clearwater Mountains (Fig. 8b) exhibits a modal wave, has a negative Tajima's *D* value, and fits a model of sudden expansion of considerable size (Table 2b). Moreover, Fu's *Fs* statistic is non-significant, as is expected of the ISR, a neutral genetic marker. These three clades are likely candidates in which a historical event of sudden expansion is genetically perceptible. The NCPA analysis did not infer a significant pattern for the Clearwater Mountain clade D-2, but agrees with the interpretation of expansion for the Salmon River Mountain clades. Continuous range expansion rather than long-distance colonization is inferred by the NCPA analysis because the D_n and/or I-t D_n values are not significantly reversed from the D_c values (Templeton 2004; Thompson et al. unpublished). Clade D-4, also from the Salmon River Mountains, has some characteristics of the shallower clades, because it has a modal wave in the distribution (Fig. 8b), but the stepwise expansion model is rejected by the sum of squared deviations test statistic (Table 2b). The recent effect of post-Pleistocene range dynamics, however, is noticeable in the consistent peak at low pairwise differences (Fig. 8). The estimated points of these peaks are identified by the unit of mutation time ($\tau = 2ut$) for the mismatch distributions that are not significantly rejected for the model of sudden expansion (Table 2).

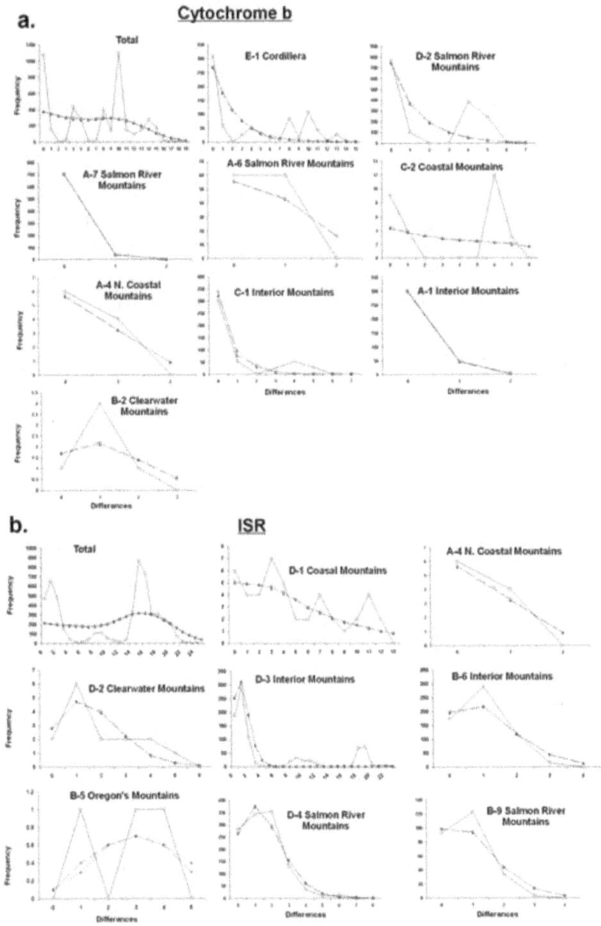

Fig. 8. Mitochondrial mismatch distribution showing the number of nucleotide site differences for phylogeographically significant clades plotted for the (a) ISR and (b) cytochrome b genealogies. The solid lines with circles represent observed values, the broken lines with squares are values obtained by coalescent simulations, and the dots with triangles are values calculated according to a model of sudden expansion (see Rogers 1995, Schneider et al. 2000, Ray et al. 2003, and Excoffier 2004 for further details). The simulated and model distributions are closely matched in each plot. Multi-modal plots tend to occur at the deeper clade levels, and uni-modal waves tend to occur at the lower clade levels

Table 2. Summary statistics of the evolutionary demographics for the (a) cytochrome b and (b) ISR markers as obtained in ARLEQUIN (Schneider et al. 2000). Values of significance are included for Fu's (1997) Fs and Tajima's (1989) neutral model D statistics, plus modeled and simulated demographic parameters from the mismatch distribution plots

Clade[a]	N	θ (S)	*D*	θ (pi)	*Fs*	P1[b]	τ	θ_0	θ_1	P2	HRI[c]	P3
						a. Cytochrome B						
Total	95	5.463	0.733	6.781	3.537	0.887	11.330	0.002	10.811	**0.001***	0.161	**0.000***
E-1	40	4.702	-0.245	4.354	1.371	0.770	10.062	0.000	1.906	0.206	0.157	0.351
D-2 SRM	55	1.311	1.150	1.915	3.225	0.912	5.455	0.016	0.998	0.061	0.304	0.290
C-2 CM	8	2.700	1.347	3.464	2.924	0.932	7.488	0.000	5.536	**0.028***	0.351	0.122
C-1 IM	31	1.502	-0.820	1.067	-0.424	0.396	3.000	0.443	0.444	0.168	0.309	0.623
B-2 CWM	4	1.091	-0.710	1.000	-0.887	0.830	1.280	0.000	2645	0.186	0.528	0.411
A-1 IM	27	0.259	-0.728	0.142	-0.349	0.167	1.280	0.000	2645	0.113	0.528	0.415
A-7 SRM	39	0.237	-1.126	0.051	-1.429	**0.038***	3.000	0.053	0.055	0.202	0.808	0.844
A-6 SRM	16	0.301	1.309	0.500	1.247	0.655	0.768	0.000	1045	0.055	0.250	0.108
A-4 NCM	5	0.480	-0.817	0.400	0.090	0.288	0.579	0.000	174.375	0.354	0.200	0.940
A-1 IM	3	0.667	0.000	0.667	0.201	0.374	0.324	0.000	0.217	0.323	0.532	0.676

Clade[a]	N	θ (S)	*D*	θ (pi)	*Fs*	P1[b]	τ	θ_0	θ_1	P2	HRI[c]	P3
b. ISR												
Total	104	8.626	0.651	10.414	-5.019	0.131	16.973	0.000	23.867	**0.006***	0.028	**0.012***
D-1 CM	10	6.363	-1.147	4.851	-0.835	0.254	2.521	4.900	8.208	0.886	0.024	0.982
D-2 CWM	6	2.190	-0.655	1.933	-0.561	0.226	1.671	0.000	1582.5	0.491	0.155	0.608
D-3 IM	49	2.467	-1.259	1.404	-7.327	**0.001***	1.230	0.000	532.188	**0.000***	0.085	0.221
D-4 SRM	42	6.275	-0.819	4.756	-0.146	0.535	1.230	0.000	532.188	**0.000***	0.085	0.219
B-5 OM	3	2.667	0.000	2.667	-0.341	0.187	3.444	0.000	2000	0.493	0.444	0.824
B-6 IM	35	0.971	-0.341	0.837	-4.440	**0.002***	1.118	0.000	1611.25	**0.029***	0.153	**0.030***
B-9 SRM	23	1.355	-1.227	0.791	-2.772	**0.014***	0.941	0.000	2310.625	0.083	0.153	0.128
A-4 NCM	5	0.480	-0.817	0.400	0.090	0.300	0.579	0.000	174.375	0.353	0.200	0.943

[a] C – Cordillera; SRM -- Salmon River Mountains; CM -- Coastal Mountains; IM -- Interior Mountains; CWM – Clearwater Mountains; NCM – Northern Coastal Mountains; OM – Oregon's Mountains.

[b] P1 = (SF_s<=OF_s), significance values for the proportion of random *Fs* statistics less or equal to the observation; P2 = (SSsd >= OSsd) / *B*, significance values for the proportion of simulated sum of square deviations greater than or equal to the observed simulated with *B* replicates; P3 = (Sr >= Or) / *B*, significance values for the proportion of simulated raggedness indices greater than or equal to the observed simulated with *B* replicates.

[c]HRI – Harpending raggedness index.

* Significant ($p<0.05$)

8.11 Conclusion

The mountainous landscape of western North America is a fruitful testing ground for examining some of the fundamental principles of phylogeography, because it presents decisive barriers to the historical and spatial geography of gene flow. There is an expectation for gene flow and range expansion occurring only after the Pleistocene, and only in places from which glaciers retreated. Molecular markers can reveal a great deal about the history of migration in these glacial settings, but the properties of genes used in the inferential process must first be examined in the context of their different modes of evolution. Our explicit method of molecular intersection gives a balanced set of phylogenetic trees (Fig. 3), where the relative degree of analytical congruence between the cytochrome b and ISR markers is identified by the different haplotype relations in each tree. For example, the longer branch length of California's haplotype (H15) in the cytochrome b tree (Fig. 3b) changes the sister relationship that is shared between H8 and H15 in the combined (Fig. 3a) and neutral ISR (Fig. 3c) trees, but the Coastal Mountain affiliation is retained (Fig. 3).

The migratory biology of ambystomatid salamanders makes them amenable subjects for phylogeographic research because their philopatric behaviour reduces the propensity for and rate of gene flow. Hence, historical events that are the cause of contemporary ranges are recorded in mutation patterns and haplotype frequencies, because the information is inherited through the evolving demographics of populations, and because station-keeping behaviors spatially fragment the genetic information into geographic races (Moritz et al. 1992; Jockusch and Wake 2002). However, *Ambystoma macrodactylum* has a cosmopolitan distribution throughout the Cordillera (Fig. 2). The extent of its range is in large part due to seasonally related station-keeping mechanisms, part of the adaptive components of the salamanders' biology. Its range is also affected by its rate of passive transportation over long distances, which was imposed by the changing conditions of the proximal climate and geography. In western North America, Pleistocene glaciers were a major limiting condition on the range of this salamander, and with glacial retreat emigrants shifted from periglacial refugia (Fig. 6) to occupy exposed and ecologically suitable regions (Fig. 7). At this time, these emigrants increased their migratory rates and distances among geographic regions (as was the case for other species as well), because glacial-fluvial systems, produced as the Pleistocene glaciers retreated, increased the amount of accidental migration relative to the occurrences of non-accidental migration.

Post-glacial history changed the mobility of long-toed salamanders and resulted in a unique genetic structure for northern populations. Their mtDNA haplotype distributions are accentuated by this history (Fig. 7). Once matters of phylogenetic congruence are considered, the more expanded data were able to be investigated via a nested cladistic phylogeographical analysis (Templeton et al. 1995). Detailed phylogeography necessitates the apportionment of data into groups of population units, or demes, which should preferably be based on their genealogical independence. Templeton et al.'s (1995) phylogeographical approach offers a convenient coalescent approach for referencing such units through explicit nesting categories. Questions regarding the cause of the genetic structure in contemporary populations can be focused on each clade, and spatial contrasts can be made between ancestors and their descendants. This approach is also useful in referencing clades for mismatch distributions. While informative, the mismatch distributions are constrained by their lack of general applicability to all species, both because the models are sensitive to assumptions regarding the migration rates of each species, and because they are used primarily to study single events of range expansion. Empirical research, simulations, and improved models will be important developments in this line of research (Ray et al. 2003; Excoffier 2004). Mismatch distributions of the long-toed salamander tend to exhibit multi-modal distributions in their deeper genealogies, but there is a shift in the more recent clades towards uni-modal patterns that are consistent with post-Pleistocene range expansion. The spatial statistics of NCPA and inferences (Templeton 2004) are greatly augmented by mismatch distributions because the plots further reduce incidences of false positives when inferring range expansion.

References

Alford R, Richards S (1999) Global amphibian declines: A problem in applied ecology. Annual Review of Ecology and Systematics 30: 133-165

Alt D, Hyndman D (1995) Northwest Exposures: A Geological Story of the North West. Mountain Press Publishing Company, Missoula, Montana

Anderson PK (1960) Ecology and evolution in Island populations of salamanders in the San Francisco Bay region. Ecological Monographs 30: 359-385

Anderson J, Graham R (1967) Vertical migration and stratification of larval *Ambystoma*. Copeia 1967: 371-374

Anderson (2004) Postglacial population history of the common shrew (*Sorex araneus*) in Fennoscandia. Molecular studies of recolonisation, sex-biased gene flow and the formation of chromosome races. Acta Universitatis Uppsaliensis. Comprehensive summaries of Uppsala dissertations from the Faculty of Science and Technology. 986. 56 pp. Uppsala, Sweden

Arbogast B, Kenagy G (2001) Comparative phylogeography as an integrative approach to historical biogeography. Journal of Biogeography 28: 819-825

Avise J (1994) Molecular Markers, Natural History and Evolution. Chapman and Hall, New York

Avise J (1996a) Introduction: The scope of conservation genetics. In Avise J, Hamrick J (eds) Conservation Genetics: Case Histories from Nature. Chapman and Hall, New York, NY 1-9

Avise J (1996b) Towards a regional conservation genetics perspective: Phylogeography of faunas in the southeastern United States. In Avise J, Hamrick J (eds) Conservation Genetics: Case Histories from Nature. Chapman and Hall, New York, NY 431-470

Avise JC (2000) Phylogeography: The History and Formation of Species. Harvard University Press, Cambridge, Massachusetts

Avise J, Ball R, Arnold J (1988) Current versus historical population sizes in vertebrate species with high gene flow: A comparison based on mitochondrial DNA lineages and inbreeding theory for neutral mutations. Molecular Biology and Evolution 5: 331–344

Baily, J (1948) Supplementary observations on the geographic variation of *Ambystoma* macrodactylum. Herpetologica 4: 171–174

Baker V (1983) Late-Pleistocene fluvial systems. In Porter SC (ed) Late-Quaternary Environments of the United States. University of Minnesota Press, Minneapolis 115–129

Ballard JWO, Whitlock MC (2004) The incomplete natural history of mitochondria. Molecular Ecology 13: 729-744

Beneski J, Zalisko E, Larsen J (1986) Demography and migratory patterns of the eastern long-toed salamander, *Ambystoma macrodactylum columbianum*. Copeia 1986: 398-408

Benkman C, Holimon W, Smith J (2001) The influence of a competitor on the geographic mosaic of coevolution between crossbills and lodgepole pine. Evolution 55: 282-294

Bermingham E, Moritz C (1998) Comparative phylogeography: Concepts and applications. Molecular Ecology 7: 367–369

Berven K, Gill D, Smith-Gill S (1979) Counter gradient selection in the green frog, Rana clamitans. Evolution 33: 609-623

Branco M, Monnerot M, Ferrand N, Templeton AR (2002) Postglacial dispersal of the European rabbit (*Oryctolagus cuniculus*) on the Iberian peninsula reconstructed from nested clade and mismatch analyses of mitochondrial DNA genetic variation. Evolution 56: 792-803

Brodie E, Ridenhour B, Brodie E (2002) The evolutionary response of predators to dangerous prey: Hotspots and coldspots in the geographic mosaic of coevolution between garter snakes and newts. Evolution 56: 2067-2082

Brosou-Aris S, Excoffier L (1996) The impact of population expansion and mutation rate heterogeneity on DNA sequence polymorphism. Molecular Biology and Evolution 13: 494-504

Brunsfeld S, Sullivan J, Soltis D, Soltis P (2001) Comparative phylogeography of north-western North America: A synthesis. In Silverton J, Antonovics J (eds) Integrating Ecology and Evolution in a Spatial Context. Fourteenth Special Symposium of the British Ecological Society. British Ecological Society, Blackwell Science Ltd. 319-339

Bury R, Pearl C (1999) Klamath-siskiyou herpetofauna: Biogeographic patterns and conservation strategies. Natural Areas Journal 19: 341–350

Byun A, Koop B, Reimchen T (1997) North American black bear mtDNA phylogeography: Implications for morphology and the Haida Gwaii glacial refugium controversy. Evolution 51: 1647–1653

Carson H, Templeton A (1984) Genetic revolutions in relation to speciation phenomena: The founding of new populations. Annual Review of Ecology and Systematics 15: 97-131

Castelloe J, Templeton A (1994) Root probabilities for intraspecific gene trees under neutral coalescent theory. Molecular Phylogenetics and Evolution 3: 102–113

Church SA, Kraus JM, Mitchell JC, Church DR, Taylor DR (2003) Evidence for multiple Pleistocene refugia in the postglacial expansion of the eastern tiger salamander, *Ambystom tigrinum tigrinum*. Evolution 57: 372-383

Clague J, James T (2002) History and isostatic effects of the last ice sheet in southern British Columbia. Quaternary Science Reviews 21: 71–87

Clayton D (1984) Transcription of the mammalian mitochondrial genome. Annual Review of Biochemistry 53: 573–594

Conroy C, Cook A (2000) Phylogeography of a post-glacial colonizer: *Microtus longicaudus* (Rodentia: Muridae). Molecular Ecology 9: 165–175

Cox A, Hebert P (2001) Colonization, extinction, and phylogeographic patterning in a freshwater crustacean. Molecular Ecology 10: 371-386

Crandall, KA (1994) Intraspecific cladogram estimation: Accuracy at higher levels of divergence. Systematic Biology 43: 222–235

Crandall KA (1996) Multiple interspecies transmissions of human and simian T-cell leukemia/lymphoma virus type I sequences. Molecular Biology and Evolution 13: 115–131

Crandall KA, Templeton AR (1993) Empirical tests of some predictions from coalescent theory with applications to intra-specific phylogeny reconstruction. Genetics 134: 959–969

Crandall KA, Templeton AR (1996) Applications of intraspecific phylogenetics. In Harvey PH, Brown AJL, Smith JM, Nee S (eds) New Uses for New Phylogenies, Oxford University Press, Oxford 81-99

Cruzan MB, Templeton AR (2000) Paleoecology and coalescence: phylogeographic analysis of hypotheses from the fossil record. Trends in Ecology and Evolution 15: 491-496

Darlington PJ (1957) Zoogeography. JC Wiley & Sons, New York.

DaSilva M, Patton J (1998) Molecular phylogeography and the evolution and conservation of Amazonian mammals. Molecular Ecology 7: 475-486

Daubenmire R (1975) Floristic plant geography of eastern Washington and northern Idaho. Journal of Biogeography 2: 1–18

Demboski J, Cook J (2001) Phylogeography of the dusky shrew, *Sorex monticolus* (Insectivora:Soricidae): Insight into deep and shallow history in northwestern North America. Molecular Ecology 10: 1227-1240

Demboski J, Stone K, Cook J (1999) Further perspectives on the Haida Gwaii glacial refugium. Evolution 53: 2008–2012

Donovan M, Semlitsch R, Routman E (2000) Biogeography of the southeastern United States: A comparison of salamander phylogeographic studies. Evolution 54: 1449–1456

Duellman WE, Sweet SS (1999) Distribution patterns of amphibians in the neararctic region of North America. In Duellman WE (ed), Patterns and Distribution of Amphibians: A Global Perspective. The Johns Hopkins University Press, Baltimore and London 31-109

Duellman WE, Trueb L (1986) Biology of Amphibians. McGraw-Hill Book Co., New York.

Dynesius M, Jansson R (2000) Evolutionary consequences of changes in species' geographical distributions driven by milankovitch climate oscillations. Proceedings of the National Academy of Sciences 97: 9115-9120

Endler J (1977) Geographic Variation, Speciation, and Clines. Princeton University Press, Princeton, New Jersey

Excoffier L (2004) Patterns of DNA sequence diversity and genetic structure after a range expansion: lessons from the infinite-island model. Molecular Ecology 13: 853-864

Farris J, Källersjö M, Kluge A, Bult C (1994) Testing significance of incongruence. Cladistics 10: 315–319

Felsenstein J (1983) Parsimony in systematics: Biological and statistical issues. Annual Review of Ecology and Systematics 14: 313–333

Ferguson DE (1961) The geographic variation of *Ambystoma macrodactylum* Baird, with the description of two new subspecies. The American Midland Naturalist 65: 311–338

Fisher R, Shaffer H (1996) The decline of amphibians in California's great central valley. Conservation Biology 10: 1387-1397

Fu YX (1997) Statistical tests of neutrality of mutations against population growth, hitchhiking and background selection. Genetics 147: 915-925

Fukumoto J, Herrero S (1998) Observations of the long-toed salamander, *Ambystoma macrodactylum*, in Waterton Lakes National Park, Alberta. The Canadian Field-Naturalist 112: 579-585

Funk W, Tallmon D, Allendorf F (1999) Small effective population size in the long-toed salamander. Molecular Ecology 8: 1633-1640

García-Parìs M, Alcobendas M, Buckley D, Wake D (2003) Dispersal of viviparity across contact zones in Iberian populations of fire salamanders (Salamandra) inferred from discordance of genetic and morphological traits. Evolution 57: 129-143

Good J, Sullivan J (2001) Phylogeography of the red-tailed chipmunk (*Tamias ruficaudus*), a northern Rocky Mountain endemic. Molecular Ecology 10: 2683–2695

Gould S, Johnston R (1972) Geographic variation. Annual Review of Ecology and Systematics 3: 457-498

Gould S, Vrba E (1982) Exaptation - a missing term in the science of form. Paleobiology 8: 4-15

Graham KL, Powell GL (1999) Status of the long-toed salamander *Ambystoma macrodactylum* in Alberta. Alberta Wildlife Status Report No. 22, Edmonton, Alberta

Green D, Campbell R (1984) The Amphibians of British Columbia. Handbook No. 45. Royal British Columbia Museum, Vancouver, British Columbia.

Green D, Sharbel T, Kearsley J, Kaiser H (1996) Postglacial range fluctuation, genetic subdivision and speciation in Western North American Spotted Frog complex, Rand pretiosa. Evolution 50: 374-390

Hamilton IM, Graham L, Powell GL, Russell AP (1996) The range of long-toed salamanders in Northwestern Alberta. Alberta Environmental Protection, Fish and Wildlife Service, Edmonton, Alberta

Hewitt G (1999) Post-glacial recolonization of European biota. Biological Journal of the Linnean Society 68: 87-112

Hewitt G, Ibrahim K (2001) Inferring glacial refugia and historical immigrations with molecular phylogenies. In Silverton J, Antonovics J (eds), Integrating Ecology and Evolution in a Spatial Context. Fourteenth Special Symposium of the British Ecological Society. British Ecological Society, Blackwell Science Ltd. 271-294

Hennig W (1966) Phylogenetic Systematics. University of Illinois Press, Urbana

Heusser C (1983) Vegetational history of the Northwestern United States, including Alaska. In Porter SC (ed), Late-Quaternary Environments of the United States, University of Minnesota Press, Minneapolis 239-258

Hillis DM, Bull JD (1993) An empirical test of bootstrapping as a method for assessing confidence in phylogenetic analysis. Systematic Biology 42: 182-192

Hipp AL, Hall HC, Systma KJ (2004) Congruence versus phylogenetic accuracy: revisiting the incongruent length difference test. Systematic Biology 53: 81-89

Hodge R (1976) Amphibians and Reptiles in Alaska, The Yukon and Northwest Territories. Alaska Northwest Publishing Company, Anchorage, Alaska

Houlahan J, Findlay C, Schmidt B, Meyer A, Kuzmin S (2000) Quantitative evidence for global amphibian population declines. Nature 404: 752-755

Howard J, Wallace R (1981) Microgeographical variation of electrophoretic loci in populations of *Ambystoma macrodactylum columbianum* (Caudata: Ambystomatidae). Copeia 1981: 466-471

Howard J. Wallace R (1983) Critical thermal maxima in populations of *Ambystoma macrodactylum* from different elevations. Journal of Herpetology 173: 400-402

Howard J, Wallace R (1984) Life history characteristics of populations of the long-toed salamander (*Ambystoma macrodactylum*) from different altitudes. The American Midland Naturalist 113: 361-373

Hudson R (1989) How often are polymorphic restriction sites due to a single mutation? Theoretical Population Biology 36: 23–33

Janzen F, Krenz J, Haselkorn T, Brodie E Jr, Brodie E III (2002) Molecular phylogeography of common garter snakes (*Thamnophis sirtalis*) in western North America: Implications for regional historical forces. Molecular Ecology 11: 1739-1751

Jarrett R, Malde H (1987) Paleodischarge of the late Pleistocene Bonneville flood, Snake River, Idaho, computed from new evidence. Geological Society of America Bulletin 99: 127-134

Jockusch E, Wake D (2002) Falling apart and merging: Diversification of slender salamanders (plethodontidae: Batrachoseps) in the American West. Biological Journal of the Linnean Society 76: 361–391

Johnson N, Marten J (1989) Evolutionary genetics of flycatchers. III. Variation in *Epidonax hammondii* (Aves: Tyrannidae). Canadian Journal of Zoology 69: 232–238

Josenhans H, Fedje D, Conway K, Barrie J (1995) Post-glacial sea levels on the western Canadian continental shelf: Evidence for rapid change, extensive subaerial exposure, and early human habitation. Marine Geology 125: 73–94

Kezer J, Farner D (1985) Life history patterns of the salamander *Ambystoma macrodactylum* in the high Cascade Mountains of southern Oregon. Copeia 1955: 127-131

Knowles L (2001) Did the Pleistocene glaciations promote divergence? Tests of explicit refugial models in montane grasshoppers. Molecular Ecology 10: 691–701

Kumar S, Tamura K, Jakobsen I, Nei M (2001) MEGA2: Molecular evolutionary genetics analysis software. Bioinformatics 17 1244–1245.

MacArthur R, Connell J (1967) The Biology of Populations. John Wiley and Sons, Inc., New York

Mandryk C, Josenhans H, Fedge D, Mathewes R (2001) Late Quaternary paleoenvironments of northwestern North America: Implications for inland versus coastal migration routes. Quaternary Science Reviews 20: 301–314

Marnell LF (1997) Herpetofauna of Glacier National Park. Northwestern Naturalist 78: 17-33

McKnight ML, Shaffer HB (1997) Large, rapidly evolving intergenic spacers in the mitochondrial DNA of the salamander family Ambystomatidae (Amphibia: Caudata). Molecular Biology and Evolution 14: 1167–1176

Meyer G, Leidecker M (1999) Fluvial terraces along the middle fork Salmon river, Idaho, and their relation to glaciation, landslide dams, and incision rates: A preliminary analysis and river-mile guide. In Hughes S, Thackray G (eds) Guidebook to the Geology of Eastern Idaho. Idaho Museum of Natural History, Pocatello 219-335

Milner AR (1983) The biogeography of salamanders in the Mesozoic and early Caenozoic: A cladistic-vicariance model. In Sims RW, Price JH, Whalley PES (eds) Evolution, Time and Space: The Emergence of the Biosphere. Academic Press, London 431-468

Moore WS (1995) Inferring phylogenies from mtDNA variation: Mitochondrialgene trees versus nuclear-gene trees. Evolution 49: 718–726

Moritz C, Schneider CJ, Wake DB (1992) Evolutionary relationships within the *Ensatina eschscholtzii* complex confirm the ring species interpretation. Systematic Biology 41: 273–291

Nelson SJ, Powell GL, Russell AP (1995) Population survey of the long-toed salamander (*Ambystoma macrodactylum*) in southwestern Alberta. Alberta Environmental Protection, Fish and Wildlife Service, Edmonton, Alberta

Nielson M, Lohman K, Sullivan J (2001) Phylogeography of the Tailed Frog (*Ascaphus truei*): Implications for the Biogeography of the Pacific Northwest. Evolution 55: 147-160

Nussbaum R, Bordie E, Storm R (1983) Amphibians and Reptiles of the Pacific Northwest. University of Idaho Press, Moscow, Idaho

Ohta T (1992) The nearly neutral theory of molecular evolution. Annual Review of Ecology and Systematics 23: 263–286

O'Reilly P, Reimchen T, Beech R, Strobeck C (1993) Mitochondrial DNA in *Gasterosteus* and a Pleistocene glacial refugium on the Queen Charlotte Islands, British Columbia. Evolution 47: 678–684

Oseen K, Powell GL, Russell AP (1995a) The distribution of the long-toed salamander (*Ambystoma macrodactylum*) in southwestern Alberta. Alberta Environmental Protection, Fish and Wildlife Service, Edmonton, Alberta

Oseen K, Powell GL, Russell AP (1995b) The distribution of the long-toed salamander (*Ambystoma macrodactylum*) in southwestern Alberta. Alberta Environmental Protection, Fish and Wildlife Service, Edmonton, Alberta

Pauken R, Metter D (1971) Geographic representation of morphologic variation among populations of *Ascaphus truei* Stejneger. Systematic Zoology 20: 434–441

Petranka J (1998) Salamanders of the United States and Canada. Smithsonian Institution Press, Washington

Pielou E (1991) After the Ice Age: The Return of Life to Glaciated North America. University of Chicago Press, Chicago

Posada D, Crandall KA (1998) MODELTEST: Testing the model of DNA substitution. Bioinformatics 14: 817–818

Posada D, Crandall KA, Templeton AR (2000) GeoDis: A program for the cladistic nested analysis of the geographical distribution of genetic haplotypes. Molecular Ecology 9: 487–488

Posada D, Crandall KA (2001) Intraspecific gene genealogies: Trees grafting into networks. Trends in Ecology and Evolution 16: 37–45

Powell G, Oseen K, Russell A (1996) Volunteer amphibian monitoring in Alberta 1992-1994: The results of the pilot project. Alberta Environmental Protection, Fish and Wildlife Division, Edmonton, Alberta

Powell G, Russell AP (1996) The long-toed salamander in the Bow corridor: A preliminary report. Alberta Environmental Protection, Fish and Wildlife Division, Edmonton, Alberta

Powell G, Russell A, James J, Nelson S, Watson S (1997a) Population biology of the long-toed salamander, *Ambystoma macrodactylum*, in the front range of Alberta. In Green D (ed) Amphibians in Decline: Canadian Studies of a Global Problem. Society for the Study of Amphibians and Reptiles 37-44

Powell GL, Russell AP, Nelson SJ, Hamilton IM, Graham KL (1997b) The status of the long-toed salamander *Ambystoma macrodactylum* in Alberta. Alberta Environmental Protection, Fish and Wildlife Service, Edmonton, Alberta

Ray N, Currat M, Excoffier L (2003) Intra-deme molecular diversity in spatially expanding populations. Molecular Biology and Evolution 20: 76-86

Richardson B, Brunsfeld S, Klopfenstein N (2002) DNA from bird-dispersed seed and wind-disseminated pollen provides insights into postglacial colonization and population genetic structure of whitebark pine (Pinus albicaulis). Molecular Ecology 11: 215-227

Rogers AR (1995) Genetic evidence for a Pleistocene population explosion. Evolution 49: 608-615

Rogers AR, Harpending H (1992) Population growth makes waves in the distribution of pairwise genetic differences. Molecular Biology and Evolution 9: 552-569

Russell R, Anderson J (1956) A disjunct population of the long-toed salamander from the coast of California. Herpetologica 12: 137–140

Russell A, Powell G, Hall D (1996) Growth and age of Alberta long-toed salamanders (*Ambystoma macrodactylum krausai*): A comparison of two methods of estimation. Canadian Journal of Zoology 74: 397-412

Schneider C, Cunningham M, Moritz C (1998) Comparative phylogeography and the history of endemic vertebrates in the wet tropic rainforests of Australia. Molecular Ecology 7: 487-498

Schneider S, Roessli D, Excoffier L (2000) ARLEQUIN. Version. 2.000: A Software for Population Genetics Data Analysis. Genetics and Biometry Laboratory, University of Geneva, Switzerland

Sheppard R (1977) The ecology and home range movements of *Ambystoma macrodactylum krausei* (Amphibia : Urodela). Master's thesis, University of Calgary, Calgary, Alberta

Soltis D, Gitzendanner M, Strenge D, Soltis P (1997) Chloroplast DNA intraspecific phylogeography of plants from the pacific northwest of North America. Plant Systematics and Evolution 206: 353–373

Stebbins R (1951) Amphibians of Western North America. University of California Press, Berkeley and Los Angeles

Stebbins R (2003) Western Reptiles and Amphibians, 3rd Edition. Houghton Mifflin, New York

Stone K, Cook J (2000) Phylogeography of black bears (*Ursus americanus*) of the Pacific Northwest. Canadian Journal of Zoology 78: 1218–1223

Stone K, Flynn R, Cook J (2002) Post-glacial colonization of northwestern North America by the forest-associated American marten (*Martes americana*, Mammalia: Carnivora: Mustelidae). Molecular Ecology 11: 2049-2063

Swofford DL (2000) PAUP*: Phylogenetic Analysis Using Parsimony (and Other Methods) Version 4.0b10. Sinauer Associates, Sunderland, Massachusetts

Tajima F (1989) Statistical method for testing the neutral mutation hypothesis by DNA polymorphism. Genetics 123: 585–595

Tallmon DA, Funk WC, Dunlap WW, Allendorf FW (2000) Genetic differentiation among long-toed salamander *Ambystoma macrodactylum* populations. Copeia 2000: 27-35

Tan A, Wake DB (1995) MtDNA phylogeography of the California newt, *Taricha torosa* (Caudata, Salamandridae). Molecular Phylogenetics and Evolution 4: 383–394

Taylor D (1985) Evolution of freshwater drainages and molluscs in western North America: Interdisciplinary studies of the Clarkia fossil beds of northern Idaho. In Smiley CJ, Leviton A, Berson M (eds) Late Cenozoic History of the Pacific Northwest. Department of Herpetology, California Academy of Sciences 265-348

Taylor D, Pollard S, Louie D (1999) Mitochondrial DNA variation in bull trout (*Salvelinus confluentus*) from northwestern North America: Implications for zoogeography and conservation. Molecular Ecology 8: 1155-1170

Templeton A, Georgiadis N (1996) A landscape approach to conservation genetics: Conserving evolutionary processes in the African bovidae. In Avise J, Hamrick J (eds), Conservation Genetics: Case Histories from Nature. Chapman and Hall, New York, New York 398-430

Templeton A, Crandall KA, Sing C (1992) A cladistic analysis of phenotypic associations with haplotypes inferred from restriction endonuclease mapping and DNA sequence data. III. Cladogram estimation. Genetics 132: 619–633

Templeton AR, Routman E, Phillips C (1995) Separating population structure from population history: A cladistic analysis of the geographical distribution of mitochondrial DNA haplotypes in the tiger salamander, *Ambystoma tigrinum*. Genetics 140: 767–782

Templeton AR (2004) Statistical phylogeography: methods of evaluating and minimizing inference errors. Molecular Ecology 13: 789-809

Thompson MD, Ray N, Russell AP (unpublished) Phylogeography of the long-toed salamander (*Ambystoma macrodactylum*): Systematics and statistical biogeography of mitochondrial genes

Titus T, Gains M (1991) Genetic variation in coastal and montane populations of Ambystoma gracile, (Caudata: Ambystomatidae). Occasional papers of the Museum of Natural History, University of Kansas, Lawrence 141: 1–12

Tzedakis P, Lawson I, Frogley M, Hewitt G, Preece R (2002) Buffered tree population changes in a quaternary refugium: evolutionary implications. Science 297: 2044-2047

Verbeylen G, De Bruyn L, Adriaensen F, Matthysen E (2003) Does matrix resistance influence Red squirrel (*Sciurus vulgaris* L. 1758) distribution in an urban landscape? Landscape Ecology 18: 791-805

Waitt R, Thorson R (1983) The Cordilleran ice sheet in Washington, Idaho, and Montana. In Porter SC (ed) Late-Quaternary Environments of the United States, University of Minnesota Press, Minneapolis 53-70

Walker IR, Pellatt MG (2003) Climate change in coastal British Columbia -- A paleoenvironmental perspective. Canadian Water Resources Journal 28: 531-566

Walsh R (1998) An extension of the known range of the long-toed salamander, *Ambystoma macrodactylum* in Alberta. Canadian Field Naturalist 112: 331-333

Walter R, Epperson B (2001) Geographic pattern of genetic variation in *Pinus resinosa*: Area of greatest diversity is not the origin of postglacial populations. Molecular Ecology 10: 103-111

Warner B, Mathewes R (1982) Ice-free conditions on the Queen Charlotte Islands, British Columbia, at the height of late Wisconsin glaciation. Science 218: 675–677

Wilke T, Duncan N (2004) Phylogeographical patterns in the American Pacific Northwest: lessons from the arionid slug *Prophysaon coeruleum*. Molecular Ecology 13: 2303-2315

Williams T, Larsen J (1986) New function for the granular skin glands of the eastern long-toed salamander, *Ambystoma macrodactylum columbianum*. Journal of Experimental Zoology 239: 329-333

Yoder A, Irwin J, Payseur B (2001) Failure of the ILD to determine data combinability for Slow Loris phylogeny. Systematic Biology 50: 408–424

Zachos J, Flower B, Paul H (1997) Orbitally paced climate oscillations across the Olicogene/Miocene boundary. Nature 388: 567-570

Zamudio K, Savage WK (2003) Historical isolation, range expansion, and secondary contact of two highly divergent mitochondrial lineages in spotted salamanders (*Ambystoma maculatum*). Evolution 57: 1631-1652

9 The geography of a faunal turnover: tracking the Vallesian Crisis

Isaac Casanovas-Vilar Salvador Moyà-Solà, Jordi Agustí and Meike Köhler

Institut de Paleontologia Miquel Crusafont (Diputació de Barcelona-Unidad Asociada CSIC), Escola Industrial, 23. 08201-Sabadell (Spain). Corresponding author's e-mail: hispanomys@yahoo.es

9.1 Abstract

The configuration of emerged land masses as well as the distribution of suitable habitats are two important constraints for migration of land mammals. The evolution of late Miocene European land mammal faunas seems to be related to climatic change, which consisted in a general cooling and an increase in seasonality starting by the late Miocene. The pre-existing subtropical evergreen forests were replaced by mixed mesophytic ones at higher latitudes, while in the peri-Mediterranean regions the forest cover was fragmented. The Vallesian Crisis, which implied the disappearance of forest-adapted taxa and a general diversity decay in Western Europe by the beginning of the late Miocene, has been related to main climatic changes. This crisis is a well-established event in the Iberian Peninsula, however the response of mammal communities in other areas is still debated. The goals of present work are to refine chronologic and geographic limits of the Vallesian Crisis. Data input consists in a series of fossil mammal sites covering most of the middle and all the late Miocene (13.8 – 4.9 Ma). Provinciality has been studied using cluster analysis, and results indicate the maintenance of three main bioprovinces (Greek-Iranian, Iberian and Central European) during most of the time span. Diversity and origination and extinction rates have been calculated for all Europe and for each main bioprovince whenever possible. The results show that diversity increased during the late Vallesian and the Turolian in the Greek-Iranian bioprovince because of the development of open-country herbivore faunas while diversity remained stable in Central Europe. A decay in diversity started in the Iberian Peninsula by the early Vallesian and genera richness kept descending during the Turolian. Although the environment was rather similar in East and West, Turolian mammal faunas from the Iberian Peninsula are very poor and show few eastern immigrants. We suggest that mixed mesophytic forests covering most of Central Europe acted as an ecological filter, preventing the migration of open-country adapted taxa from Anatolia.

Keywords: Europe, Miocene, mammals, biogeography, paleoclimatology, faunal turnover, migration.

9.2 Climatic change and mammalian faunal dynamics

The influence of climatic change and its relation to major ecological and evolutionary transitions on fossil mammal faunas has been an extensively studied subject during the past decades. Detailed works have linked changes in species composition with climatic events (see for example Vrba 1985, 1995; Van der Meulen and Daams 1992; De Bonis et al. 1992; Fortelius et al. 1996; Agustí et al. 1999; Bobe and Eck 2001). However, as Alroy et al. (2000) have shown in the particular case of North American mammals, sometimes climatic shifts are correlated with ecological and evolutionary changes, while often other such transitions occur at time of little global climate change. The effect of global climatic change in mammal communities seems clear, but we should put more emphasis on the particular environmental conditions of the areas affected. Considering distribution of landmasses and past ecosystems can help us to understand better mammal turnovers and why sometimes they occur at moments of moderate climatic shifts.

Vrba (1985) proposed two mechanisms in order to explain turnover pulses as recorded in East African ecosystems: habitat tracking and ecologically induced genealogical events. Habitat tracking simply implies the expansion of the geographic ranges of pre-existing species. Otherwise climatic change may cause local speciation because habitat alteration and degradation leads to a more heterogeneous "patchy" environment. Nevertheless, Vrba (1985) notes that speciation does not occur in response to new empty niches, but fragmentation of species population. Distribution of emerged lands and particular ecosystems will evidently determine the principal kind of mechanism involved in a turnover event.

9.3 The Vallesian Crisis and the late Miocene faunal turnover

Within the Neogene the late Miocene is a complex moment which appears punctuated by several climatic and phisiographic events. This is the case, for instance, of the progressive closing of Eastern Mediterranean gateway during the late Serravallian – early Tortonian (Rögl 1999, 2001) and the uplift of the Tibetan plateau (Amano and Taira 1992), which led to a restructuration of atmospheric and marine circulation. These events resulted in the onset or intensification of monsoonal circulation in Asia and the appearance of a characteristic dry season in circum-Mediterranean regions (Kutzbach et al. 1993). At high latitudes temperatures decreased because of changes in position of polar ice fronts.

In the terrestrial realm, these events must have influenced the pattern of dispersal and turnover among the mammalian faunas, according to the ideas of turnover pulses exposed above. In fact, the late Miocene, that is the period covered by continental mammal stages Vallesian and Turolian, coincides with several faunal dispersals, induced either by climatic events, intercontinental relationships or interplay between both causes. Particularly, the Vallesian stage (between 11.1 Ma and 8 Ma) is characterized by the overall dispersal at 11.1 Ma of three-toed hipparionine horses, which entered Eurasia from North America via the Bering Strait (Bernor et al. 1993; Garcés et al. 1997). This event was possible because of an important sea level fall of about 100 m (Haq et al. 1987) coinciding with a major shift in δ^{18} O (Mi 5) (Miller et al. 1991) indicating general cooling. However, this climatic event seems to have had few effects over land mammal communities, and in fact diversity increased continuously during the Vallesian (at least in Spain, see Agustí et al. 1999).

This scenario changed abruptly at 9.7 Ma, when a significant mammal turnover, known as the Vallesian Crisis, deeply altered the composition of the Western and Central European terrestrial ecosystems. This point marks the boundary between the early and late Vallesian at 9.7 Ma (Garcés et al. 1996). The Vallesian Crisis (sometimes also cited as Mid or Late Vallesian Crisis) was recognized for the first time in the Vallès-Penedès Basin (northeastern Spain) and involved the disappearance of most of the middle Miocene elements adapted to the warm wet-subtropical environments (Agustí and Moyà-Solà 1990; Moyà-Solà and Agustí 1990). The diversity and number of forest-adapted rodents decreased significantly coinciding with the entry and widespread of murids that completely outnumbered the cricetids, major elements of the rodent communities until the late Vallesian (Agustí 1982; Van Dam 1997). Among the large mammals the crisis specially affected perissodactyls (tapirids and the rhinoceroses *Lartetotherium sansaniense* and "*Dicerorhinus*" *steinheimensis*), artiodactyls (the suid *Conohyus*, the cervid *Amphiprox*, the moschid *Hispanomeryx*, the bovids *Miotragocerus* and *Protragocerus*), hominoids (*Dryopithecus*) and the large carnivores of the families Nimravidae and Amphycionidae (Moyà-Solà and Agustí 1990; Agustí et al. 2003). It has also been suggested that the Vallesian Crisis extended to the Central European mammal faunas (Fortelius et al. 1996; Franzen and Storch 1999). The effects of the crisis there would have been less marked, and some middle Miocene relics persisted in Germany, Austria and Hungary during the late Vallesian. On the other hand the Vallesian Crisis is not observed in the Rhône Valley (Mein 1999), while in southeastern Europe coincides with an increase in diversity and turnover. The crisis coincides with a minor isotopic event, Mi 7 (Miller et al. 1991), even weaker than previous Mi 6 although this time the shift in δ^{18} O is accompanied of a noticeable sea level lowering (Haq et al. 1987). The Vallesian Crisis has been related to this event (Agustí et al. 1999), which would have implied the onset of a late Miocene glaciation in the Arctic zones (Zubakov and Borzemkova 1990). This scenario of high latitude cooling and low latitude drying continued during the Turolian. This stage records the progressive extension to the west of the bovid and giraffid dominated faunas that characterized Anatolian communities (Fortelius et al. 1996; Koufos 2003).

As exposed above, after decades of work, a general picture of the climatic changes taking place during the late Miocene and their effect over European mammal faunas is emerging. However, problems appear when focusing in concrete events, areas, or time spans. For example, the effects of the Vallesian Crisis and the existence of such event in some areas are still debated. The absence of a reasonably complete record in many countries may obscure the effects of the crisis in these areas, or simply the crisis did not occur there. There seems to be a diversity loss in Spain and Central Europe, but in Anatolia diversity increased during the crisis as result of the entry of new taxa. To what extend are we dealing with an extinction or a turnover event? And why the response of mammal communities is different depending on the area? The aim of this work is to find the temporal and geographic limits of the Vallesian Crisis, and to construct a late Miocene picture of Europe consistent with them.

9.4 Material and methods

9.4.1 The database

We have focused in the western part of the Old World, covering a latitudinal range from 35 to 55 degrees and from –10 to 40 degrees of longitude. Northern Africa and the British Islands are not included. We have selected a time span covering nearly all the late Miocene, ranging from the late Aragonian (middle Miocene) until the end of the epoch (13.8-4.9 Ma, datings after Agustí et al. 2001). This time span includes seven MN zones ("Mammal Neogene" biozones): MN 6 (late Aragonian, 13.8-12.5 Ma) MN 7 + 8 (late Aragonian, 12.5-11.1 Ma); MN 9 (early Vallesian, 11.1-9.7 Ma); MN 10 (late Vallesian, 9.7-8.7 Ma); MN 11 (early Turolian, 8.7-7.5 Ma); MN 12 (middle Turolian, 7.5-6.8 Ma); and MN 13 (late Turolian, 6.8-4.9 Ma) (datings are after Agustí et al. 2001).

Data source for this chapter is the version of the NOW (Neogene Old World) database, which is available at the following site: http://www.helsinki.fi/science/now/data.html, realized in July 2003. This database has been completed with data from many Spanish localities. We have included localities dated at the MN zone level of resolution at least and ignored those inaccurately dated. Finally a genera occurrence by localities database has been written using the data compiled. For this second database we have selected only the localities from the first one that included at least five genera. If more than one species of the same genus is present in the locality the genus is counted only once. Taxa identified to a suprageneric level are counted as an additional genus only if there are no identified genera of the same suprageneric level. For instance, if in a locality the genus *Deinotherium* is present and we have also recorded Proboscidea indet., we only count one genus: *Deinotherium*. If in another site only Proboscidea indet. is recorded, it counts as a genus in our database. Macromammals and micromammals have been recorded in separate databases because they are recovered using different field methods. As a result we have obtained a database that includes a total of 341 sites: 145 macromammal sites and 196 micromammal sites. A total of 442 taxa are recorded in the studied time span: 253 macromammals and 189 micromammals. In order to study the geographic trends in mammal assemblages through time we have divided the western Old World in 40 squares of the same area. These squares are defined by the intersection of latitude and longitude lines each one equally separated 5° from their nearest neighbors. Thus the side of the squares is also 5°. Each site is assigned the to one square numbered from 1 until 40 (see Fig. 1 and Table

1). Of course, some squares will include land and sea, and the ratio land to sea will not always be the same. Although we consider that this fact will not affect the results in an important way, it must not be forgotten. Appendix shows the localities used in the cluster analysis of Figs 2-8 (see below) as well as the square in which each one is included.

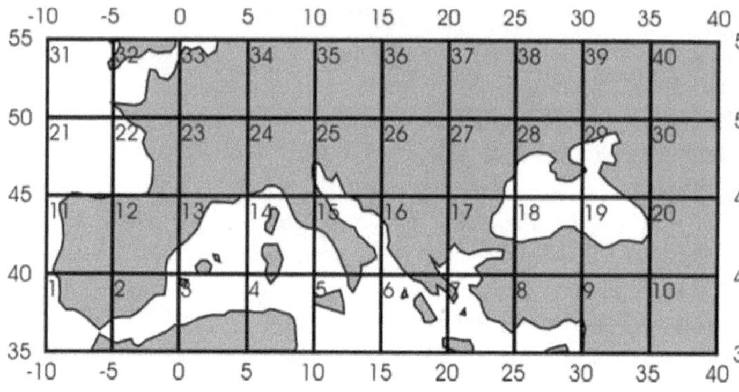

Fig. 1. Map of the study area. The western Old World is divided in 40 squares of the same area defined by the intersection of latitude and longitude lines, each one equally separated 5° from their nearest neighbors. Each square is numbered from 1 until 40

9.4.2 Provinciality

We have considered sites in each MN zone and constructed taxa occurrence by locality matrices. Macromammals and micromammals are considered separately resulting in two different matrices for each MN zone, thus fourteen matrices. Cluster analysis has been performed on each matrix. We have used unweighted pair-group average algorithm (UPGMA) to join clusters, while distance matrix has been computed using Raup-Crick index for presence / absence data (Raup and Crick 1979). This index uses randomization procedure, comparing the observed number of genera co-occurring in two associations with the distribution of co-occurrences of 200 random replicates. Calculation was carried out using the program PAST 1.27 (Hammer et al. 2001).

Table 1. Per-square (see Fig. 1) percent of the total number of macromammal and micromammal sites recorded in each MN zone. The higher proportion of both micro and macromammal localities comes from central and northeastern Spain, so it is expected that Spanish results will clearly bias calculations when Europe is considered as a whole

	MACROMAMMALS (% of total number of localities in each MN)							MICROMAMMALS (% of total number of localities in each MN)						
square	MN 6	MN 7 + 8	MN 9	MN 10	MN 11	MN 12	MN 13	MN 6	MN 7 + 8	MN 9	MN 10	MN 11	MN 12	MN 13
1	0	0	0	0	0	0	0	8	0	0	7	0	0	0
2	0	0	0	0	7	7	11	0	0	0	0	4	4	15
3	0	0	0	0	0	0	0	0	0	0	0	0	0	0
4	0	0	0	0	0	0	0	0	0	0	0	0	0	0
5	0	0	0	0	0	0	0	0	0	0	0	0	0	0
6	0	0	0	0	0	0	5	0	0	0	0	0	0	0
7	0	0	0	0	0	10	0	0	4	0	0	0	0	0
8	0	6	3	0	14	13	0	0	4	0	0	0	0	0
9	11	6	0	6	7	3	0	4	4	0	0	0	0	0
10	0	0	0	0	0	0	0	0	4	0	0	0	0	0
11	0	0	0	0	0	0	0	0	0	0	3	0	0	4
12	22	0	7	18	21	27	37	50	23	43	64	46	81	56
13	17	56	31	24	7	3	5	4	35	26	18	12	0	7
14	0	0	0	0	0	3	0	0	0	0	0	0	4	4
15	0	0	0	0	0	10	16	0	0	0	0	4	4	7
16	0	0	0	0	0	0	0	0	0	0	0	0	0	0
17	6	0	0	24	21	0	21	0	4	0	0	4	0	7
18	0	0	0	0	0	0	0	0	0	0	0	0	0	0
19	0	0	0	0	7	3	0	0	0	0	0	0	4	0
20	0	0	0	6	0	0	0	0	0	0	0	0	0	0
21	0	0	0	0	0	0	0	0	0	0	0	0	0	0
22	0	0	0	0	0	0	0	0	0	0	0	0	0	0
23	0	0	0	0	0	0	0	0	0	0	5	8	0	0
24	11	6	31	6	7	0	0	17	8	4	5	15	0	0
25	11	13	3	0	0	0	0	4	4	0	0	0	0	0
26	17	0	7	0	7	0	5	4	4	9	3	8	0	0
27	0	6	0	0	0	0	0	0	4	4	0	0	0	0
28	0	0	17	18	0	13	0	0	0	9	0	0	4	0
29	0	0	0	0	0	3	0	0	0	0	0	0	0	0
30	0	0	0	0	0	0	0	0	0	0	0	0	0	0
31	0	0	0	0	0	0	0	0	0	0	0	0	0	0
32	0	0	0	0	0	0	0	0	0	0	0	0	0	0
33	0	0	0	0	0	0	0	0	0	0	0	0	0	0
34	6	0	0	0	0	0	0	8	0	0	0	0	0	0
35	0	0	0	0	0	0	0	0	0	0	0	0	0	0
36	0	6	0	0	0	0	0	0	4	0	0	0	0	0
37	0	0	0	0	0	0	0	0	0	4	0	0	0	0
38	0	0	0	0	0	0	0	0	0	0	0	0	0	0
39	0	0	0	0	0	3	0	0	0	0	0	0	0	0

However, there is an important pitfall in our provinciality analysis: all sites included in an MN zone are not of the same absolute age. This would not be a great problem if the MN zones consisted in relatively short time spans, but some zones, such as MN 7 + 8 and MN 9 last nearly 1.5 Ma. Then, differences expressed by clusters could not correspond to real ecological or biogeographical differences between sites at a specific time just because the localities compared are not necessarily contemporary. Unfortunately accurate datings are not available for most of the sites, so an analysis at a finer level than an MN zone as a whole is not possible with our data set. The cases when differences expressed by cluster analysis seem to be explained by a temporal factor will be discussed.

9.4.3 Diversity

We have computed diversity for each MN zone in all the study area and in each latitude and longitude rank in order to explore geographic trends in diversity through time. We have calculated total diversity in each interval (N_{tot}) as:

$$N_{tot} = N_{FL} + N_{bL} + N_{Ft} + N_{bt} \tag{10.1}$$

Where N_{FL} is the number of genera confined to the interval (singletons sensu Foote 2000); N_{bL} is the number that cross the bottom boundary only; N_{Ft} is the number that cross the top boundary only; and N_{bt} is the number that cross both boundaries. N_{bt} also includes taxa known before and after the time span but not during. These quantities are calculated for each latitude and longitude rank, considering only the genera present in this single rank and discarding those present in adjacent ranks. Thus they are different in each latitude and longitude rank. N_{tot} is the total number of taxa that existed during the interval. This diversity measure is strongly affected by the duration of the time span considered: the longer time interval the higher number of N_{FL}. Diversity measures that estimate standing diversity at a point in time express better the number of taxa susceptible to origination or extinction at an instant in time (Van Valen, 1984). One of this second kind of diversity measures is the estimated mean standing diversity (N_{st}):

$$N_{st} = (N_b + N_t) / 2 \tag{10.2}$$

Where,

$$N_b = N_{bL} + N_{bt} \tag{10.3}$$

and similarly,

$$N_t = N_{Ft} + N_{bt} \tag{10.4}$$

N_b and N_t are the bottom-boundary and top-boundary crossers respectively. Thus, N_{st} estimates the diversity as their average. Because its definition N_{st} cannot be calculated for our first and last intervals (i. e. MN 6 and MN 13). Foote (2000) has shown the advantages and pitfalls of this and other diversity measures. Modeling shows that this measure becomes progressively worse as the length of the interval increases overestimating mean standing diversity if origination rate is different to extinction rate. The advantage of this measure is that singletons are simply irrelevant. Singletons are especially sensitive to variation in preservation and interval length, so basing diversity measures in estimated number of taxa partly removes this undesirable effects. Unfortunately there is not any measure of absolute diversity independent of temporal variation in preservation and interval length.

We have also estimated changes in diversity using the ratio (Foote, 2000):

$$\ln (N_t / N_b) \tag{10.5}$$

Normalized by interval length (Δt), which gives the proportional change in diversity through an interval. This measure is directly derived from the difference between origination and extinction rates (see below) and is particularly useful because all differences in species richness that may exist between areas (many of them resulting from the quality of their record) are leveled of. The proportional change in diversity is calculated for the study area as a whole and for selected squares. As the previous measure it cannot be calculated for the first and the last interval in the studied time span.

9.4.4 Origination and extinction rates

We have calculated estimated per-capita origination ($\hat{p}$) and extinction ($\hat{q}$) rates (Foote, 2000) for the study area as a whole and for selected squares:

$$\hat{p} = -\ln (N_{bt} / N_t) / \Delta t \tag{10.6}$$

$$\hat{q} = -\ln (N_{bt} / N_b) / \Delta t \tag{10.7}$$

As modeling has shown (Foote 2000) these estimates are unaffected by interval length. The ratios N_{bt} / N_t and N_{bt} / N_b decay exponentially with time if rates are constant within the interval; thus the logarithm of each ratio declines linearly with time. Even if true origination and extinction rates (o and e respectively) are not constant, $\hat{p}$ and $\hat{q}$ provide unbiased estimates of the mean rate within an interval. In contrast to other rate this rates do not rely on counting events within the interval and normalizing by a diversity measure. Thus $\hat{p}$ is not affected by extinction rate, while $\hat{q}$ is unaffected by origination rate. Singletons are irrelevant for these rate measures, so they are expected to be less sensitive to variation in preservation rates. The ratio ln (N_t / N_b) used here to estimate relative changes in diversity derives from substracting $\hat{q}$ from $\hat{p}$. These measures were intended for analyzing the whole fossil record of animal groups, rather than studying small areas and concrete time intervals. So in our case it is accurate to refer to $\hat{p}$ and $\hat{q}$ as per-capita entry and exit rates respectively because true origination and extinction may have taken place outside the area considered. Then $\hat{p}$ and $\hat{q}$ include not only true originations and extinctions but also immigration and regional disappearance of taxa.

9.4.5 Distinguishing variation in origination and extinction rates from variation in rates of preservation

A singular increase in preservation rate in one interval will cause an increase in both rates in this interval (see Foote 2000, Fig. 15). The number of genera that would have made their last appearance in preceding intervals is reduced because they now appear last in the interval with better preservation. The same is true for first appearances in succeeding intervals. Metrics underestimate extinction before the pulse in preservation, and underestimate origination afterwards. The measures are also affected by a singular decrease in preservation in an opposite way. A pulse in extinction will not affect origination, and a pulse in origination will not affect extinction (see Foote 2000, Fig. 12). Then we will be able to distinguish variation in taxonomic rates from variation in preservation rates but we will have to analyze the variation in apparent taxonomic rates that is potentially attributable to variation in preservation rates.

Preservation probability (R_i) can be estimated as:

$$R_i = X_{bt,\, samp} / X_{bt} \qquad (10.8)$$

Where X_{bt} is the number of genera known both before and after the interval and $X_{bt,\, samp}$ is the number of these genera actually sampled during the interval. Preservation rate r_i is estimated as (Foote 2000):

$$r_i = -\ln (1 - R_i) / \Delta t \qquad (10.9)$$

If variation in apparent taxonomic rates were dominated by variation in preservation rates, then the correlation between both kinds of rates would be large and positive. We will calculate r_i for the study area as a whole and for selected squares. We will test for correlation between two variables by calculating Kendall's τ. As in previous calculations, preservation rates cannot be calculated for our first and last time intervals.

9.5 Results

9.5.1 Provinciality

Cluster analysis (see Figs. 2 to 8) show the existence of a high degree of provinciality. Although there are some incongruences (MN 6, see Fig. 2 and MN 7+8, see Fig. 3) the same bioprovinces can be recognized in the case of micro and macromammals. Results clearly show the maintenance of three main bioprovinces during most of the time span. Those are an eastern province (also mentioned as Greek-Iranian province), a Central European province and an Iberian (including southern France) province. The existence of two distinct provinces (Eastern and Western Europe) occurring synchronously in the late Miocene was first recognized by Tobien (1967), while the distinction of an Iberian province by the same time (although never recognized) seems evident in the results presented by Fortelius et al. (1996, Fig. 31. 6 p. 422).

In MN 6 three main bioprovinces can be recognized: the Greek-Iranian province, the Central European province and the Iberian province (Fig. 2). The distinction between Central European province and Iberian province is clear considering the macromammals, while in the case of micromammals, localities from Germany, Slovakia and Switzerland appear close to Spanish and Portuguese ones. These localities are characterized by a high diversity of insectivores, dormice and mid-Miocene cricetids. Three Swiss sites cluster apart, due to a higher diversity of flying squirrels and Eomyidae. This fact may reflect the existence of a certain differentiation between Iberia and Central Europe starting at MN 6, although both provinces are very similar.

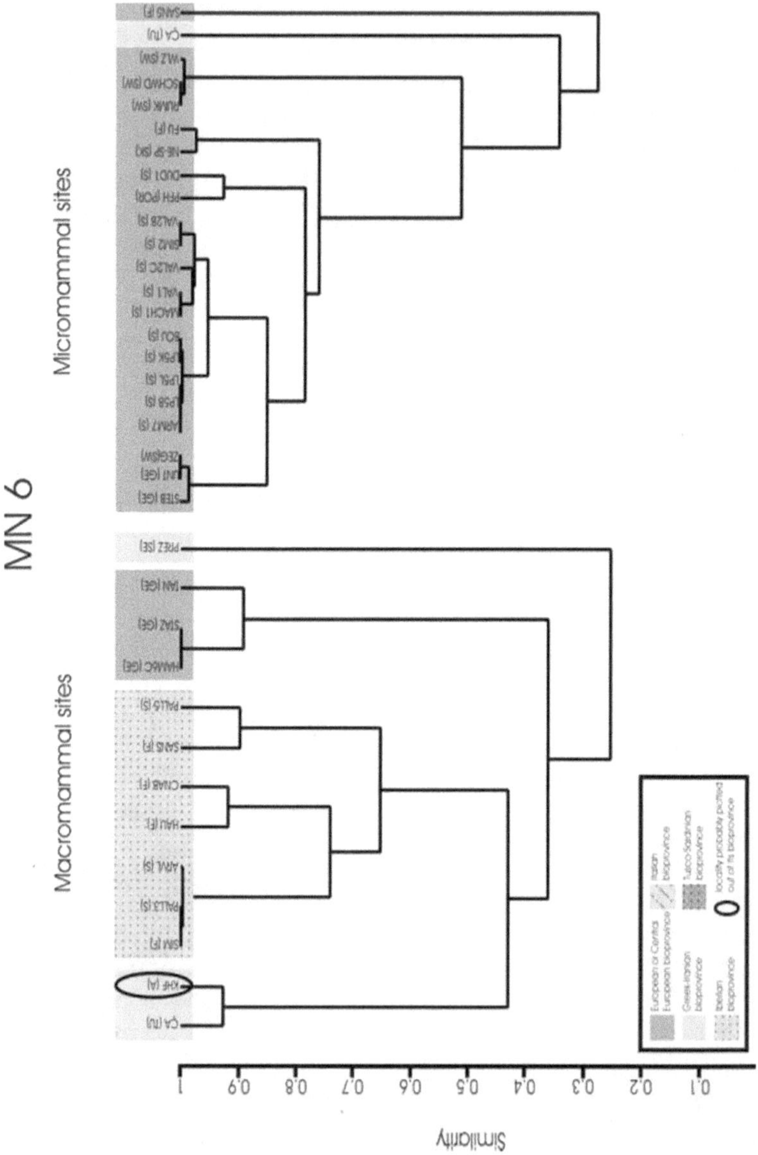

Fig. 2. Dendrogram displaying the results of cluster analysis for MN 6 micro and macromammal sites. Key to locality codes is given in the appendix, whereas key to country codes is given in Table 2

Table 2. Key to country codes used in clusters from Figs. 2 to 8

Country	Country code	Country	Country code
Austria	A	Portugal	POR
France	F	Romania	RO
Germany	GE	Serbia	SE
Greece	GR	Slovakia	SK
Hungary	HU	Spain	S
Italy	IT	Switzerland	SW
Moldova	MO	Turkey	TU
Poland	PO	Ukraine	U

The locality of Sansan in France (SANS (FR) in Fig. 2) appears as very different of the other sites because of its exceptionally rich fauna, but it should be included in the Central European-Iberian province. The single micromammal fauna from the Greek-Iranian province (Çandir, Turkey; ÇA (TU) in Fig. 2) also appears isolated from the rest of sites. This fauna is characterized by the absence of many western cricetid genera, and by the presence of Spalacidae. Considering the macromammals, eastern faunas (Turkey, Serbia) are clearly separated from western ones, except from the site of Kohfidisch (Austria), which seems close to Çandir (Turkey). The eastern faunas are characterized by a higher bovid diversity including many exclusively eastern genera, although others reached Central Europe (*Turkoceras*, *Hypsodontus*). This fact results in a higher similarity between southeastern Europe and some Central European sites. Central European faunas are characterized by the presence of the rhinos *Alicornops*, *Plesiaceratherium* and *Hoploaceratherium*, together with more diverse Suidae and Cervidae. Macromammal faunas in Spain included less Cervidae genera and appear somewhat different from Central European ones.

In MN 7 + 8 the situation remains unchanged (Fig. 3), and again the three bioprovinces are clearly different if we attain to macromammals, while some confusion between Iberian and Central European provinces occurs in the micromammal sites cluster. Considering micromammals, the Greek-Iranian province is clearly recognized and it is characterized by a relatively lower diversity of dormice and squirrels. Cricetids also differ, so different forms lived in East and West of Europe. An illustrative example is the different lineages of hypsodont cricetids (Cricetodontini) which are represented by the genus *Byzantinia* in the East and *Hispanomys* in the West. Both genera arose from *Cricetodon* (although from different species) and followed a parallel evolution during the late Miocene (for a recent review of the evolutionary history of the Cricetodontini see De Bruijn et al. 1993; De Bruijn and Ünay 1996). Some differentiation can be seen between Central European micromammal faunas and those from inner Spanish basins. The differences expressed by the analysis are due to the presence of less dormice genera in assemblages dominated by few species of cricetids. Other Spanish sites, placed in the northeastern margin of the peninsula (Castell de Barberà and Sant Quirze (CB (S) and SQ (S) in Fig. 3), both from the Vallès-Penedès Basin), cluster with Central European

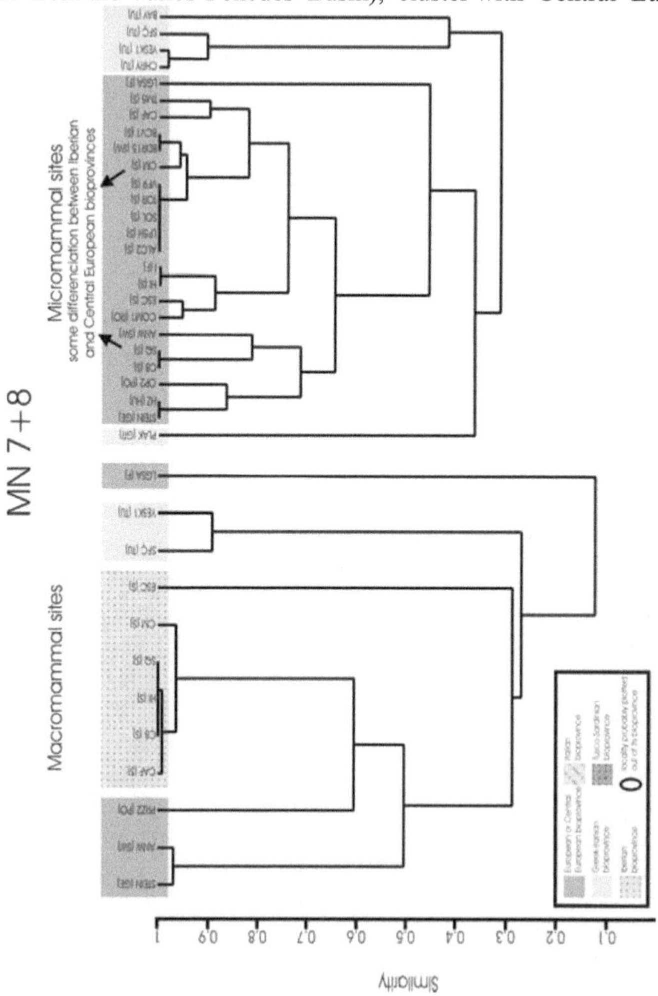

Fig. 3. Dendrogram displaying the results of cluster analysis for MN 7 + 8 micro and macromammal sites. Key to locality codes is given in the appendix, whereas key to country codes is given in Table 2

because of their more diverse rodent faunas including many dormice species andalso some beavers and Eomyidae, very rare in the inner peninsula. However, other sites from the same basin (Hostalets Inferior (HI (S)), Can Missert (CM (S)) and Barranc de Can Vila 1 (BCV1 (S))) appear closer to the remaining Spanish sites. This fact may reflect that the boundary between both bioprovinces may fluctuate in time due to climatic changes (it must not be forgotten that the time span covered by MN 7 + 8 lasts nearly 1.5 Ma). An important bovid radiation occurred in the Eastern Mediterranean by this time, so Turkish sites include more than six bovid species. The occurrence of these elements explains why the sites appear so different from the rest of the European ones in the macromammal cluster. La Grive Saint Alban (France, LGSA (FR) in Fig. 3) also seems very different (a situation also observed in the micromammal cluster), but this is an undesirable taphonomical effect, because it is a karstic site (as well as Escobosa (ESC (S)), which appears as very different too). La Grive Saint Alban must be part of the Central European bioprovince, characterized by the less diverse bovid fauna and the presence of certain rhinos (*Dicerorhinus steinheimensis*), suids (*Albanohyus*) and primates. The sites from Spain cluster all together because of the presence of Boselaphini, while this bovid tribe is absent from many Central European sites. However, these sites do not seem very different from Central European ones and similarity between the two provinces (or should we say subprovinces?) is higher than in MN 6.

The MN 9 recorded the dispersal of hipparionine horses in the Old World. Some other taxa (the sabertoothed cat *Machairodus* and the hyaenid *Thalassictis* for instance, the giraffid *Palaeotragus* forms part of an older migration wave by the end of MN 7 + 8) extended their range from the East and accompanied the hipparions in their dispersal into Western Europe. This event coincided with a rhino turnover in Eastern Europe (appearance of *Ceratotherium* and *Chilotherium*) while bovid diversity continued rising. There seems not to have been any turnover in macromammal faunas in central and southwestern Europe, and the new immigrants are simply incorporated to the pre-existing faunas. The degree of similarity between the Central European and the Iberian bioprovinces (Fig. 4) is close to the one recorded in MN 7 + 8, but in MN 9 the localities from each province do not cluster in separate (although closely related) groups, and in Fig. 4 the Spanish sites appear altogether with German, Swiss and Hungarian ones showing few differences. The locality of Los Valles de Fuentidueña (VFÑ (S) in Fig. 4) appears close to Moldavian sites, clustered out of its province. Other sites such as Rudabánya (Hungary) and Nombrevilla (RUD (HU) and NOM (S) in Fig. 4) may also cluster out of their respective provinces. The Moldavian sites are clearly different from European ones, thus belonging to the Greek-Iranian province, but they also differ a little from the Turkish site Esme Akçaköy (EAKY (TU) in Fig. 4) because bovid diversity in Moldova was lower. The micromammal cluster shows the three main provinces. The Greek-Iranian province (represented only by Moldavian sites) is completely split from a great cluster including Central Europe and the Iberian Peninsula. These later provinces are more dissimilar than they were in previous analysis. Southwestern Europe (Iberian Peninsula and Southern France) is characterized by cricetid-dominated assemblages, while rest of rodent families are less diverse. The genus *Cricetulodon* appears in Spain at the beginning of the Vallesian and will be a major component of the MN 9 rodent faunas. This genus is an eastern immigrant (recorded in the Turkish MN 7 + 8 site Bayraktepe 1 (BAY (TU) in Fig. 3)), which is absent from Central European sites. Another eastern immigrant is the ground dormouse *Myomimus*, which are the most abundant dormice in the faunas of the inner Peninsula. Central European faunas were similar to those of the MN 7 + 8, with high levels of diversity and several genera of Eomyidae, dormice, ground and flying squirrels and beavers. The site of Vösendorf (Austria, VOS (A) in Fig. 4) appears closer to the Spanish ones because the poor fauna recovered lacks Gliridae. The late Vallesian (MN 10) seems to be a time of increasing similarity between East and West (see macromammal cluster of Fig. 5). For the first time Greek sites cluster with French and Spanish ones, although the differences are still important. The Greek-Iranian province (maybe the gentilice "Greek" should not be used to define the name of the bioprovince during MN 10) includes Moldavian and Turkish sites characterized by open-country Bovidae and Giraffidae plus a high diversity of the Bovoidea genera that are going to characterize the Turolian. Furthermore, there was a vicariant evolution of hipparionine horses that led to the origin of *Cremohipparion* in the Eastern Mediterranean, and to *Hipparion* s. s. in the West (Bernor and Armour-Chelo 1999). Western faunas include less Bovoidea and a higher diversity of Cervidae, Suidae and Tragulidae plus some eastern immigrants such as the hyaena *Adcrocuta* and the Boselaphini *Tragoportax*. The clustering of Greek localities with western ones may reflect the diachronous dispersal of open-country eastern herbivores. In the micromammal cluster we can only detect two groups: the Iberian province and the Central European province, which is mainly integrated by French sites. The presence of a still diverse dormice and squirrel fauna in Central Europe characterizes these sites. Another major component of the rodent faunas is *Rotundomys* and hypsodont cricetid that appears in few Spanish sites and many French ones.

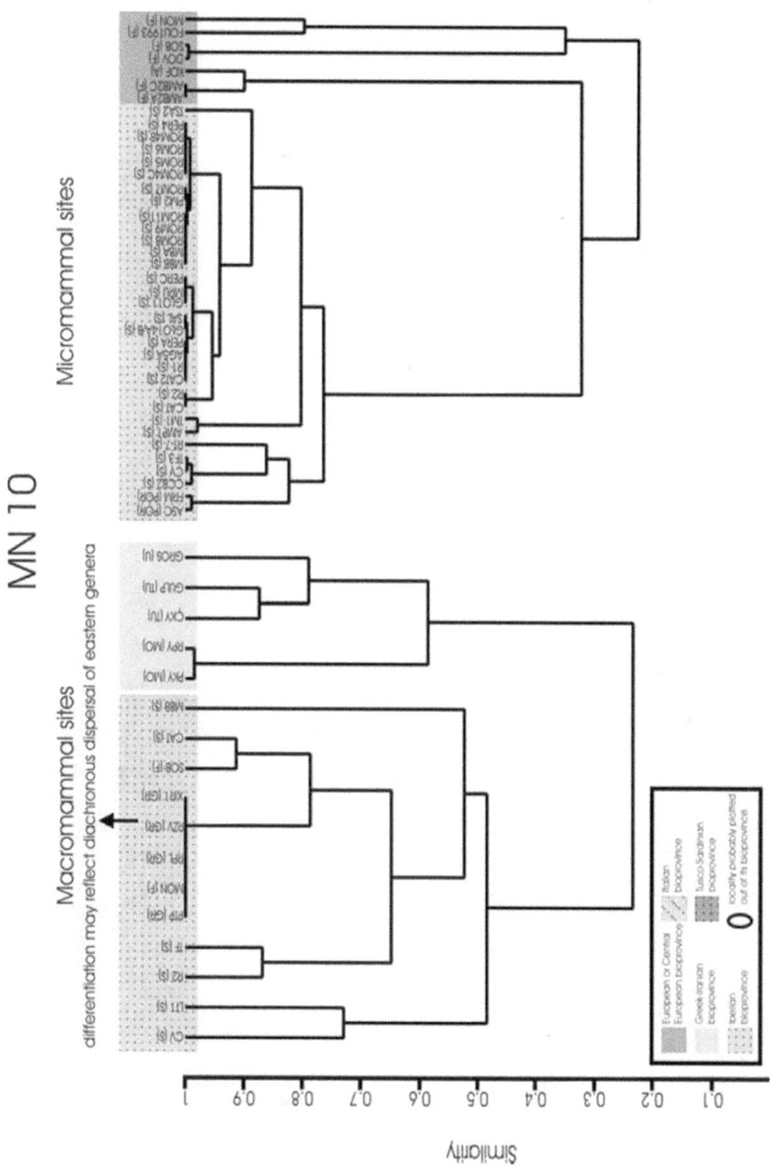

Fig. 4. Dendrogram displaying the results of cluster analysis for MN 9 micro and macromammal sites. Key to locality codes is given in the appendix, whereas key to country codes is given in Table 2

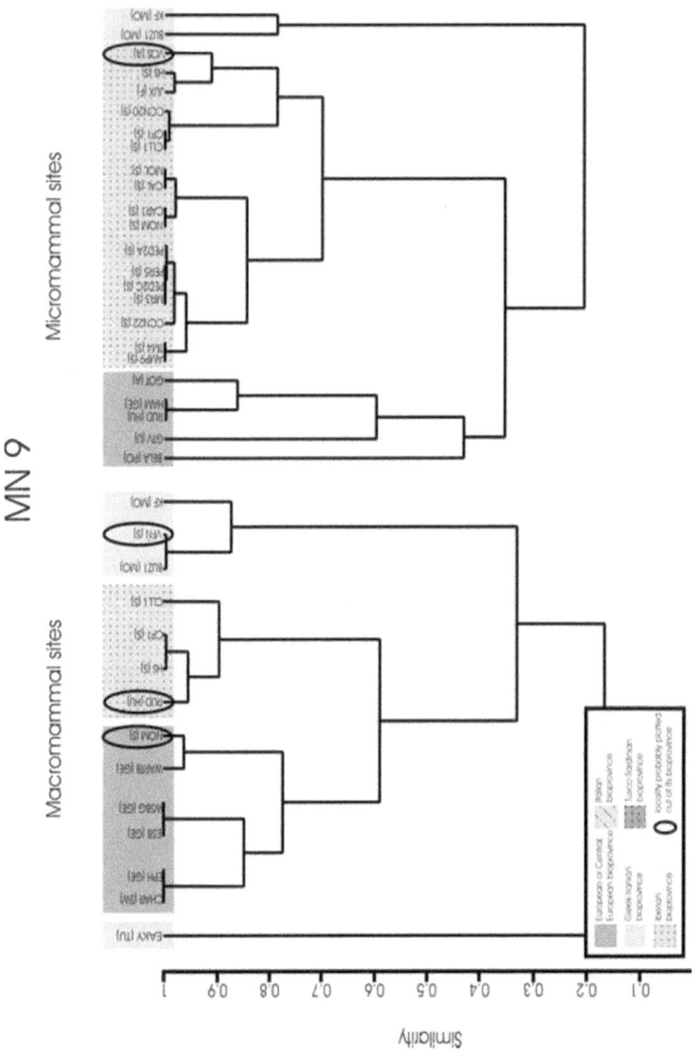

Fig. 5. Dendrogram displaying the results of cluster analysis for MN 10 micro and macromammal sites. Key to locality codes is given in the appendix, whereas key to country codes is given in Table 2

Rodent faunas in the Iberian province are very poor and mice (Muridae) have replaced cricetids being the most abundant rodents in the majority of the sites. The Muridae evolved during the mid-Miocene in Asia, and entered Europe by MN 9. The first occurrence of the Muridae *Progonomys* in Turkey is dated at 10.135 Ma (Kappelman et al. 2003), but it needed about 0.4 Ma to reach Spain (it appears at 9.7 Ma in the Vallès-Penedès Basin (Garcés et al., 1996)). After their arrival in Spain the mice will characterize the rodent assemblage and evolve into several endemic genera during the Vallesian and the Turolian.

Similarity between Greek and Turkish macromammal assemblages increased during MN 11 (Fig. 6) with the arrival into Greece of open-country Bovoidea and Giraffidae. The cercopithecid monkeys (*Mesopithecus*) and the proboscidean *Choerolophodon* accompanied these artyodactyls in their dispersal. These open-country faunas will be characteristic of the Greek-Iranian province during the whole Turolian. The analysis clearly distinguishes an Iberian province with impoverished macromammal faunas. There is a third province very different from the remaining, the Central European one. The persistence of many forest-adapted families (tragulids, tapirids) and the absence of open-country Bovoidea is the main criterion to distinguish its faunas. Considering micromammals, the Greek-Iranian province is represented by the poor fauna of Lefkon (Greece, LK in Fig. 6) including exclusively eastern genera. The faunas of Central Europe appear as clearly different from the Iberian ones and are characterized by a still diverse fauna including several Gliridae and Sciuridae, plus some mice, beavers and cricetids with prismatic teeth. The faunas from Dorn Dürkeheim (Germany) and Sümeg (Hungary) (DDURK (GE) and SUG (HU) in Fig. 6) may appear so different because of their very diverse faunas including many forest-adapted genera. The rodent faunas of the Iberian province are very poor and characterized by the murids *Parapodemus*, *Occitanomys* and *Huerzelerimys*. Minor components of the fauna include modern cricetids (Cricetinae of the genus *Kowalskia*) and the Cricetodontini (*Hispanomys*). A new fourth group appears, the Tusco-Sardinian province, which presents bizarre insular faunas (for recent synthesis see Rook et al. 1999; Moyà-Solà et al. 1999a; Bernor et al. 2001) and cluster apart of all other sites.

The highest degree of similarity in macromammal faunas from the Greek-Iranian and the Iberian provinces is attained at MN 12 (Fig. 7), when eastern faunas attain their maximum range. However, some differences still exist, and the bovid diversity in Western Europe is much lower. The insular fauna from the Tusco-Sardinian province continued evolving isolated from the continent. The micromammal sites cluster shows the same provinces recognized in MN 11, Iberian rodent faunas being still very poor. The Central European record disappears at MN 11, so the persistence of a Central European province cannot be confirmed.

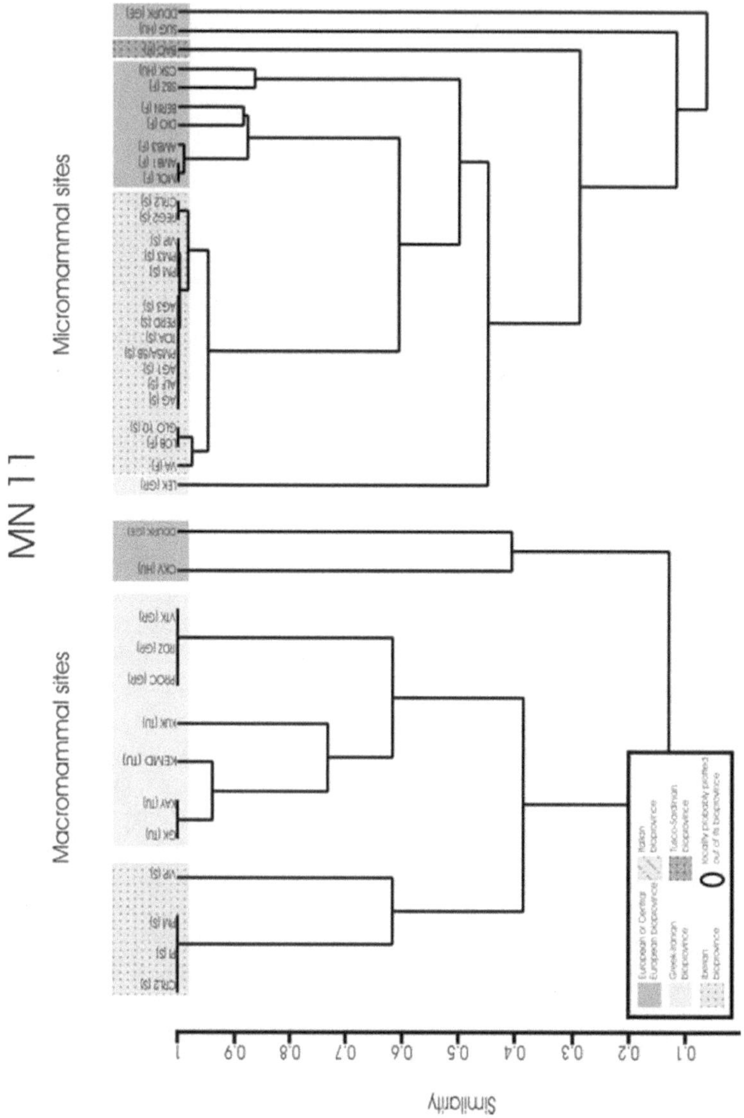

Fig. 6. Dendrogram displaying the results of cluster analysis for MN 11 micro and macromammal sites. Key to locality codes is given in the appendix, whereas key to country codes is given in Table 2

The situation is going to change in the Messinian (MN 13, Fig. 8) and provinces will increase their differences. The macromammal cluster clearly shows three provinces: Iberian, Greek-Iranian and a new Italian province. The Iberian province is characterized by its low diversity, except in the case of the locality of Venta del Moro (VMOR (S) in Fig. 8) which is slightly differenced from the others. This site records the dispersal into Spain of the canids of the genus *Nyctereutes* and camelids (*Paracamelus*) from North America. The relatively poor fauna from Gravitelli (Italy, GRA (IT) in Fig. 4) may explain well why it appears so close to Spanish sites. The Greek-Iranian province is only represented by Greek sites with a diverse bovid and giraffid fauna inherited from the Turolian. The Italian faunas appear rather different from the ones of the remaining sites. The Tusco-Sardinian province and its bizarre insular faunas disappeared when Italy became part of the continent during the Messinian, so the differences observed are not an insularity effect. The Italian province includes three sites (four if we consider that Gravitelli is not plotted in its province) showing a clear Central European influence, so they include many taxa present in Central Europe during MN 11, but absent from rest of provinces since MN 10 (*Tapirus*, *Euprox*, Moschidae) as well as the exclusively Central European proboscidean *Zygolophodon*. The macromammal community is completed with few immigrants from other areas such as *Mesopithecus* and elements that are going to characterize the Pliocene (*Parabos*, *Korynochoerus* (= *Propotamochoerus*), *Nyctereutes*). The presence of many MN 11 relicts may support the existence of a Central European province at least until the end of the Miocene. Many genera from Central Europe would have arrived into Italy when it became part of main land. The micromammal faunas also show the existence of an Italian province, characterized by the presence of some Gliridae (*Muscardinus*), beavers and mice. By MN 13 the rabbits and hares (Leporidae) appear in all micromammal faunas of Europe. The Leporidae evolved in North America and dispersed into Eurasia during the Messinian, but they did not replace the Ochotonidae (the Lagomorph family including pikas) since the Pleistocene. It should be noted that the French micromammal fauna of Lissieu (LIS (FR) in Fig. 8) includes a considerable diversity of cricetids with prismatic teeth, Muridae, Eomyidae, Zapodidae and dormice, that seems closer to the Italian sites than Spanish ones. However the assignment of this site to the Italian province is doubtful. A Greek-Iranian province is also recognized by the cluster, characterized by the absence of some Western mice genera and by the presence of gerbils, beavers, ground and flying squirrels, absent in most of Western sites.

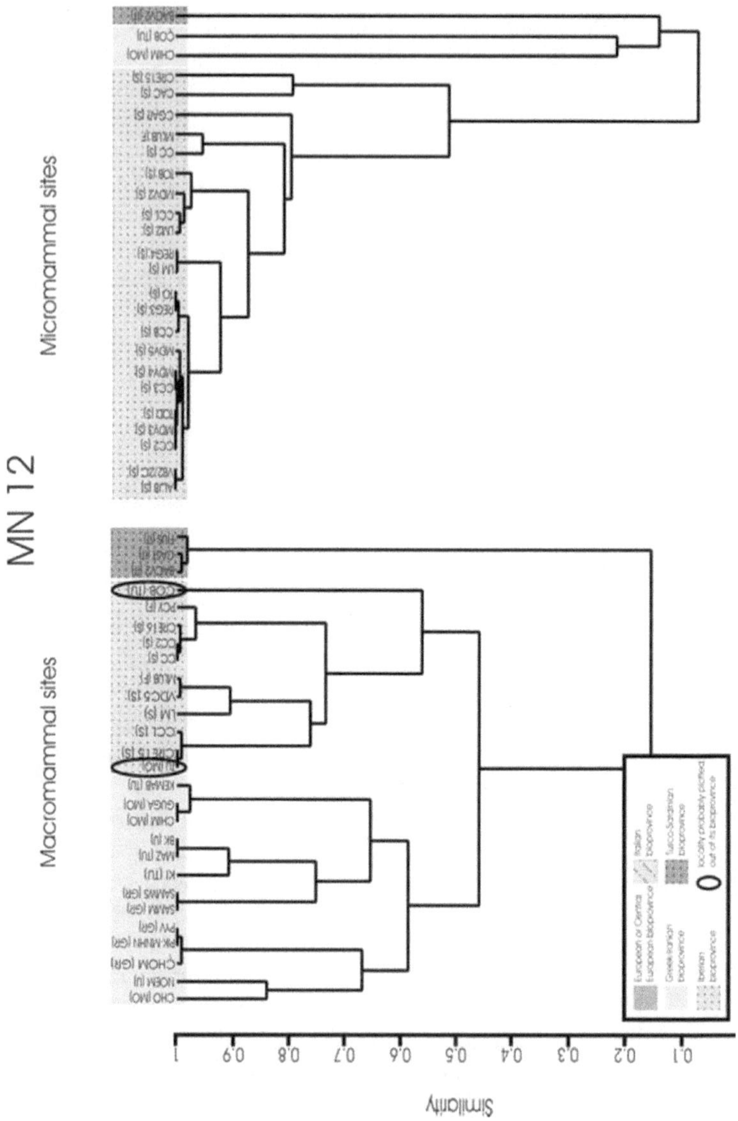

Fig. 7. Dendrogram displaying the results of cluster analysis for MN 12 micro and macromammal sites. Key to locality codes is given in the appendix, whereas key to country codes is given in Table 2

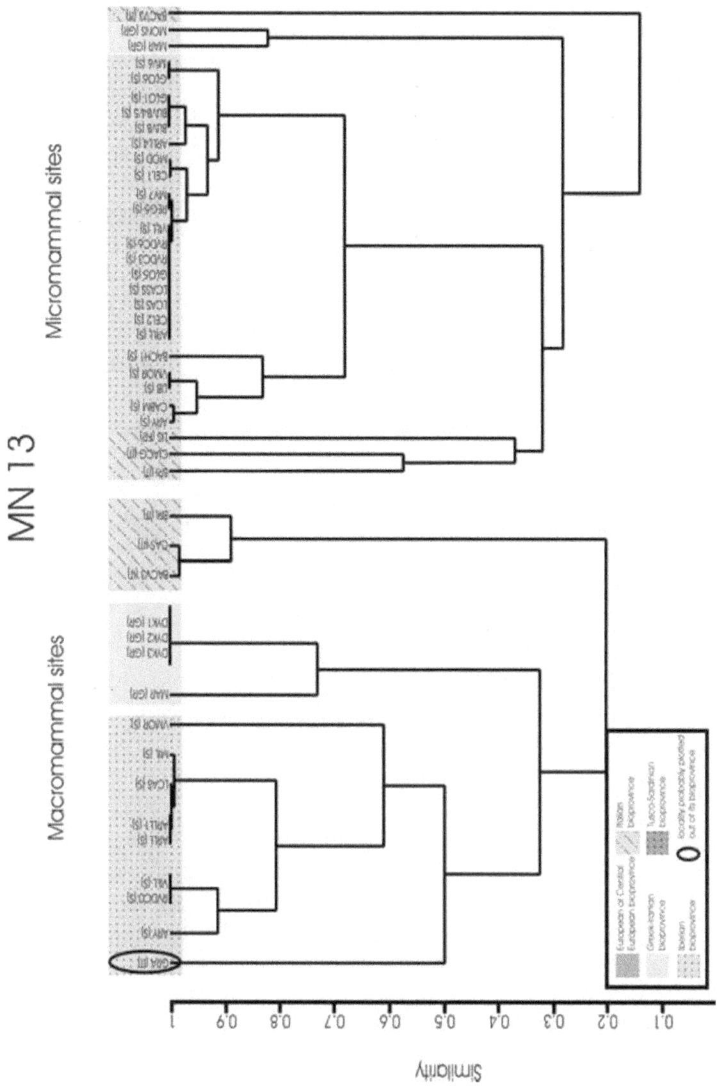

Fig. 8. Dendrogram displaying the results of cluster analysis for MN 13 micro and macromammal sites. Key to locality codes is given in the appendix, whereas key to country codes is given in Table 2

The Iberian bioprovince still persists but it is divided in two biogeographic regions (Levant and Central Iberia). Levant sites (Venta del

Moro, Casablanca M, Librilla, etc.) exhibit a diverse fauna including several Muridae, Gerbillidae (of Asian or African origin, see Agustí 1989; Geraads 1998) ground squirrels and (in a few cases) beavers. The sites from central Spain are characterized by poor faunas dominated by mice (*Castillomys*, *Stephanomys*, *Apodemus*, *Occitanomys*, *Castromys*) and very hypsodont endemic Cricetodontini (*Ruscinomys*), while beavers and squirrels are very rare and gerbils are absent.

9.5.2 Diversity

We have estimated diversity as well as origination and extinction rates for the whole study area and for selected squares. We have only selected squares containing a moderately rich mammal record covering at least five MN zones without temporal gaps in the range. Only three squares suit the criteria (see Fig. 1 and Table 1): 12, 13 and 24. Thus we could only consider Northern Spain plus Southern France and Central Europe. In order to include SE Europe we have summed squares 8 and 9 to obtain a moderately rich macromammal record ranging from MN 6 to MN 12. However, we would like to compare areas of the same surface, so if we take two squares for SE Europe, we would have to take also two squares for the other areas compared. We have summed squares 12 and 13 and obtained a continuous record including both macromammals and micromammals covering the whole time span. Note that squares 12 and 13 include most of European localities in every MN zone. We have also considered square 24, and added the adjacent square 25. This results in a rather complete Central European micro and macromammal record that lasts until the MN 11.

Total number of genera (N_{TOT}) in Europe is shown in Fig. 9A. Macromammals show two diversity peaks: MN 9 and MN 12. These peaks are also associated with an abrupt increase in the number of singletons (N_{FL}), especially in the case of the MN 12. MN 13 shows a moderate decrease although N_{FL} keeps increasing. The lowest values of diversity are attained at MN 6 and MN 10 representing the loss of about 30 genera. MN 10 minimum coincides with a decrease in N_{FL}, whereas this is not the case in the MN 6 lowest. Micromammals increase diversity until their optimum at MN 7 + 8 and afterwards a fall in diversity starts. This loss is moderate until MN 11 but at MN 12 N_{TOT} abruptly falls to its minimum values. Micromammals would have lost close to 30 genera from MN 7 + 8 to MN 12. A moderate recover takes place at MN 13, and N_{TOT} reaches values similar to those of MN 11. Decrease in N_{TOT} coincides with a progressive decrease in N_{FL} and the abrupt fall observed in MN 12 is associated with the absence of singletons in this zone.

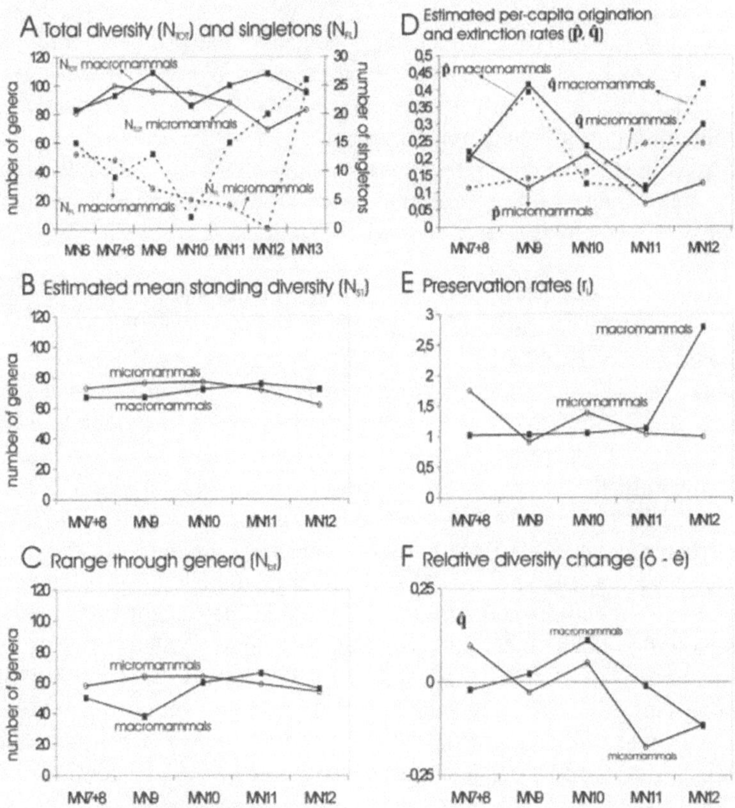

Fig. 9. Diversity and taxonomic and preservation rates plots for the whole study area (see text for details). A) Total diversity (N_{TOT}) and singletons (N_{FL}); B) Estimated mean standing diversity (N_{ST}); C) Range through genera (N_{bt}); D) Estimated per-capita origination and extinction rates ($\hat{p}$, $\hat{q}$); E) Preservation rates (r_i); F) Relative diversity change ($\hat{p}$ minus $\hat{q}$). Plots display results for macromammals and micromammals. For results of correlation test between N_{ST}, $\hat{p}$, $\hat{q}$ and r_i see Table 2

The apparent recovery in MN 13 coincides with a spectacular increase in N_{FL}. If we exclude singletons from our calculations and estimate mean standing diversity, the situation changes considerably (see Fig. 9B). The apparent diversity crisis that affected macromammals at MN 10 disappears, and their diversity gradually increases until a peak in MN 11. Afterwards N_{ST} remains approximately the same. The loss of genera concerning to micromammals appears less marked but still exists. Micromammal diversity reaches its maximum at MN 9 – MN 10, and afterwards decreases until a minimum at MN 12. Thus, a diversity crisis seems to have affected micromammal genera in Europe although it occurred in the Turolian rather than in the Vallesian. Nevertheless, as we have seen, Spanish sites represent most of the European micromammal record, so we should consider to what extent these results are affected by this fact.

Figure 10A show N_{TOT} and N_{FL} of macromammals and micromammals respectively for the selected squares 8 + 9 (labeled as Turkey), 12 + 13 (labeled as SW Europe), and 24 + 25 (labeled as Central Europe). In SW Europe, N_{TOT} of micro and macromammals has its higher values at MN 7 + 8 and MN 9 respectively. Afterwards diversity starts a moderate decrease in the case of macromammals while micromammals loose genera faster. Macromammal N_{TOT} keeps falling until MN 13 with a slight recover at MN 12. Micromammals had lost about 30 genera since MN 7 + 8, but an important recover took place at MN 13 and N_{TOT} attained values similar to those of MN 9. Increases in N_{TOT} in micromammals are associated with increases in the number of singletons, while in the case of macromammals this pattern is not so clear and only the extreme low values of N_{TOT} at MN 11 coincide with extreme low values of N_{FL}. When considering N_{ST} in SW Europe (see Fig. 10B) the picture changes only a little. Macromammals loose diversity gradually from MN 7 + 8 until they reach a minimum at MN 10, afterwards there is a slight recover at MN 11 to continue falling during MN 12. Micromammals were very diverse at MN 9 and afterwards diversity decreased gradually at higher rate than macromammals during the whole time span.

The existence of a diversity crisis starting at MN 10 (or maybe earlier in the case of macromammals) seems clear in SW Europe. This crisis affected all land mammal communities but its effects were probably more severe on micromammals. Central Europe and Turkey show lower values of diversity than SW Europe, but this is almost certainly and effect of the quality of the record, which is clearly best in SW Europe. Central European record unfortunately disappears in MN 11, but if the Vallesian Crisis at MN 10 affected all Europe it should be detected. This crisis is apparent, and N_{TOT} of macromammals shows an abrupt decrease in MN 10, while micromammals reduced their diversity earlier, at MN 9. In both cases there is a quick recover afterwards.

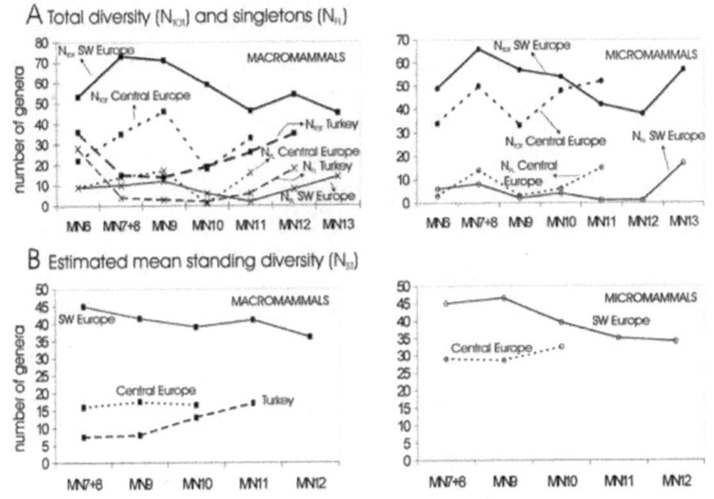

Fig. 10. Diversity calculated for selected squares: SW Europe (squares 12 + 13), Central Europe (squares 24 + 25) and Turkey (squares 8 + 9) (see text for details). A) Total diversity (N_{TOT}) and singletons (N_{FL}); B) Estimated mean standing diversity (N_{ST}). Plots display results for macromammals and micromammals. For results of correlation test between N_{ST} and r_i see Table 2

Changes in N_{TOT} are clearly associated with changes in N_{FL} during the same zones. Moreover the decrease in the number of singletons observed at MN 9 and MN 10 is related with a decrease in the number of localities relative to previous time spans. Thus, these apparent falls are most probably an effect of completeness and should be removed if singletons are excluded from the analysis. This is what we have done by computing N_{ST} (see Fig. 10B), and results evidence these undesirable effects of a poor record.

Macromammal diversity remains more or less stable, while micromammals even increase their diversity during MN 10, showing opposite trends than SW Europe. We could only consider macromammals in the case of Turkey, but the trends observed in this area are very different from those seen in SW and Central Europe. N_{TOT} has high values at MN 6 but abruptly falls to its minimum at MN 7 + 8, afterwards diversity gradually recovers until a new maximum at MN 12. Note that the trends show by N_{TOT} clearly follow those of N_{FL} (Fig. 10A) but in this case they are not associated with a significant decrease in the number of sites. Removing the singletons we obtain the same pattern: diversity in Turkey is low during MN 7 + 8 and MN 9 and afterwards gradually increases. It seems that a diversity crisis could have occurred in Turkey, but it happened during the late Aragonian, about 3 Ma earlier than in SW Europe.

We have tested for correlation between N_{ST} and preservation rate (r_i) in Europe and the selected areas (see Figs. 9E and 11C respectively) using Kendall's τ (see Table 3). In every situation the effects of r_i over N_{ST} are not statistically significant but the probability of correlation is high (about 0.4 in all cases). When considering the whole macromammal record in Europe this probability increases to 0.9. So in any case effects of preservation and quality of the record should be taken on account.

9.5.3 Origination and Extinction rates

Figure 9C shows the number of genera that range through the entire MN zone (N_{bt}) for all Europe. Macromammal N_{bt} is surprisingly low during MN 9, and quickly recovers afterwards. Micromammal N_{bt} decay gradually from MN 10 onwards. These patterns are also observed in all squares (results not presented), and this fact points to the existence of an important macromammal turnover at MN 9 in contrast to a more gradual micromammal turnover starting at MN 10. Figure 9D shows origination and extinction rates for Europe, and both rates increase spectacularly during MN 9 in the case of macromammals. Then an important decrease in both rates follows and they rise again in MN 12, although this time extinctions outnumber originations and reach their higher values. Micromammal originations reach their maximum values at MN 7 + 8 and MN 10, but there is a first minimum at MN 9. Afterwards a gradual increase in extinctions occurred until MN 11, when there is an abrupt rise associated with a decrease in originations. In MN 12 extinctions stopped their increase and the origination rate rose. We have tested for correlation between rates and r_i (see Fig. 9E and Table 3) in every situation and results show that both variables are not correlated in any case. For macromammals in both situations Kendall's τ is 0, so variables are totally uncorrelated.

Table 3. Results of correlation test between standing diversity (N_{ST}), origination ($\hat{p}$) and extinction ($\hat{q}$) rates with preservation rate (r_i). Kendall's τ and probability of being uncorrelated is given for macromammals and micromammals in each area and the whole continent (see text for details)

Area		N_{ST}		ô		ê	
		τ	p(unc.)	τ	p(unc.)	τ	p(unc.)
Europe	macromammals	0.67	0.10	0.00	1.00	0.00	1.00
	micromammals	0.10	0.79	0.60	0.14	-0.20	0.62
SW Europe	macromammals	-0.20	0.62	0.20	0.62	0.40	0.33
	micromammals	-0.20	0.62	0.60	0.14	0.00	1.00
Central Europe	macromammals	-0.33	0.60	0.33	0.60	0.33	0.60
	micromammals	0.00	1.00	0.33	0.60	1.00	0.00
Turkey	macromammals	0.18	0.71	-1.00	0.00	0.33	0.50

However, these surprisingly good results are probably an effect of the few points included in the calculations: the fewer points the test finds perfect no correlation if the two variables are not very correlated. On the contrary if variables are strongly correlated, few data points will result on perfect positive or negative correlations and an absolute value of 1 for τ. In our particular situation we assume that our variables are nearly uncorrelated. In the case of micromammals there is no statistically significant correlation between $\hat{p}$ and r_i but results are not so good and probably origination peaks of MN 7 + 8 and MN 10 are partly due to an increase in r_i. In the case of micromammal $\hat{q}$ correlation is negative ($\tau = -0.2$). If variation in $\hat{q}$ was dominated by changes in r_i, τ should be large and positive, so in this last situation we also conclude that micromammal preservation and extinction rates are nearly uncorrelated.

Origination rates have their maximum values at MN 10 for SW Europe and Turkey in the case of macromammals (Fig. 11A) and decrease to a minimum in MN 11. A moderate increase is seen in SW Europe during MN 12. Central Europe shows a peak in originations at MN 9 coinciding with the peak seen when considering Europe as a whole, which otherwise is not evident in Turkey and SW Europe. The existence of this peak of macromammal $\hat{p}$ in the European record is due to the existence of high rates in Central Europe plus moderately high rates in Turkey and SW Europe. The decrease seen in MN 11 is mostly due to the decay in $\hat{p}$ seen in these two areas during this zone. Micromammal $\hat{p}$ reached its maximum values at MN 7 + 8 in SW Europe, and decrease significantly during MN 9. Finally it recovered at MN 10. Central Europe shows the same trends than lower latitudes but the increase in originations in MN 10 was more marked. The Central European record disappears at MN 11 but in Spain data show that $\hat{p}$ decreased during the following intervals.

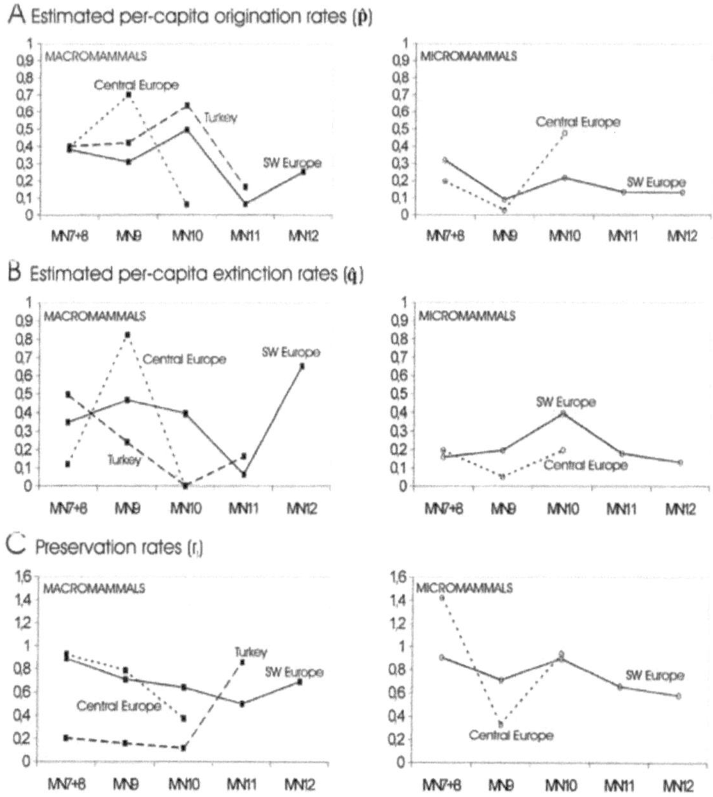

Fig. 11. Taxonomic and preservation rates calculated for selected squares: SW Europe (squares 12 + 13), Central Europe (squares 24 + 25) and Turkey (squares 8 + 9) (see text for details). A) Estimated per-capita origination rate ($\hat{p}$); B) Estimated per-capita extinction rate ($\hat{q}$); C) Preservation rates (r_i). Plots display results for macromammals and micromammals. For results of correlation test between $\hat{p}$, $\hat{q}$ and r_i see Table 2

The pulse in macromammal $\hat{p}$ seen in Europe in MN 9 (Fig. 9D) is associated with a pulse in $\hat{q}$. As Fig. 11B shows, this pulse in $\hat{q}$ only occurred in Central Europe, and only a moderate increase in $\hat{q}$ took place in SW Europe. $\hat{q}$ decreased in Turkey. However, the values of this rate were rather high in all regions, so they result in a peak when the whole of Europe is considered. $\hat{q}$ fell to 0 in Turkey and Central Europe during MN 10, while in SW Europe there was only a slight decrease. In SW Europe extinctions have their minimum at MN 11, when Turkish macromammal $\hat{q}$ increase again. In MN 12 $\hat{q}$ increase abruptly in Spain having a maximum of 0.65. Note that most of macromammal data from MN 12 and MN 13 are from central Spain, so the increase in $\hat{q}$ seen in Europe in MN 12 partly results from the local increase in $\hat{q}$ in Spain during this interval. In the case of micromammals (Fig. 11B) plots for Central Europe $\hat{q}$ are depressed at MN 9, and in MN 10 rise to approximately the same value they had before. In SW Europe there was an abrupt increase in $\hat{q}$ in MN 10 and a decrease immediately afterwards. The increase in $\hat{q}$ seen in Europe during MN 11 and MN 12 may be attributable to the absence of other micromammal records in other areas than SW Europe. As in previous calculations we have tested for correlation between $\hat{p}$, $\hat{q}$ and r_i using Kendall's τ (see Table 3). Correlation is not statistically significant in any case except for macromammal $\hat{p}$ in Turkey (0.04) and micromammal $\hat{q}$ in Central Europe where test detects perfect positive correlation ($\tau = 1$). In Turkey correlation is negative ($\tau = -1$), and if changes in $\hat{p}$ where related to variations in r_i these value should be high and positive, so significant correlation has to be ruled out.

9.5.4 Biogeographical stability and its consequences during the late Miocene mammal turnover

The Spanish record is the most complete mammal record for the late Miocene (Alba et al. 2001, calculated the completeness of the Spanish Neogene mammal record in the means of Foote and Raup (1996) and showed that completeness was even higher than some marine records), so this fact is going to influence the whole European record. The graphs for diversity and rates in Europe (Fig. 9) resemble those of southwestern Europe (Figs. 10 to 12), especially in the case of micromammals. Thus, events taking place only in the Iberian Peninsula may also be expressed when considering Europe as a whole.

There are three main bioprovinces maintained during most of the time span: Iberian (plus southern France), Central European and Greek-Iranian provinces (see Figs. 2-8). The existence of the Greek-Iranian province is well known (Tobien 1967), but the division of western Europe in two bioprovinces is a new point of view (althoug Fortelius et al. 1996) recognized some degree of differentiation in southwestern Europe during the late Miocene). During the middle Miocene the two western provinces were not very different and they even may constitute a single province characterized by a high diversity of forest-adapted mammals. Levels of diversity during MN 7 + 8 (see Fig. 10B) are the highest of the considered period (the differences of diversity between central and southwestern Europe are probably an artifact of completeness). In contrast the faunas from Turkey exhibit low levels of diversity (Fig. 10B).

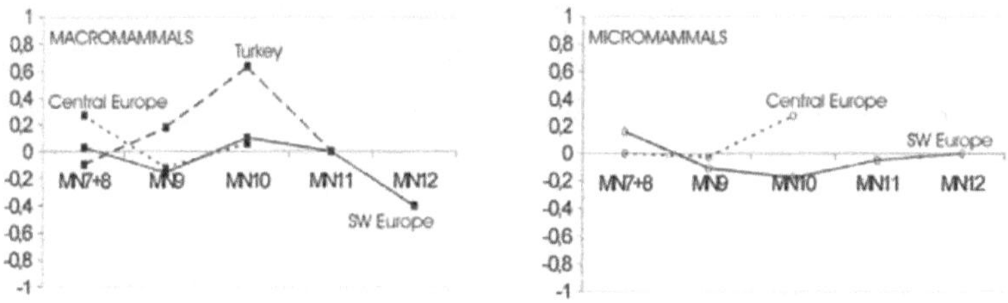

Fig. 12. Relative diversity change ($\hat{p} - \hat{q}$) calculated for selected squares: SW Europe (squares 12 + 13), Central Europe (squares 24 + 25) and Turkey (squares 8 + 9) (see text for details)

The Iberian province is hardly distinguished from the Central European one in the macromammal cluster (Fig. 2), evidencing a similar environment in both areas. During the late middle Miocene the Central European paleovegetation record shows a progressive decrease in evergreen woody taxa and an increase in deciduous ones, resulting in associations comparable to modern mixed mesophytic forests (Kovar-Eder 2003). The progressive cooling during the early Middle Miocene led to the disappearance of megathermic taxa and the *Avicennia* mangrooves of southwestern Europe (Bessedik et al. 1984). Palynological data from the Duero Basin (Spain) dated as MN 7 + 8 (after García-Moreno 1987) indicates the presence of a moderately open woodland and also wet areas including many hydrophilous elements (*Typha*, *Sparganium*, *Magnolia*) (Rivas-Carballo and Valle 1986). Thus, the environment was relatively warmer and more opened in Spain than in Central Europe. The vegetation of the Greek-Iranian province was more similar to Central Europe during the early / early-middle Miocene (on the Balkan Peninsula and Evia Island, see Kovar-Eder 2003). As the Western Mediterranean it included some sclerophyllous taxa together with some evergreen genera present in Central Europe in preceeding periods (*Platanus neptuni*, *Calocedrus*). So the biogeographic pattern shown by mammals (see biogeographical map of the Serravallian in Fig. 13) is also observed when the vegetation is considered. Fortelius et al. (2002, 2003) used mean hypsodonty in Eurasian macromammal sites as a proxy of mean annual precipitation. Their maps for the Middle Miocene show the existence of two areas of different mean hypsodonty corresponding to the two well known Greek-Iranian and Western European province, while the Iberian province cannot be distinguished (Fortelius et al. 2003).

During MN 7 + 8 taxonomic rates in the case of macromammals were moderate in Western Europe ($\hat{q}$ in Central Europe is very low, see Fig. 11B), while in the case of micromammals were rather high. Originations exceeded extinctions resulting in an increase in micromammal diversity (Fig. 10B). This fact may be explained because of the minor environmental change affecting forest environments all over Europe. These small changes would have created new microhabitats and fragmented the populations favoring ecologically induced genealogical events (sensu Vrba 1985). Regarding the macromammals, habitat heterogenity was not so important as to induce any turnover. The macromammal extinction rates in Turkey are considerably high (Fig. 11B), while originations (Fig. 11A) equal those of the other provinces. This fact explains the relatively low diversity recorded in the area, and may reflect some opening of the mid Miocene forests, which led to an important turnover in herbivore faunas.

The beginning of the late Miocene (MN 9) is characterized by a progressive decrease in the $\hat{p}$ rates in all the provinces (except for macromammals in Central Europe, see Fig. 11 A), while extinctions increased in central and southwestern Europe (Fig. 11 B). The moderate decrease in $\hat{p}$ and the increase in $\hat{q}$ in southwestern Europe (mainly Spain) resulted in a decline in macromammal diversity. This did not occur in the case of micromammals, which maintained their diversity (althoug extinctions were also high). The macrofloral remains recovered in the Vallès-Penedès Basin and adjacent areas (Spain) records the appearance of broad-leafed deciduous taxa such as *Acer tricuspidatum* and *Platanus leucophylla* (Sanz de Siria 1993, 1994). However some subtropical genera (*Ficus*, *Sabal*) survived during the latest MN 9 in the same area. The environment in Spain may have been less forested during the lowermost Vallesian (Fortelius et al. 1996; Casanovas-Vilar and Agustí submitted). A brief recovery in humidity is seen prior to the early/late Vallesian boundary (Van der Meulen and Daams 1992; Casanovas-Vilar and Agustí submitted).

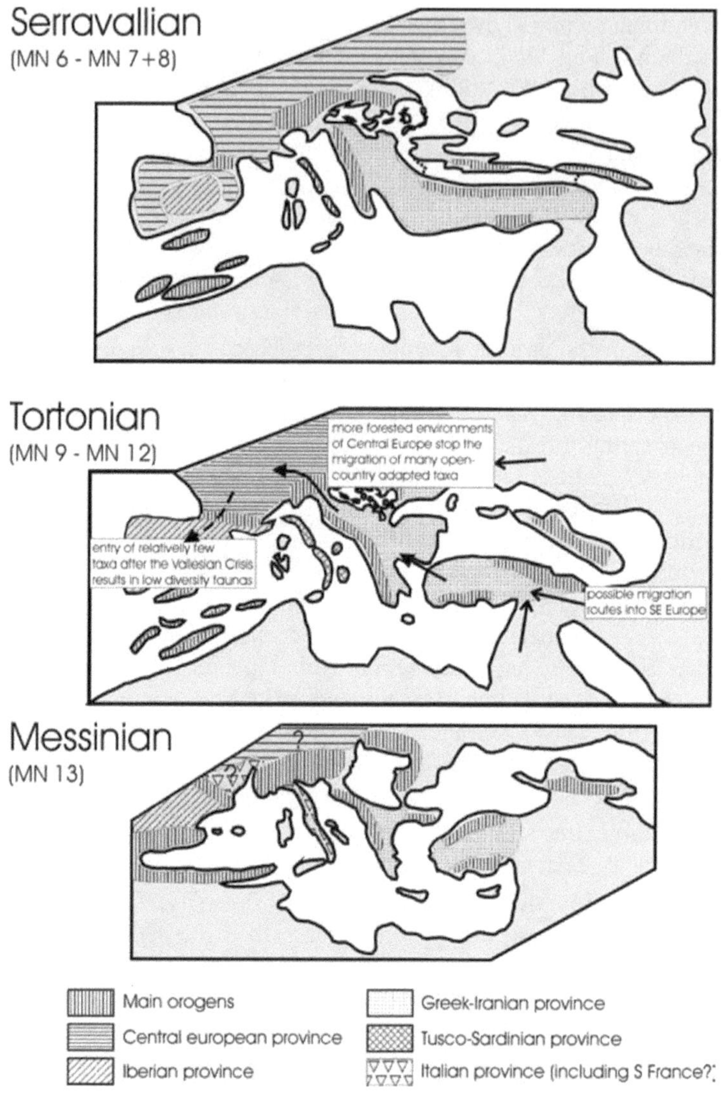

Fig. 13. Paleogeographical maps for the Serravallian, Tortonian (after Rögl 1999, 2001; modified) and Messinian (after Ghetti et al. 2002; modified) showing evolution of the Mediterranean Basin during the studied time-span. Distribution of main orogens and the area covered by identified bioprovinces is also shown (see text for details)

A clear effect of the reduction of the forest cover during the earliest Vallesian is the disappearance from the Spanish record of many forest dwelling rodents, while they are still present in Central Europe where dense mixed mesophytic forests persist (Bernor et al. 1987) (many genera will temporally re-appear by the latest MN 9 in the Vallès-Penedès Basin). The beginning of the fragmentation of forest cover in Spain started at MN 9, coinciding with a major isotopic event (Mi 5) (Miller et al. 1991; also recorded in the Mediterranean by Turco et al. 2001) and a sea level fall of nearly 100 m (Haq et al. 1987). These events, as well as the Tibetan uplift would have produced a high-latitude cooling and a mid-latitude drying, with an increase in wet / dry seasonality in the Mediterranean regions. The increase of seasonality would have favored the development of opened areas, especially in the inner Iberian Peninsula (Casanovas-Vilar and Agustí submitted). The habitat of some micromammals would have been reduced to "patches" of the old middle Miocene forests, while other genera inhabited the new opened areas. In Turkey extinction rates kept lowering while originations remain stable. Bovid diversity continued increasing with both browsing and grazing adaptations being represented, indicating the beginning or the break-up of the forest environments in the region (Bernor et al. 1996).

The Vallesian Crisis is though to coincide with early / late Vallesian boundary (or the boundary between MN 9 and MN 10), but the results (Fig. 10 B) do not show a decrease in diversity in all areas. In fact, diversity only decreased in the Iberian province, especially in the case of the small mammals. In Central Europe remained stable (with an slight increase in the case of micromammals) and in Turkey diversity start increasing. In southwestern Europe the important micromammal turnover implies a rising in $\hat{p}$ and $\hat{q}$, but the increase in the number of extinctions was higher (Fig. 11 A and B). The per-capita macromammal origination rates in this region rose following the pattern seen in Turkey, so large mammal diversity stopped decreasing. The Vallesian Crisis also coincides with an isotopic event (Mi 7 in Miller et al. 1991), although it is not so marked as the preceding Mi 5 and Mi 6. However, during the late Vallesian global temperature kept decreasing while the changes in atmospheric circulation due to the Tibetan uplift (Kutzbach et al. 1993) intensified wet / dry seasonality at low latitudes. The rodent record from different basins of Spain shows this increase in seasonality that would have led to the onset of the characteristic dry season of the Mediterranean climate (Van Dam and Weltje 1999; Casanovas-Vilar and Agustí submitted). This change in seasonality would have favored the replacement of cricetids by murids in the inner Peninsula (Van Dam 1997).

However, in the northern basins rodent communities were rather different and this seasonality in humidity was accompanied by an increase in the seasonality in temperature (Casanovas-Vilar and Agustí submitted). The late Vallesian floras from the Vallès-Penedès Basin (NE Spain) (Sanz de Siria 1997; Agustí et al. 2003) include a 45 % of deciduous trees (*Acer, Alnus, Fraxinus* and others) and a 15 % of drier, sclerophyllous pre-Mediterranean elements (*Quercus ilex, Quercus praecursor*) confirming the presence of the two kinds of seasonality. However, these floras record the persistence of a "hard core" of evergreen trees (close to 33 %). Pollen from deciduous trees is also the most abundant in the palynological record of the Duero Basin (NW Spain) (Rivas-Carballo and Valle 1986). The "patches" of the Middle Miocene forest would have reduced even more, and the composition of this forest would have changed, evergreen trees being outnumbered by deciduous ones. The forest-adapted mammals would have been confined to this "patches", but finally those became too small and the mammals could not maintain their populations, thus being locally extinct. The effects would have been greater in the small mammals, because of their reduced migration capability, while many large mammals migrated to more suitable environments. The fauna of Dorn Dürkheim (MN 11, Germany) includes a number of forest-adapted forms that migrated from the Greek-Iranian and the Iberian provinces (Franzen and Storch 1999) Figures 10B, 12 show a sharp decrease in micromammal diversity and a moderate decrease and recovery in macromammals. Some macromammals, unable to tolerate a seasonality in temperature would have disappeared. This is the case of the hominoid primate *Dryopithecus* and the pliopithecids. These primates needed whole-year availability of fruits, so they could not migrate to higher latitudes and were trapped in the "patches" of southern forests (Andrews 1992; Fortelius and Hokkanen 2001; Agustí et al. 2003), the combinated effect of the increasing reduction of those and the progressive substitution of evergreen by deciduous trees brought hominoids to the extinction in Europe. The survival of the hominoid *Oreopithecus bambolii* in the Turolian faunas of Tuscany and Sardinia is consequence of the insular conditions of this geographic area at that time. The peculiar trophic adaptations of this primate, consequence of the peculiar selective pressures of insular ecosystems, allowed this hominoid to tolerate the effect on vegetation trophic resources of seasonality (Köhler and Moyà-Solà 1996; Moyà-Solà et al. 1999b; Köhler and Moyà-Solà 2003).

The same events that caused the Vallesian Crisis in southwestern Europe affected extensive areas of the Old World, but their effects on the land mammal communities were different. In Turkey diversity started increasing as a result of a minimum in extinction rates (Fig. 11B) and a maximum in originations (Fig. 11A). This fact reflects the radiation of open-country bovids and giraffids that is going to characterize the Turolian in the Greek-Iranian province. The reduction of forest in southeastern Europe, had opposite effects than in the Iberian Peninsula, allowing the migration of many taxa of Asian or African origin. Why did many of those genera not arrive into Spain? Migrations of some large mammals into all European regions, such as the hipparionine horses or the bovid *Tragoportax*, were quick events taking no more than 1 million years length. In contrast, many eastern open-country mammal taxa failed to disperse into Western Europe (Koufos 2003). Large areas of southwestern and Eastern Europe (De Bonis et al. 1992) were covered by open woodlands, then why we do not find the same open-country mammals in both areas? One hypothesis might be that some orogens such as the Alps or the Pyrenees acted as important barriers for migration. However, palynologic data have shown that the high of these reliefs was much lower during the Miocene (Pérez-vila et al. 2001), thus they would not have constituted effective barriers. It seems more logical that the persistence of humid and forested environments in Central Europe (Fortelius et al. 2002, 2003) may have acted as an ecological filter, so only ubiquist taxa (for instance *Tragoportax*, *Microstonyx*) could cross it and successfully arrive into the Iberian Peninsula (see the biogeographical map for the Tortonian on Fig. 13). This situation would persist during the Turolian.

By the Turolian, mammal diversity in the Iberian province keeps declining (Figs. 10, 12), this was first caused by a decay in $\hat{p}$ and $\hat{q}$, but in MN 12 extinctions exceed originations (Fig. 11A, B). Again this is an effect of the impossibility of eastern faunas to disperse to southwestern Europe (only few taxa as for example sivatherine giraffids, some species of *Gazella* and *Protoryx*, probably ubiquist taxa, reached the Iberian Peninsula during the Turolian), plus the extinction of the last forest-dwelling relicts (*Dorcatherium*, *Micromeryx*, *Lucentia*). It is surprising that an endemic open-country fauna did not develop in the Iberian Peninsula during the Turolian. This could be because macromammal populations were not deeply fragmented, so ecologically induced genealogical events (as exposed by Vrba 1985) did not occur. However, the situation was not the same in the case of micromammals, and a characteristic fauna of mice and hypsodont cricetids developed in the inner basins (although it was not very diverse). The macromammal clusters on Figs. 6 and 7 show an increasing degree of similarity between the Iberian and the Greek-Iranian province during the Turolian. The eastern open-country faunas, covered their maximum geographic range during MN 12. The high similarity results may reflect some degree of aridification of

Central Europe, allowing the dispersal of more eastern herbivores than in preceding periods. Cerling et al. (1997) have provided impressive isotopic evidence for a global vegetational change starting at 8-7 Ma (thus coinciding with the Turolian) and given by the transition from C3 to C4 vegetation. This event presumably occurred earlier at lower latitudes and, although there are no data, C3 vegetation might have appeared in Central Europe by the mid Turolian.

Restoration of the level of provincial diversity is observed again by the end of the Turolian (MN 13, see Fig. 8) when an Iberian, Greek-Iranian and a new Italian province is observed. Italy was a small archipelago during the lower and middle Turolian inhabited by bizarre insular faunas. The connection to main land during the MN 13 led to the extinction of these faunas, and the migration into the new peninsula of faunas showing clear Central European affinities (see previous section). The Messinian (MN 13) records an important recovery of the environmental conditions in the Mediterranean. Floras from the lower part of Komnina Formation (Ptolemais Basin, Greece) generally indicate continuously humid and warm climatic conditions (Kloosterboer-van Hoeve et al. 2000). The early Messinian deposits of the Velona Basin (central Italy) have yielded a palynological record interpreted as unstable shallow lacustrine/marshy environment, rich in aquatic vegetation, surrounded by swamps dominated by Taxodiaceae and uplands dominated by warm temperate deciduous forest (Ghetti et al. 2002). However, the Iberian Peninsula is still characterized by a low diversity of macromammals (including few immigrants, such as *Paracamelus* and few African forms such as *Hexaprotodon* and *Macaca*), suggesting that the Central European forest filter kept working until the Pliocene (see the biogeographical map for the Messinian on Fig. 13). Fortelius et al. (2002, 2003) have shown the persistence of a high mean annual precipitation zone in Central Europe during the Pliocene, which would explain the maintenance of the forest filter. However, Iberian micromammal fauna was enriched with the entry of some immigrants of Asian or African origin, during the "Messinian Salinity Crisis". This event led to the dessecation of the Mediterranean because of the tectonic closing of its western gateway during the latest Miocene.

Although open, savanna-like environments persisted in southeastern Europe, no new occurrences are recorded in Greek sites by MN 13. Furthermore, many of the taxa that characterized the Turolian in the province (the giraffids, *Adcrocuta*, *Choerolophodon*, *Microstonyx*, *Tragoportax*, *Prostrepsiceros*) will not cross the Miocene / Pliocene boundary and will be replaced by a new macromammal comunity including *Parabos*, *Sus*, *Paracamelus* and *Korynochoerus* (= *Propotamochoerus*) amongst others (Koufos 2003).

9.5.5 Epilogue: This view of the European late Miocene

The several climatic and phisiographic events that punctuated the late Miocene (Mi 6, Mi 7 peaks, Himalayan and Tibetan uplift, intensification of seasonality, decrease of mean global temperatures because of first Artic glaciations), had different effects depending on the area. For instance, the Vallesian Crisis (as defined by Agustí and Moyà-Solà 1990) was not a continent-wide event, and it only affected southwestern Europe. In the Iberian bioprovince we see an evident decrease of mammal diversity from MN 9 to MN 10, which is especially evident in the case of micromammals (Fig. 10B). However mammal diversity started to decline earlier (see Fig. 12), in MN9, when extinctions exceed originations (Fig. 11A, B). Then the crisis is not an abrupt event marking a boundary between early (MN 9) and late Vallesian (MN 10), instead it reflects a more gradual process of environmental deterioration. The main climatic change was the onset of seasonality in both humidity and temperature. This crisis coincided with an increase in diversity in Turkey, with the entry of many open-country Bovoidea and Giraffidae, while in Central Europe is little changed, and micromammals even increased their diversity (Fig. 10B). Fortelius et al. (1996) and Franzen and Storch (1999) identified the Vallesian Crisis in Central Europe, but there is no evidence of it when mean standing diversity is considered, so we think that this identification is an undesirable effect of using absolute countings of taxa in their calculations (as we have shown, the number of records in Central European MN 10 is very low compared with MN 9 and MN 11).

The only phenomenon we detect is a diversity crisis in the Iberian province product of its geographical position at the southwestern margin of Europe, and the persistence of forested environments in Central Europe acted as an ecological filter for eastern open-country mammal faunas (which would have prospered in the Iberian environments, probably fairly close to Greek and Turkish ones, that are situated in similar latitudes, if they had arrived). The Iberian Peninsula did not become a "small species factory" and did not developed an autochthonous open-country herbivore fauna. This situation persisted during the Turolian, thus explaining the low diversity of the Iberian faunas (Fig. 10B). Diversity increased in southeastern Europe as a result of the entry of many eastern taxa. The eastern immigrants attained their maximum geographical range to the west during the Turolian, coinciding with the transition from C3 to C4 vegetation (Cerling et al. 1997), which could have resulted in some opening of the canopy in Central Europe. Provinciality increased again in the Messinian, with a higher segregation of the three provinces. Italy, which had been a small archipelago during most of the Turolian, was joined to the continent and a mammal community with clear Central European affinities replaced the bizarre insular faunas.

As we have shown, climatic changes, correlate well with main faunal changes in Europe. This was not the case in the North American mammal record (Alroy et al. 2000), where some climatic changes are correlated with mammal turnovers while others are not. On the other side of the Atlantic, Central Europe is characterized by overall stability, while on the Mediterranean region climatic changes especially affected mammal faunas. This difference can be explained because of the different geography of both continents, the complex distribution of emerged lands in Europe prevented habitat tracking. The presence of the Paratethys at the East, and the Mediterranean at the South, increased the probability for taxa of being trapped in one of the Mediterranean peninsulas. The only connection with Africa was through Anatolia (except in the MN 13 interval), so when climate changed, many taxa preferring warm forest environments could not migrate to lower latitudes because they had to cross Central Europe which stopped their dispersal. By its side, Anatolia was invaded for new Asian and African immigrants, resulting in an increase in diversity in that area. The situation in North America is very different, because habitat tracking is not stopped by barriers such as the Mediterranean. The connections of North America with other continents during the late Miocene occurred during important climatic changes (such as the sea level lowering of 100 m at the mid / late Miocene boundary), but were brief and most of the time the continent was isolated, implying the entry of fewer immigrants in the area.

9.5.6 Data availability

The whole database used in this work is available from the first author under request. Additional tables and calculations (diversity, taxonomic rates) are also available.

9.6 Acknowledgements

We thank A. Elewa for giving us the occasion for taking part in this book. I. C.-V. acknowledges the support from Generalitat de Catalunya in the form of a predoctoral grant by DURSI (Departament d'Universitats, Recerca i Societat de la Informació) 2003 FI 00083. This paper has also been supported by project BOS2001-1004 of the Spanish Ministry of Science and Technology, Grup de Recerca de Qualitat SGR2001-00077 of DURSI and RHOI-HOMINID-NSF-BCS-0321893 of the Researching Hominid Origins Initiative. We would like to thank M. Fortelius and his team for developing and maintaining NOW, which was the main source for the database used in this paper. We have used the PAST program in many of our calculations, so we thank Ø. Hammer the effort done to create and improve it. The idea of writing this work arose after the conversation of the first author with Drs. Hans de De Bruijn and A. J. van der Van der Meulen at Utrecht University and I. C.-V. is very grateful to them for indirectly suggesting it. This work does not result from the single work by four people, but from the work of many paleontologists who compiled the data during decades, we wish to thank them all for making it possible.

References

Agustí J (1982) Biozonación del Neógeno continental de Catalunya mediante roedores (Mammalia). Acta Geológica Hispánica, Spain 17 (1-2): 221-226

Agustí J (1989) On the peculiar distribution of some Muroid taxa in the Western Mediterranean. Bolletino della Società Paleontologica Italiana, Italy 28 (2-3): 147-154

Agustí J, Moyà-Solà S (1990) Mammal extinctions in the Vallesian (Upper Miocene). Lecture Notes in Earth Sciences, Springer-Verlag, Germany 30: 425-432

Agustí J, Cabrera L, Garcés M, Llenas M (1999) Mammal turnover and Global climate change in the late Miocene terrestrial record of the Vallès-Penedès basin (NE Spain). In Agustí J, Rook L, Andrews P (eds) Hominoid evolution and climatic change in Europe. Vol. 1: The evolution of Neogene terrestrial ecosystems in Europe. Cambridge University Press, UK: 390-397

Agustí J Cabrera L, Garcés M, Krijgsman W, Oms O, Parés JM (2001) A calibrated mammal scale for the Neogene of Western Europe. State of art. Earth-Science Reviews, Elsevier, The Netherlands 52: 247-260

Agustí J, Sanz de Siria A, Garcés M (2003) Explaining the end of hominoid experiment in Europe. Journal of Human Evolution, Elsevier, The Netherlands 45: 145-153

Alba DM, Agustí J, Moyà-Solà S (2001) Completeness of the mammalian fossil record in the Iberian Neogene. Paleobiology, The Paleontological Society, USA 27 (1): 79-83

Alroy J, Koch PL, Zachos JC (2000) Global climate change and North American mammalian evolution. In Erwin HD, Wing SL (eds) Deep Time:

Paleobiology's Perspective. Supplement to Paleobiology, The Paleontological Society, USA 26 (4): 259-288

Amano K, Taira A (1992) Two-phase uplift of Higher Himalaya since 17 Ma. Geology, The Geological Society of America, USA 20: 391-394

Andrews P (1992) Evolution and environment in the Hominoidea. Nature, Nature Publishing Group, UK 360: 641-646

Bernor RL, Armour-Chelu M (1999) Family Equidae. In Rössner GE, Heissig K (eds) The Miocene land mammals of Europe. K Verlag Dr Friedrich Pfeil, Germany: 193-202

Bernor RL Kovar-Eder J Lipscomb D Rögl F Sen S, Tobien H (1987) Systematic, stratigraphic and paleoenvironmental contexts of first-appearing Hipparion in the Vienna Basin, Austria. Journal of Vertebrate Paleontology, Society of Vertebrate Paleontology, USA 8 (4): 427-452

Bernor RL, Mittmans HW, Rögl F (1993) Systematics and chronology of the Götzendorf "Hipparion" (Late Miocene, Pannonian F, Vienna Basin). Annales Naturhistorisches Museum Wien, Austria 95 (A): 101-120

Bernor RL, Fahlbusch V, Andrews P, De Bruijn H, Fortelius M, Rögl F, Steininger FF, Werdelin L (1996) The evolution of Western Eurasian Neogene mammal faunas: a chronologic, systematic, biogeographic and paleoenvironmental synthesis. In Bernor RL Fahlbusch V, Mittmann H-W (eds) The Evolution of Western Eurasian Neogene Mammal Faunas, Columbia University Press, USA: 449-469

Bernor RL, Fortelius M, Rook L (2001) Evolutionary biogeography and paleoecology of the Oreopithecus bambolii "Faunal Zone" (late Miocene, Tusco-Sardinian Province). Bollettino della Società Paleontologica Italiana, Italy 40(2):139-148

Bessedik M, Aguilar J-P, Cappeta H, Michaux J (1984) Le climat du Néogène dans le sud de la France (Provence, Languedoc, Rousillon), d'après l'analyse des faunes (rongeurs, sélaciens) et des flores polliniques. Paléobiologie continentale, France 14 (2): 181-190

Bobe R, Eck G (2001) Responses of African bovids to Pliocene climatic change. Paleobiology Memoirs, The Paleontological Society, USA supplement to 27 (2): 1-48

Casanovas-Vilar I, Agustí J (submitted) Ecogeographical stability and climate forcing in the late Miocene (Vallesian) rodent record of Spain.

Cerling TE, Harris JR, MacFadden BJ, Leakey MG, Quade J, Eisenmann V, Ehleringer JR (1997) Global vegetation change through the Miocene / Pliocene boundary. Nature, Nature Publishing Group, UK 389: 153-158

De Bonis L, Bouvrain G, Geraads D, Koufos G (1992) Diversity and paleoecology of Greek late Miocene mammalian faunas. Paleogeography, Paleoclimatology, Paleoecology, Elsevier, The Netherlands 91: 99-121

De Bruijn H, Ünay E (1996) On the evolutionary history or the Cricetodontini from Europe and Asia Minor and its bearing on the reconstruction of migrations and the continental biotope during the Neogene. In Bernor RL Fahlbusch V, Mittmann H-W (eds) The Evolution of Western Eurasian Neogene Mammal Faunas, Columbia University Press, USA: 227-234

De Bruijn H Fahlbusch V Saraç G, Ünay E (1993) Early Miocene rodent faunas from the Eastern Mediterranean area. Part III: The genera Deperetomys and Cricetodon, with a discussion of the evolutionary history of the Cricetodontini. Proceedings Koninklije Nederlandse Akademie van Wetenschappen, The Netherlands B 96: 151-216

Foote M (2000) Origination and extinction components of taxonomic diversity: general problems. In Erwin HD, Wing SL (eds) Deep Time: Paleobiology's Perspective. Supplement to Paleobiology, The Paleontological Society, USA 26 (4): 74-103

Foote M, Raup DM (1996) Fossil preservation and the stratigraphic ranges of taxa. Paleobiology, The Paleontological Society, USA 22: 121-140

Fortelius M, Hokkanen A (2001) The trophic context of hominoid occurence in the later Miocene of Western Eurasia – a primate-free view. In de Bonis L Koufos G, Andrews P (eds) Hominoid evolution and climatic change in Europe. Vol. 2: Phylogeny for the Neogene hominoid primates of Eurasia Cambridge University Press, UK: 19-47

Fortelius M, Werdelin L, Andrews P, Bernor RL, Gentry A, Humphrey L, Mittmann H-W, Viratana S (1996) Provinciality, Diversity, Turnover, and Paleoecology in Land Mammal Faunas of the Later Miocene of Western Eurasia. In Bernor RL Fahlbusch V, Mittmann H-W (eds) The Evolution of Western Eurasian Neogene Mammal Faunas, Columbia University Press, USA: 414-448

Fortelius M, Eronen J, Jernvall J, Liu L, Pushkina D, Tesakov A, Vislobokova I, Zhang Z, Zhou L (2002) Fossil mammals resolve regional patterns of Eurasian climate change during 20 million years. Evolutionary Ecology Research, USA 4: 1005-1016

Fortelius M, Eronen J, Liu L, Pushkina D, Tesakov A, Vislobokova I, Zhang Z (2003) Continental-scale hypsodonty patterns, climatic paleobiogeography and dispersal of Eurasian Neogene large mammal herbivores. In Reumer JWF, Wessels W (eds) Distribution and Migration of Tertiary Mammals in Eurasia. A volume in honour of Hans de Bruijn. Deinsea, Backhuys Publishers, The Netherlands 10: 1-11

Franzen JL, Storch G (1999) Late Miocene mammals from Central Europe. In Agustí J Rook L, Andrews P (eds) Hominoid evolution and climatic change in Europe. Vol. 1: The evolution of Neogene terrestrial ecosystems in Europe. Cambridge University Press, UK: 165-190

Garcés M, Agustí J, Cabrera L, Parés JM (1996) Magnetostratigraphy of the Vallesian (late Miocene) in the Vallès-Penedès Basin (northeast Spain). Earth and Planetary Science Letters, Elsevier, The Netherlands 142: 381-396

Garcés M, Cabrera L, Agustí J, Parés JM (1997) Old World first appearance datum of Hipparion-horses: Late Miocene large-mammal dispersal and global events. Geology, The Geological Society of America, USA 25 (1): 19-22

García-Moreno E (1987) Roedores y lagomorfos del Mioceno de la zona central de la Cuenca del Duero. Sistemática, bioestratigrafía y paleoecología. Tesis Doctoral Universidad Complutense Madrid, Spain unpublished: 1-259

Geraads D (1998) Biogeography of circum-Mediterranean Miocene-Pliocene rodents; a revision using factor analysis and parsimony analysis of endemicity. Paleogeography, Paleoclimatology, Paleoecology, Elsevier, The Netherlands 137: 273-288

Ghetti P, Anadón P, Bertini A, Esu D, Gliozzi E, Rook L, Soulié-Märsche I (2002) The early Messinian Velona basin (Siena, central Italy): paleoenvironmental and paleobiogeographical reconstructions. Paleogeography, Paleoclimatology, Paleoecology, Elsevier, The Netherlands 187: 1-33

Hammer Ø, Harper DAT, Ryan P (2001) PAST: Paleontological statistics software package for education and data analysis. Paleontologia Electronica 4 (1, 4): 1-9

Haq BU, Hardenbol J, Vail PR (1987). Chronology of fluctuating sea levels since the Triassic. Science, American Association for the Advancement of Science, USA 235: 1156-1167

Kappelman J, Duncam A, Feshea M, Lunkka J-P, Ekart D, McDowell F, Ryan T, Swisher III CC (2003) Chronology of the Sinap Formation. In Bernor RL Kappelman J Sen S, Fortelius M (eds) Monograph of the Sinap Formation, Anatolia, Turkey, Columbia University Press, USA: 41-69

Kloosterboer-van Hoeve M, Steenbrink J, Vught Nvan, Hilgen FJ (2000) Refinement of the Messinian APTS from cycle patterns in the lacustrine Lava section. Earth and Planetary Science Letters, Elsevier, The Netherlands 181:161-173

Köhler M, Moyà-Solà S (1997) Ape-like or hominid-like? The positional behavior of Oreopithecus bambolii reconsidered. Proceedings National Academy of Sciences of the USA 94: 11747-11750

Köhler M, Moyà-Solà S (2003) Understanding the enigmatic ape Oreopithecus bambolii. Cour Forsch-Inst Senckenberg, Germany 243: 111-123

Koufos GD (2003) Late Miocene mammal events and biostratigraphy in the Eastern Mediterranean. In Reumer JWF, Wessels W (eds) Distribution and Migration of Tertiary Mammals in Eurasia. A volume in honour of Hans de Bruijn. Deinsea, Backhuys Publishers, The Netherlands 10: 343-372

Kovar-Eder J (2003) Vegetation dynamics in Europe during the Neogene. In Reumer JWF, Wessels W (eds) Distribution and Migration of Tertiary Mammals in Eurasia. A volume in honour of Hans de Bruijn. Deinsea, Backhuys Publishers, The Netherlands 10: 373-392

Kutzbach JE, Prell L, Rudiman WF (1993) Sensitivity of Eurasian climate to surface uplift of Tibetan Plateau. Journal of Geology, University of Chicago Press, USA 101: 177-190

Mein P (1999) The late Miocene small mammal succession from France, with emphasis on the Rhône Valley localities. In Agustí J Rook L, Andrews P (eds) Hominoid evolution and climatic change in Europe. Vol. 1: The evolution of Neogene terrestrial ecosystems in Europe. Cambridge University Press, UK: 140-165

Miller KG, Wright JD, Fairbanks RG (1991) Unlocking the icehouse, Oligocene-Miocene oxygen isotopes, eustacis and margin erosion. Journal of

Geophysical Research, American Geophysical Union, USA 96 (B4): 6829-6848
Moyà-Solà S, Agustí J (1990) Bioevents and mammal successions in the Spanish Miocene. In Lindsay EH Fahlbusch V, Mein P (eds) European Neogene Mammal Chronology, Plenum Press, USA: 357-373
Moyà-Solà S, Quintana J, Alcover JA, Köhler M (1999a) Endemic Island Faunas of the Medirerranean Miocene. In Rössner GE, Heissig K (eds) The Miocene land mammals of Europe. K Verlag Dr Friedrich Pfeil, Germany: 435-442
Moyà-Solà S, Köhler M, Rook L (1999b) Evidence of hominid-like adaptations to precision grip in the hand of Oreopithecus. Proceedings National Academy of Sciences of the USA 96: 56-59
Pérez-Vila M-J, Fauquette S, Suc J-P, Bessedik M (2001) Palynological contribution to estimation of Mio-Pliocene altitude of eastern Pyrenees. In Agustí J, Oms O (eds) Late Miocene to early Pliocene environments and ecosystems. Abstracts of the 2nd EEDEN Plenary Workshop Sabadell (Spain) 15-17 November 2001: 52-54
Raup DM, Crick RE (1979) Measurement of faunal similarity in paleontology. Journal of Paleontology, The Paleontological Society, USA 53 (5): 1213-1227
Rivas-Carballo R, Valle MF (1986) Nuevas aportaciones a la palinología del Terciario de la Cuenca del Duero. Torremormojón (Palencia). Stvdia Geologica Salmanticensia, Ediciones Universidade de Salamanca, Spain 22: 133-145
Rögl F (1999) Mediterranean and Paratethys paleogeography during the Oligocene and Miocene. In Agustí J Rook L, Andrews P (eds) Hominoid evolution and climatic change in Europe. Vol. 1: The evolution of Neogene terrestrial ecosystems in Europe. Cambridge University Press, UK: 8-23
Rögl F (2001) Mid-Miocene Circum-Mediterranean paleogeography. In Latal C, Piller WE (eds) Environment and Ecosystem Dynamics of the Eurasian Neogene. Stratigraphy & Paleogeography, Workshop Graz 15-18.3.2001. Berichte des Institutes für Geologie und Paläontologie der Karl-Franzens-Universität Graz 4: 49-59
Rook L Abazzi L, Engesser B (1999) An overview on the Italian Miocene land mammal faunas. In Agustí J Rook L, Andrews P (eds) Hominoid evolution and climatic change in Europe. Vol. 1: The evolution of Neogene terrestrial ecosystems in Europe. Cambridge University Press, UK: 191-205
Sanz de Siria A (1993) Datos sobre la paleoclimatología y paleocología del Neógeno del Vallès-Penedès según las macrofloras halladas en la cuenca y zonas próximas. Paleontologia i Evolució, Spain 26-27: 281-289
Sanz de Siria A (1994) La evolución de las paleofloras en las cuencas cenozoicas catalanas. Acta Geologica Hispánica, Spain 29 (2-4): 169-189
Sanz de Siria A (1997) La macroflora del Vallesiense superior de Terrassa (Barcelona). Paleontologia i evolució, Spain 30-31: 247-268
Tobien H (1967) Subdivision of Pontian mammal faunas. Comitee Mediterranean Neogene Stratigraphy, Proceedings IV Session, Bologna 1967. Giornale di Geologia, Italy (2) 35: 1-5

Turco E, Hilgen FJ, Lourens LJ, Shackleton NJ, Zachariesche WJ (2001) Punctuated evolution of global climate during the late Middle to early Late Miocene: High-resolution planktonik foraminiferal and oxygen isotopes records from the Mediterranean. Paleoceanography, American Geophysical Union, USA 16 (4): 405-423
Van Dam JA (1997) The small mammals from the upper Miocene of the Teruel-Alfambra region (Spain): Paleobiology and paleoclimatic reconstructions. Geologica Ultraiectina, Mededelingen van de Faculteit Aardwetenschappen Universiteit Utrecht, The Netherlands 156: 1-204
Van Dam JA, Weltje GJ (1999) Reconstruction of the late Miocene climate of Spain using rodent paleocommunity succesions: an application of end-member modelling. Paleogeography, Paleoclimatology, Paleoecology, Elsevier, The Netherlands 151: 267-305
Van der Meulen AJ, Daams R (1992) Evolution of Early-Middle Miocene rodent faunas in relation to long-term paleoenviromental changes. Paleogeography, Paleoclimatology, Paleoecology, Elsevier, The Netherlands 93: 227-253
Van Valen LM (1984) A resetting of Phanerozoic community evolution. Nature, Nature Publishing Group, UK 307: 50-52
Vrba ES (1985) Environment and evolution: Alternative causes of the temporal distribution of evolutionary events. South African Journal of Science, National Research Foundation, South Africa 81: 229-236
Vrba ES (1995) The fossil record of African antelopes (Mammalia, Bovidae) in relation to human evolution and paleoclimate. In Vrba ES, Denton GH Partridge TC, Burckle LH (eds) Paleoclimate and evolution, with emphasis on human origins. Yale University Press, USA 385-424
Zubakov VA, Borzenkova II (1990) Global Paleoclimate of the Late Cenozoic. Elsevier, The Netherlands 1-456

Appendix

Key to localities included in the clusters from Figs. 2 to 8. Most of the localities and their respective ages are taken from the NOW database. Presence of micro or macromammals is indicated by "X". Each locality is assigned to one of the squares that appear in Fig.1.

Locality code	Locality	Country	Age	Macrom.	Microm.	Square
AG(S)	Los Aguanaces	Spain	MN11		X	12
AG1(S)	Los Aguanaces 1	Spain	MN11		X	12
AG5(S)	Los Aguanaces 5	Spain	MN11		X	12
AG5A(S)	Los Aguanaces 5A	Spain	MN10		X	12
ALC2(S)	Alcocer 2	Spain	MN7+8		X	12

ALF(S)	Alfambra	Spain	MN11		X	12
AMB1(F)	Ambérieu 1	France	MN11		X	23
AMB2A(F)	Ambérieu 2A	France	MN10		X	23
AMB2C(F)	Ambérieu 2C	France	MN10		X	23
AMB3(F)	Ambérieu 3	France	MN11		X	23
AMP1(S)	Ampudia 1	Spain	MN10		X	12
AMP9(S)	Ampudia 9	Spain	MN9		X	12
ANW(SW)	Anwil	Switzerland	MN7+8	X	X	24
ARLL(S)	Arquillo	Spain	MN13	X		12
ARLL1(S)	Arquillo 1	Spain	MN13	X	X	12
ARLL4(S)	Arquillo 4	Spain	MN13		X	12
ARM7(S)	Armantes 7	Spain	MN6		X	12
ARVL(S)	Arroyo del Val	Spain	MN6	X		12
ARY(S)	Arenas del Rey	Spain	MN13	X	X	2
ATVK(MO)	Atavaska	Moldova	MN9	X		28
BAC(IT)	Baccinello V0	Italy	MN11		X	15
BACH1(S)	Bacochas 1	Spain	MN13		X	2
BACV2(IT)	Baccinello V2	Italy	MN12	X	X	15
BACV3(IT)	Baccinello V3	Italy	MN13	X	X	15
BAL(HU)	Baltavas	Hungary	MN13	X		26
BALL(S)	Ballestar	Spain	MN9	X		13
BAY(TU)	Bayraktepe 1	Turkey	MN7+8		X	10
BCV1(S)	Barranc de Can Vila 1	Spain	MN7+8		X	13
BDR15(SW)	Bois de Raube 15	Switzerland	MN7+8		X	24
BELA(PO)	Belchatow A	Poland	MN9		X	37
BERN(F)	Bernardière	France	MN11		X	24
BK(U)	Belka	Ukraine	MN12	X		39
BOJ(S)	Borjas 1	Spain	MN6		X	12
BRI(IT)	Brisighella	Italy	MN13	X	X	15
BUV4/5(S)	Bunker de Valdecebro 4/5	Spain	MN13		X	12
BUVB(S)	Bunker de Valdecebro B	Spain	MN13		X	12
BUZ1(MO)	Buzhor 1	Moldova	MN9	X	X	28
ÇOB(TU)	Çobanpinar	Turkey	MN12	X	X	19

ÇA(TU)	Çandir	Turkey	MN6	X	X	9
CABM(S)	Casablanca M	Spain	MN13	X	X	13
CAC(S)	Casa del Acero	Spain	MN12		X	12
CAF(S)	Can Feliu	Spain	MN7+8	X	X	13
CAL(S)	Casas Altas	Spain	MN9		X	12
CAR1(S)	Carrilanga 1	Spain	MN9		X	12
CAS(IT)	Casino	Italy	MN13	X		15
CAST(IT)	Casteani	Italy	MN12	X		15
CAT(S)	La Cantera	Spain	MN10	X	X	12
CAT2(S)	La Cantera 2	Spain	MN10		X	12
CB(S)	Castell de Barberà	Spain	MN7+8	X	X	13
CC(S)	Concud	Spain	MN12	X	X	12
CC2(S)	Concud 2	Spain	MN12	X	X	12
CC3(S)	Concud 3	Spain	MN12		X	12
CCB(S)	Concud B	Spain	MN12		X	12
CCB2(S)	Can Casablanques 2	Spain	MN10		X	13
CCL(S)	Concud Barranco	Spain	MN12	X	X	12
CCN20(S)	Creu Conill 20	Spain	MN9	X	X	13
CCN22(S)	Creu Conill 22	Spain	MN9		X	13
CEL1(S)	Celadas 1	Spain	MN13		X	12
CEL2(S)	Celadas 2	Spain	MN13		X	12
CGAR(S)	Cerro de la Garita	Spain	MN12	X	X	12
CHAR(SW)	Charmoille	Switzerland	MN9	X		24
CHIM(MO)	Chimishlija	Moldova	MN12	X	X	28
CHO(MO)	Chobruchi	Moldova	MN12	X		28
CHOM(GR)	Chomateres	Greece	MN12	X		7
CHRY(GR)	Chrysavgi	Greece	MN7+8		X	17
CIACG(IT)	Ciabot Cagna	Italy	MN13		X	14
CKV(HU)	Csakvar	Hungary	MN11	X	X	26
ÇKY(TU)	Çorak Yerler	Turkey	MN10	X		20
CLL1(S)	Can Llobateres 1	Spain	MN9	X	X	13
CM(S)	Can Missert	Spain	MN7+8	X	X	13
CNAB(F)	Castelnau-d'Arbieu	France	MN6	X		13
COM1(RO)	Comanesti 1	Romania	MN7+8		X	27
CP1(S)	Can Ponsic 1	Spain	MN9	X	X	13

CRE15(S)	Crevillente 15	Spain	MN12	X	X	2
CRE16(S)	Crevillente 16	Spain	MN12	X		2
CRL2(S)	Crevillente 2	Spain	MN11	X	X	2
CV(S)	Ceràmiques Viladecavalls	Spain	MN10	X	X	13
DDURK(GE)	Dorn Dürkheim	Germany	MN11	X	X	24
DIN(GE)	Dinotheriensande	Germany	MN9	X		24
DIO(F)	Dionay	France	MN11		X	24
DOV(F)	Douvre	France	MN10		X	24
DUD1(S)	Duredos 1	Spain	MN6		X	12
DYK1(GR)	Dytiko 1	Greece	MN13	X		17
DYK2(GR)	Dytiko 2	Greece	MN13	X		17
DYK3(GR)	Dytiko 3	Greece	MN13	X		17
EAKY(TU)	Esme Akçaköy	Turkey	MN9	X		8
EPH(GE)	Eppelsheim	Germany	MN9	X		24
ESB(GE)	Esselborn	Germany	MN9	X		24
ESC(S)	Escobosa	Spain	MN7+8	X	X	13
FIUS(IT)	Fiume Santo	Italy	MN12	X		15
FOU1993(F)	Lo Fournas 1993	France	MN10		X	13
FRM(POR)	Freiria do Rio Maior	Portugal	MN10		X	11
FU(F)	Four	France	MN6	X	X	13
GBG(A)	Gaiselberg	Austria	MN9	X		26
GK(TU)	Garkin	Turkey	MN11	X		8
GLO1(S)	La Gloria 1	Spain	MN13		X	11
GLO10(S)	La Gloria 10	Spain	MN11		X	12
GLO11(S)	La Gloria 11	Spain	MN10		X	12
GLO14A/B(S)	La Gloria 14A/B	Spain	MN10		X	12
GLO5(S)	La Gloria 5	Spain	MN13		X	11
GLO6(S)	La Gloria 6	Spain	MN13		X	11
GOT(A)	Götzendorf	Austria	MN9	X	X	26
GRA(IT)	Gravitelli	Italy	MN13	X		6
GROS(U)	Grossulovo	Ukraine	MN10	X		28
GTV(U)	Gritsev	Ukraine	MN9		X	28
GUGA(MO)	Gura-Galben	Moldova	MN12	X		28

GULP(TU)	Gülpinar	Turkey	MN10	X		9
HAM(GE)	Hammerschmiede	Germany	MN9	X	X	24
HAM6C(GE)	Hambach 6C	Germany	MN6	X	X	34
HAU(F)	Haulies	France	MN6	X		13
HI(S)	Hostalets de Pierola inferior	Spain	MN7+8	X	X	13
HOW(GE)	Höwenegg	Germany	MN9	X		24
HS(S)	Hostalets de Pierola superior	Spain	MN9	X	X	13
HZ(HU)	Hasznos	Hungary	MN7+8		X	26
I(F)	Isere	France	MN7+8		X	13
JUX(F)	Jujurieux	France	MN9		X	14
KAY(TU)	Kayadibi	Turkey	MN11	X		8
KEMAB(TU)	Kemiklitepe A-B	Turkey	MN12	X		8
KEMD(TU)	Kemiklitepe D	Turkey	MN11	X		9
KF(MO)	Kalfa	Moldova	MN9	X	X	28
KHF(A)	Klein Hadersdorf	Austria	MN6	X		26
KI(TU)	Kinik	Turkey	MN12	X		9
KOF(A)	Kohfidisch	Austria	MN10		X	26
KUK(TU)	Küçükçekmece	Turkey	MN11	X		19
LCAS(S)	Las Casiones	Spain	MN13	X	X	12
LCASS(S)	Las Casiones superior	Spain	MN13		X	12
LCAT(S)	La Cantera	Spain	MN11	X		12
LEK(GR)	Lefkon	Greece	MN11		X	17
LGSA(F)	La Grive St. Alban	France	MN7+8	X	X	13
LIB(S)	Librilla	Spain	MN13		X	2
LIS(F)	Lissieu	France	MN13		X	13
LM(S)	Los Mansuetos	Spain	MN12	X	X	12
LM2(S)	Los Mansuetos 2	Spain	MN12		X	12
LOB(F)	Lobrieu	France	MN11		X	13
LP5B(S)	Las Planas 5B	Spain	MN6		X	12
LP5H(S)	Las Planas 5H	Spain	MN7+8		X	12
LP5K(S)	Las Planas 5K	Spain	MN6		X	12
LP5L(S)	Las Planas 5L	Spain	MN6		X	12
LT1(S)	La Tarumba 1	Spain	MN10	X		13

MACH1(S)	Manchones 1	Spain	MN6		X	12
MAR(GR)	Maramena	Greece	MN13	X	X	17
MAZ(TU)	Mahmutgazi	Turkey	MN12	X		8
MBA(S)	Masía del Barbo A	Spain	MN10		X	12
MBB(S)	Masía del Barbo B	Spain	MN10	X	X	12
MDV2(S)	Masada del Valle 2	Spain	MN12		X	12
MDV3(S)	Masada del Valle 3	Spain	MN12		X	12
MDV4(S)	Masada del Valle 4	Spain	MN12		X	12
MDV5(S)	Masada del Valle 5	Spain	MN12		X	12
MIL(S)	Milagros	Spain	MN13	X		12
MLUB(F)	Mt. Luberon	France	MN12	X	X	14
MOD(S)	Modorras	Spain	MN13		X	12
MOL(F)	Mollon	France	MN11		X	24
MOL(S)	Molina de Aragón	Spain	MN9		X	12
MON(F)	Montredon	France	MN10	X	X	13
MONS(GR)	Monasteri	Greece	MN13		X	17
MR3(S)	Masía la Roma 3	Spain	MN9		X	12
MRU(S)	Masada Ruea	Spain	MN10		X	12
MV6(S)	Masada del Valle 6	Spain	MN13		X	12
MV7(S)	Masada del Valle 7	Spain	MN13		X	12
NE-SP(SK)	Neudorf-Spalte	Slovakia	MN6	X		26
NOEM(U)	Novaja Emetovka	Ukraine	MN12	X		29
NOM(S)	Nombrevilla	Spain	MN9	X	X	12
OP2(PO)	Opole 2	Poland	MN7+8		X	36
PALL3(S)	Paracuellos 3	Spain	MN6	X		12
PALL5(S)	Paracuellos 5	Spain	MN6	X		12
PCY(F)	Puy Courny	France	MN12	X		13
PED2A(S)	Pedregueras 2A	Spain	MN9		X	12
PED2C(S)	Pedregueras 2C	Spain	MN9		X	12
PER4(S)	Peralejos 4	Spain	MN10		X	12
PER5(S)	Peralejos 5	Spain	MN9		X	12
PERA(S)	Peralejos A	Spain	MN10		X	12
PERC(S)	Peralejos C	Spain	MN10		X	12

PERD(S)	Peralejos D	Spain	MN11		X	12
PFH(POR)	Pero Filho	Portugal	MN6		X	1
PI(S)	Piera	Spain	MN11	X		13
PIK-MNHN(GR)	Pikermi-MNHN	Greece	MN12	X		7
PKY(MO)	Poksheshty	Moldova	MN10	X		28
PLAK(GR)	Plakia	Greece	MN7+8		X	7
PM(S)	Puente Minero	Spain	MN11	X	X	12
PM2(S)	Puente Minero 2	Spain	MN10		X	12
PM3(S)	Puente Minero 3	Spain	MN11		X	12
PM5A/5B(S)	Puente Minero 5A/5B	Spain	MN11		X	12
POSAT(POR)	Póvoa de Satarém	Portugal	MN6		X	1
POU-CAY(F)	Poudenas-Cayron	France	MN7+8	X		13
PREZ(SE)	Prebreza	Serbia	MN6	X		17
PROC(GR)	Prochoma	Greece	MN11	X		17
PRZ2(PO)	Przeworno 2	Poland	MN7+8	X		36
PTP(GR)	Pentalophos	Greece	MN10	X		17
PYV(GR)	Pyrgos Vassilissis	Greece	MN12	X		7
R1(S)	La Roma 1	Spain	MN10		X	12
R2(S)	La Roma 2	Spain	MN11	X	X	12
RDZ5(GR)	Ravin des Zouaves 5	Greece	MN11	X		17
REG2(S)	Regajo 2	Spain	MN11		X	12
REG3(S)	Regajo 3	Spain	MN12		X	12
REG4(S)	Regajo 4	Spain	MN12		X	12
REG5(S)	Regajo 5	Spain	MN13		X	12
ROM11(S)	Masía de la Roma 11	Spain	MN10		X	12
ROM4B(S)	Masía de la Roma 4B	Spain	MN10		X	12
ROM4C(S)	Masía de la Roma 4C	Spain	MN10		X	12
ROM5(S)	Masía de la Roma 5	Spain	MN10		X	12
ROM6(S)	Masía de la Roma 6	Spain	MN10		X	12
ROM7(S)	Masía de la Roma 7	Spain	MN10		X	12
ROM8(S)	Masía de la Roma 8	Spain	MN10		X	12
ROM9(S)	Masía de la Roma 9	Spain	MN10		X	12

RPL(GR)	Ravin de la Pluie	Greece	MN10	X		17
RPY(MO)	Respopeny	Moldova	MN10	X		28
RT-7(S)	Rubí-Terrassa 7	Spain	MN10		X	13
RUD(HU)	Rudabánya	Hungary	MN9	X	X	27
RUMK(SW)	Rümikon	Switzerland	MN6	X	X	24
RVDC0(S)	Rambla de Valcecebro 0	Spain	MN13	X		12
RVDC3(S)	Rambla de Valcecebro 3	Spain	MN13		X	12
RVDC6(S)	Rambla de Valcecebro 6	Spain	MN13		X	12
RZV(GR)	Ravin des Zouaves 1	Greece	MN10	X		17
SAL(S)	La Salle	Spain	MN10		X	12
SAMM(GR)	Samos Main Bone Beds	Greece	MN12	X		8
SAMWS(GR)	Samos White Sands	Greece	MN12	X		8
SAN(S)	Santiga	Spain	MN9	X		13
SANS(F)	Sansan	France	MN6	X	X	12
SBZ(F)	Saint Bauzille	France	MN11		X	13
SCHWD(SW)	Schwamendingen	Switzerland	MN6		X	24
SFÇ(TU)	Sofça	Turkey	MN7+8	X	X	9
SGAUVAL(F)	Saint-Gaudens(Valentine)	France	MN7+8	X		13
SIM(F)	Simorre	France	MN6	X		13
SIM2(S)	Simancas 2	Spain	MN6		X	12
SOB(F)	Soblay	France	MN10	X	X	24
SOL(S)	Solera	Spain	MN7+8		X	12
SQ(S)	Sant Quirze	Spain	MN7+8	X	X	13
STAZ(GE)	Stätzling	Germany	MN6	X		25
STEB(GE)	Steinberg	Germany	MN6		X	34
STEIN(GE)	Steinheim	Germany	MN7+8	X	X	25
STSPH(A)	St. Stephan im Lavanttal	Austria	MN7+8	X		25
SUG(HU)	Sümeg	Hungary	MN11		X	26
TF(S)	Torrent de Febulines	Spain	MN10	X		13
TF3(S)	Torrent de Febulines 3	Spain	MN10		X	13
THAN(GE)	Thannhausen	Germany	MN6	X		25
TM1(S)	Torremormojón 1	Spain	MN10		X	12
TM4(S)	Torremormojón 4	Spain	MN9		X	12

TM5(S)	Torremormojón 5	Spain	MN7+8		X	12
TO(S)	Tortajada	Spain	MN12		X	12
TOA(S)	Tortajada A	Spain	MN11		X	12
TOB(S)	Tortajada B	Spain	MN12		X	12
TOC(S)	Tortajada C	Spain	MN12		X	12
TOD(S)	Tortajada D	Spain	MN12		X	12
TOR(S)	Toril	Spain	MN7+8		X	12
TSA2(S)	Tal·lús Sud Autopista 2	Spain	MN10		X	13
TU(MO)	Tudorovo	Moldova	MN12	X		28
UNT(GE)	Unterneul	Germany	MN6		X	25
VA(F)	Valréas	France	MN11		X	13
VAL1(S)	Valalto 1	Spain	MN6		X	12
VAL2B(S)	Valalto 2B	Spain	MN6		X	12
VAL2C(S)	Valalto 2C	Spain	MN6		X	12
VAR(MO)	Varnitsa	Moldova	MN9	X		28
VB2/2C(S)	Villalba Baja 2/2C	Spain	MN12		X	12
VDC5(S)	Valdecebro 5	Spain	MN12	X		12
VF9(S)	Vilafeliche 9	Spain	MN7+8		X	12
VFÑ(S)	Valles de Fuentidueña	Spain	MN9	X		12
VILL(S)	Villastar	Spain	MN13	X	X	12
VIP(S)	Vivero de Pinos	Spain	MN10	X	X	12
VMOR(S)	Venta del Moro	Spain	MN13	X	X	2
VOS(A)	Vösendorf	Austria	MN9	X	X	26
VTK(GR)	Vathylakkos	Greece	MN11	X		17
WARTB(GE)	Wartenberg	Germany	MN9	X		25
WHF(GE)	Westhofen	Germany	MN9	X		24
WLZ(SW)	Wiesholz	Switzerland	MN6		X	24
WSBG(GE)	Wissberg	Germany	MN9	X		24
XIR1(GR)	Xirochori 1	Greece	MN10	X		17
YESK1(TU)	Yeni Eskihisar 1	Turkey	MN7+8	X	X	8
ZEG(SW)	Zeglingen	Switzerland	MN6		X	24

10 Evaluating the roles of connectivity and environment on faunal turnover: patterns in recent and fossil Iberian mammals

Joaquín Hortal[1,2], Manuel Nieto[3,4], Jesús Rodríguez[3], Jorge M. Lobo[1]

[1]Departamento de Biodiversidad y Biología Evolutiva. Museo Nacional de Ciencias Naturales (CSIC). C/José Gutiérrez Abascal, 2, Madrid – 28006, Spain, jhortal@mncn.csic.es, [2]Departamento de Ciências Agrárias. Universidade dos Açores. Campus de Angra, Terra-Chã; Angra do Heroísmo - 9701-851 - Terceira (Açores), Portugal, [3]Departamento de Paleobiología. Museo Nacional de Ciencias Naturales (CSIC). C/José Gutiérrez Abascal, 2, Madrid – 28006, Spain, [4]Instituto de Neurobiología Ramón y Cajal (CSIC). Avda. del Doctor Arce, Madrid – 28002, Spain

10.1 Abstract

The role of dispersal events in shaping patterns of geographic differentiation of biotas has been often studied regarding to invasive events occurring in short periods of time. However, migrations of species from one area to another are occurring continuously. Thus, patterns on differentiation diversity may be affected by the connectivity among areas. We provide evidence of this fact using the recent mammal faunas at the Iberian Peninsula. When compared with differences in environment and habitat structure, connectivity-mediated distances where the most important factor affecting β-diversity patterns at 100x100 km scale. Moreover, most of the explanatory capacity of environmental differences could not be separated from that attributable to connectivity. We developed a model that links patterns in β-diversity with geomorphologically-based connectivity using these recent data, and tested it on the patterns of macromammal variation among sedimentary basins during the Iberian Neogene. β-diversity values where much higher than those observed for recent faunas, and our model was unable to predict Neogene patterns. A critical discussion on the causes of this lack of agreement among recent patterns and those observed from fossil data is provided.

Keywords: geographic connectivity, Iberian mammals, geomorphology, Neogene and recent faunas, beta diversity, faunistic similarity, environmental differences.

10.2 Introduction

Most times, biological invasions are regarded as discrete events in time. In a few years, or decades, single species or complete faunas are able to colonize broad territories, competing with, and even displacing, native species and communities (see some examples at, e.g., Vitousek et al. 1987, Niemela and Spence 1991, and reviews at Hengeveld 1989, Shigesada and Kawasaki 1997, Levine and D'Antonio 1999, or Sakai et al. 2001). Most of these invasions are continuously taking place in a silent and gradual way. As Kozlowski (1999) states, species produce numerous offspring, which migrate in a density-dependent regulatory mechanism. Thus, the world is flooded with migrants of different species, which can suffer microevolutionary changes in the newly-established populations (e.g. Hendry and Kinnison 2001). This mixture of dispersal and local microevolutionary processes is one of the causes of biodiversity, assuring the immense, permanent variability of nature, and regulating the genetic and ecological structure of populations via permanent gene flow in heterogeneous landscapes (Kozlowski, 1999), as well as the structure of local and regional communities (see, e.g., Cornell and Lawton 1992).

Although implicitly accepted as a necessary part of the faunistic differences among different areas, the importance of these 'shadow invasions' in shaping the variation in community composition remains frequently unnoticed. Other factors, such as environmental variability or differences in habitat structure have been claimed to play significant roles in the structure of ecological communities. However, recent works have found significant relationships between spatial distance among local communities, and community dissimilarity (e.g. Duivenvoorden et al. 2002). These differences could be attributed to increasing differences in environmental conditions between increasingly distant sites due to environmental gradients (see, e.g., Pereira and Pires da Fonseca 2003). However, spatial distance and, most accurately, connectivity among areas, together with dispersal ability, mediates the success of species in reaching a new locality. Here, connectivity is defined as the facility of movement between two areas, taking into account geomorphologic and/or landscape characteristics.

In this framework, differences in the connectivity among areas could give rise to differences in community composition. To carry out a preliminary test of this hypothesis at a regional scale, in the first part of this work we use data on the community composition of recent Iberian mammals to evaluate the relative importance of differences in connectivity, environment (climate and geology) and habitat over differentiation diversity, measured as β-diversity.

Most times, the relationships of the above mentioned factors (environment and/or habitat) with the ecological responses of several recent mammal species have been used to extrapolate past conditions from fossil faunas, a point that we have criticized extensively (Rodríguez 1999, Rodriguez and Nieto 2003, Hortal et al. in revision; see discussion). However, to test if the present relationship between connectivity among areas and β-diversity could be extrapolated to the past, we have developed a function relating both variables using recent data. Then, we have tested the performance of our model on the mammal fossil record of the Iberian Neogene. Finally, we assess the possible effects of well known invasive and/or macroevolutionary events over the β-diversity patterns throughout this age.

10.3 Assessing the role of environment, habitat and connectivity on recent mammal communities

As we commented before, we carry out a preliminary test on how the differences among areas in i) environment and ii) habitat structure, and iii) connectivity, shape the patterns of faunistic differentiation of recent Iberian mammals. To avoid effects of area size and/or data coverage, we have used the cells from the UTM 100x100 km grid where more than 90% of their total area is occupied by land surface at the Iberian Peninsula. Grid cell size was selected to approximate to the area of the sedimentary basins that we use as units of analysis for the fossil data. 52 Iberian 100x100 km resolution grid squares fulfilled these criteria, so we use them to develop three distance matrices for the abovementioned three factors (see Figure 1). The methodological limitations of the GIS and the community analysis package used, prevented us from developing the high number of possible pairwise comparisons among these grid squares (1300). Thus, we compare the distance values for 50 pairs of squares selected at random with the β-

diversity scores obtained from the pairwise comparisons of the mammal faunas of these squares.

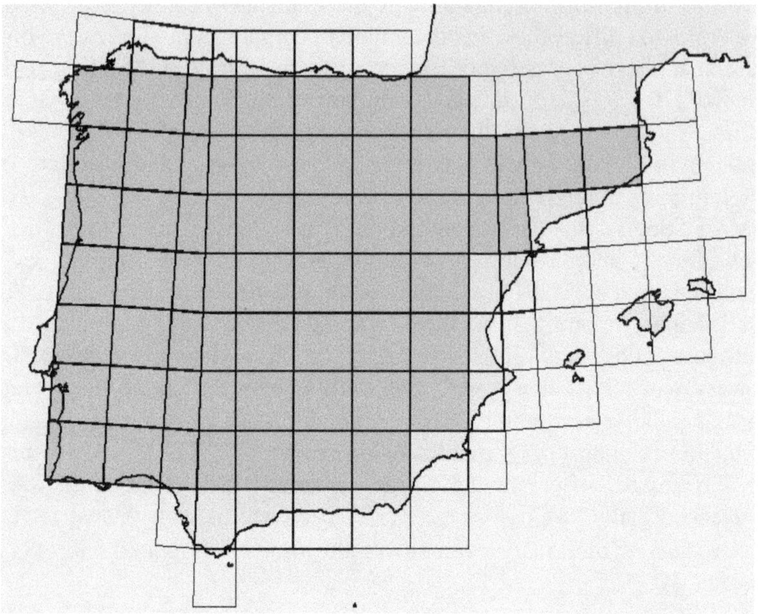

Fig. 1. Spatial location of the 52 squares (dark grey) from the UTM 100 km grid used for the recent fauna analyses (see text)

10.3.1 Estimation of present environmental and habitat differences between areas

To assess the influence of variation in environmental conditions and habitat structure on recent mammal communities, we developed two matrices of distances among the UTM 100x100 grid squares:
*The first one, d_{env}, accounts for environmental variation. It was built using the information from six climatic variables (minimum, maximum and mean annual temperature, mean annual and summer precipitation, and percentage of sunny hours per year), and three bedrock geologic (calcareous rock, acid rock and clay area). Climate data extracted from the 50x50 km resolution data provided by W. Cramer (CLIMATE database version 2; http://www.pik-potsdam.de/~cramer/climate.html). To obtain bedrock data, we used a previously digitised three-category map (IGN 1995). All these variables were successfully used in other analyses on the distribution of species richness and composition of several groups of

Iberian plants and insects (e.g., Lobo et al. 2001, Lobo and Martín-Piera 2002, or Hortal et al. 2004). These analyses usually include geomorphologic and structural variables, such as altitudes, slopes, aspects, etc. However, we use such information to develop friction surfaces and connectivity analyses (see below). Thus, we have excluded them from the calculation of the environmental matrix, in order to avoid self-replication and to separate more clearly both sources of faunistic variability.

*The second one, d_{lulc}, accounts for the differences in the land use/land cover of each area, that is, for differences in habitat structure among areas. To develop this matrix, the original CORINE Land Use/Land Cover map (EEA 2000) were reclassified into 13 categories (urban areas, dry cultures, irrigated cultures, patchy cultures, olive, vineyard and fruit plantations, anthropogenic pastures, natural pastures, broadleaf forests, conifer forests, mixed forests, shrubland, bare rock and land masses). The percentage of land coverage devoted to each category was extracted using GIS techniques. Similar variables were also successfully used in several of the formerly related studies.

Following the recommendations of Legendre and Legendre (1998), after the standardization of the truly continuous variables (those related to climate), we used the Gower Similarity Index (Gower 1971) to measure the environmental and habitat proximities among grid squares. In the case of the development of the d_{env} matrix, different weights were assigned to bedrock and climate variables, to account for the differential importance of each group of environmental variables in the distribution of the species (see discussion at Hortal and Lobo 2005). Although bedrock geology is not a direct determinant of variations in mammal communities, it does for plants (see, e.g., Lobo et al. 2001 for the Iberian Peninsula), a well known determinant of mammal diversity (see, e.g., Olff et al. 2002 or Cristoffer and Peres 2003). Thus, substrate composition may have an indirect effect over mammal communities. Although no analysis is available on the different weights of geology and climate in the configuration of mammal communities (see Hortal and Lobo 2005 for an example of such analysis on Iberian dung beetles), we have assumed the latter having greater importance. In absence of any hypotheses of the magnitude of such difference, and following the empirical evidence obtained by Hortal and Lobo (2005), climate was assumed to be three times more important than geology. Thus, every climatic variable had a weight of 1/2 (3/6 variables), giving a total weight of 3 to climate (0.5 x 6 climatic variables). In the same way, every geologic category had a weight of 1/3 (1/3 variables), and geology a total weight of 1 (0.33 x 3 variables). The resulting similarity matrices (calculations made with Clustan Graphics software; Clustan Ltd.

2002) were transformed into distance matrices by calculating the inverse of the index value (that is, $d = 1 -$ Gower value).

10.3.2 Estimation of connectivity between areas

Both habitat structure and topography play a well known role in the connectivity among areas. Our main objective is to analyze the relationship between connectivity and β-diversity in both present times and during the Neogene. Since we lack accurate data to estimate connectivity during the Neogene, we will use recent data to extrapolate past conditions in order to estimate past connectivity. We assume that environmental conditions strongly varied during the Neogene, so we have chosen geomorphology as the factor that remained more constant during this period.

In spite of the differential effects on climate changes throughout the Iberian Peninsula (see, e.g., maps in Adams 1998), the spatial location of the main geographic accidents (mountain chains, plateaus, valleys, etc.) has not suffered dramatic changes. It is true that topographic variations due to the Alpine Orogeny, and sea level variations due to glaciation processes, have leaded to great temporal differences in the spatial configuration of habitats in the Iberian Peninsula. However, most current valleys, plateaus and mountain chains have been valleys (or sea intrusions) plateaus and mountain chains for a long period of time. Thus, it can be assumed that, if point A is best connected to point B than to point C in the present, it remained so throughout the last MYs. Thus, we have assumed as a working hypothesis, that their effect on broad scale differences in mammal communities could have remained constant. This assumption should not be hold for true, as we are going to see below.

The spatial configuration of the physical matrix where dispersal occurs, limits or facilitates the migrations of species (see, e.g., Lobo and Halffter 2000, or Spector 2002), thus determining the connectivity among areas. For example, the direction of frictional effects to movement (that is, factors that increase the cost of moving through a territory) is important in mountain environments. The direction of slopes affects the efforts needed to cross an area depending upon the direction of movement (i.e., up, down, or along a slope). Areas with the same slope values, but at different directions (aspect), might have very different frictions for the dispersal in a given direction. On the one hand, longitudinal mountains (that is, those having an east-west configuration) can act as barriers to latitudinal (north-south) dispersal. The harsh conditions appearing at high altitudes impede the communication between both sides (see, e.g. Brown and Lomolino 1998, or Bhattarai and Vetaas 2003), thus diminishing the relationship between northwards and southwards faunas. On the other, latitudinal mountain chains (those placed in a north-to-south direction), in spite of being barriers for the longitudinal (east-west) dispersal, can also act as corridors for latitudinal movements. The climatic gradients they produce may permit to the populations or individuals of a given species to reach suitable conditions just by rapid shifts in altitude. Thus, latitudinal mountain chains facilitate latitudinal movements by providing corridors of suitable habitat conditions, allowing communication between faunas of sites located at different latitudes.

We have taken this issue into account to calculate the distances between areas, by means of a connectivity analysis in a GIS environment. 1 km^2 resolution altitude data was obtained from a global Digital Elevation Model (Clark Labs 2000). Aspect and slope maps where derived from this DEM using Idrisi software (Clark Labs 2001). Then, we developed connectivity distance images, where the friction-mediated cost of moving to each spatial object studied (in this case, 100x100 km squares) from each 1 km^2 pixel is stored. In the case of recent data, these cost distances are calculated from the centroids of each 100x100 km square. In the case of the fossil faunas, several centroids where located at each basin, and the cost distances to each pixel where calculated from the nearest one. Anisotropic cost algorithms implemented in Idrisi GIS software calculate such connectivity distance images from three different maps (see Clark Labs 2001):

*A magnitude of friction surface, where different slope values are reclassified into friction values. A sigmoid-shape relationship between slope and friction was assumed, with low increments of cost of movement at low slope values, and progressively higher increments. Friction was assumed to be minimum at 0º, and maximum at slopes of 50º or higher.

*A map of the direction of friction (anisotropic costs; see Clark Labs 2001). A new GIS coverage was obtained from the aspect map, where longitudinal aspects were assigned to frictions in latitudinal directions, and latitudinal aspects did to longitudinal directions. This image is used to weight the cost of movement through each pixel. Thus, movements are restricted by transversal mountains, under the assumption that mountain chains act as barriers to the movement of most species. This may be especially true during cold periods, where the extreme conditions happening at high altitudes may difficult migrations of many of the temperate, mediterranean and semi-tropical species present in the Iberian Peninsula.

*An isotropic friction surface, representing the non-directional frictions to the movement. In this case, we used just altitude, as it constitutes a proxy for the environmental variability that may be similar effects through time.

In this case, a sigmoid-shape relationship was also assumed, taking current altitudes of 1000 m.a.s.l. as completely suitable to movements, and those higher to 5000 (non-existent in the Iberian Peninsula) to be completely unsuitable. These values were selected to provide a realistic simulation of the effects of altitude on dispersal; many mammal species can disperse easily up to an intermediate altitude, and the difficulty of movement increases exponentially from these point, up to a ceiling sited at extreme altitudes, were environmental conditions do not permit the survival of migrants. This way, moving through high altitude mountains is possible, but at high costs.

Using these three cost images, a connectivity distance image is calculated for each 100x100 km grid cell (or sedimentary basin, see below). Then, the distance of connectivity (d_{con}) from each cell to another is calculated using a least-cost pathway algorithm (see Clark Labs 2001). This analysis calculates the shortest way from a given point to another using the connectivity distance image of one of them, developing a pathway that minimizes the costs of moving through the friction surface. The spatial location of the 50 pathways developed for recent faunas is shown in Figure 2 (see another example in Figure 6 below). d_{con} values were calculated as the sum of the costs of moving through 1 km^2 each pixel in the pathway. The higher the value of d_{con}, the higher the difficulty to move from a point to the other. This way, we produced a triangular distance matrix among grid cells.

10.3.3 β-diversity estimation

Recent mammal community data comes from the current atlas of the distribution of Spanish terrestrial mammals (Palomo and Gisbert 2002). We have extracted checklists from the previously defined grid cells. To measure differences in composition from such presence/absence data, we have chosen β_w, the classic measure from Whittaker (1960). Although a few newly-developed estimators present better properties (e.g., additivity or usability with nested areas; see Koleff et al. 2003), this measure has been extensively used in the literature, and has the crucial properties of symmetry and independence of sample size (see Wilson and Shmida 1984, Magurran 1988, and Koleff et al. 2003). This index relates the total number of species recorded in the two areas being compared (in this case, grid squares), with the average number of species found within them. According to the formulation of Koleff et al. (2003), this index is calculated as:

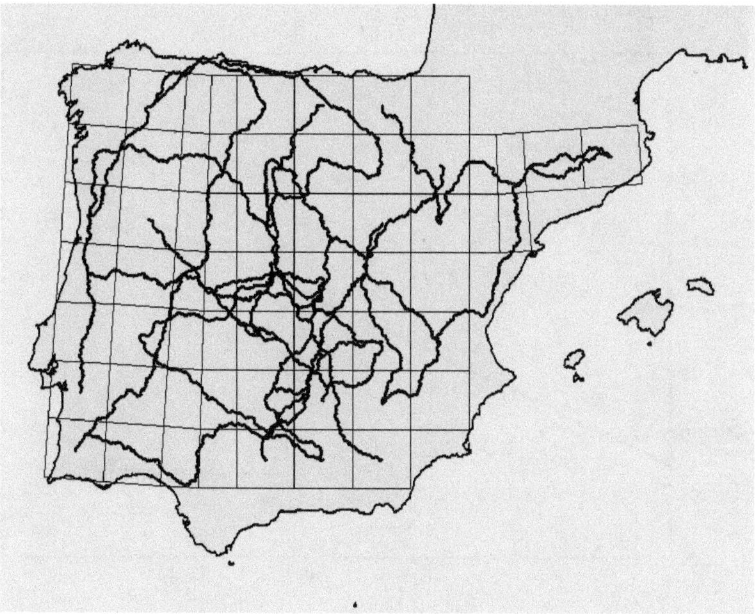

Fig. 2. Pathways obtained in the recent fauna connectivity analyses. Each line shows the spatial location of the least-cost pathway between the centroids of one of the pairs of grid cells analyzed (thin squares), using a geomorphologically-based friction surface (see text)

$$\beta_w = \frac{a+b+c}{(2a+b+c)/2} - 1 \qquad (10.1)$$

where a is the number of species found in both samples, b is the number of species found in the first grid cell but not in the second, and c is the number of species found in the second grid cell but not in the first one. As mentioned before, we used the 50 pairs of 100x100 km grid squares selected at random, and calculated β_w scores using Species Diversity and Richness 3.0 software (Henderson and Seaby 2002). These species replacement scores were mainly low, ranging from 0.0423 to 0.3628, with a mean β_w value of 0.2037 (±0.0805 SD). Interestingly, the resulting values showed a clear relationship with d_{con} scores (Figure 3; see also Table 2 below).

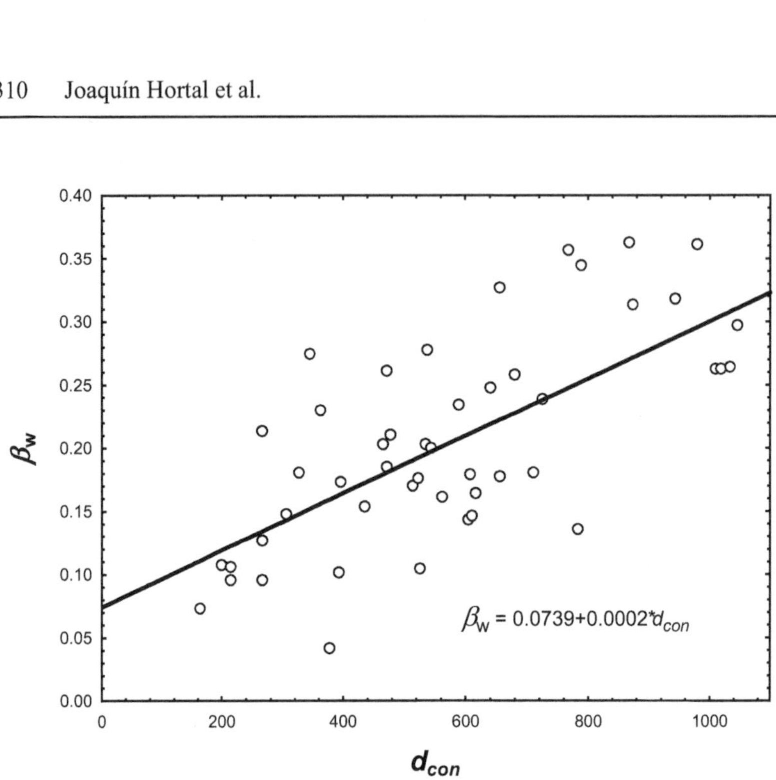

Fig. 3. Relationship between mammal differentiation diversity (β_w; see text) and connectivity-mediated distance (d_{con}). Circles correspond to the 50 pairwise comparisons between pairs of 100x100 km UTM cells, and the line and function correspond to the linear model developed using common regression techniques (see text)

10.3.4 Determinants of β-diversity

Once the distance matrices accounting for differences in environment, habitat and connectivity are developed, we have carried out a preliminary analysis of their relative importance in shaping the variation in mammal faunistic composition between pairs of areas, using the 50 previously defined pairwise comparisons.

After standardizing these variables, and assessing their normality, their single importance was calculated by means of the regression techniques included in the General Regression Models (GRM; see, e.g., Dobson 1999, or StatSoft 2001). Adjusted R^2 statistic has been used to estimate the amount of variance explained by each factor (see Table 1), applying the correction for negative Adjusted R^2 values suggested by Legendre and Legendre (1998). Interestingly, whilst the effect of environmental distances was small (around 10%), and differences in habitat were non-significant, connectivity between pairs of areas explained 45.9% of their β_w scores. A full model, based on the three kinds of distances, explained up to 47.1% of total variance.

Table 1. Results from the General Regression Models (that is, ANOVA results; see StatSoft 2001) using environmental, habitat structure, and connectivity distances over β_w. *df* is the number of degrees of freedom of the model; F is the Snedecor's F statistic; Mult. *R*, Mult. R^2, and Adj. R^2 are the multiple and adjusted estimates the amount of variance explained by each model. Adj. R^2 scores were corrected for negative values according to the method suggested by Legendre and Legendre (1998). Predictor names as in text. ** is $p<0.01$, * is $p<0.05$, and n.s. is non significant

Predictor	Df	F	Mult. R	Mult. R^2	Adj. R^2
Full regression models					
d_{env}	48	5.366*	0.317	0.101	0.100
d_{lulc}	48	2.041 n.s.	0.202	0.041	0.039
d_{con}	48	39.621**	0.672	0.452	0.459
$d_{env} + d_{lulc}$	47	2.817 n.s.	0.327	0.107	0.087
$d_{env} + d_{con}$	47	22.050**	0.696	0.484	0.481
$d_{lulc} + d_{con}$	47	20.444**	0.682	0.465	0.461
$d_{env} + d_{lulc} + d_{con}$	46	14.524**	0.697	0.486	0.471
Partial regression models					
rd_{env}	48	1.041 n.s.	0.146	0.021	0.019
rd_{lulc}	48	0.113 n.s.	0.049	0.002	0.000
rd_{con}	48	29.344**	0.616	0.379	0.385

To partition the effects of these factors into their independent and joint components (e.g., the amount of β_w variation that can be attributed to the single effect of d_{con}, independently from the variations in d_{env} or d_{lulc}, or the amount of variation that is due to the joint effect of d_{con} and d_{env}), we used partial regression analyses (see Legendre and Legendre 1998, or Lichstein et al. 2002). In these analyses, each predictor (or group of predictors) is regressed against all the others. The resulting residuals are retained, thus representing the variability in each predictor that is independent from the others. This way, we used GRM analyses to extract three new variables:

rd_{env} (the residuals of regressing d_{env} over d_{lulc} and d_{con}), rd_{lulc} (from d_{lulc} over d_{env} and d_{con}), and rd_{con} (from *dcon* over d_{env} and d_{lulc}). We then regressed β_w values over these new variables, thus obtaining the independent effects of the three factors (see Table 1). Here, it is remarkable that the effect of the environmental differences falls out of significance, as most of its explanatory capability can not be separated from that of the connectivity.

Table 2. Significance and parameter estimation of the model relating the scores of β_w between UTM 100x100 grid squares and connectivity distances (β_w = Intercept + *estimate**d_{con}; see also Figure 3), developed from the data of the 50 pairwise comparisons of recent Iberian mammals. Estimate shows the parameter scores in the model, *Std. Err.* does for the standard error of these estimates, *t* for the *t* statistic, and *cl*-95% and *cl*+95% are the lower and upper confidence limits for the estimates. The rest of the abbreviations as in text and Table 1

	df	F	Mult. *R*	Mult. R^2	Adj. R^2
Model significance	48	39.621**	0.672	0.452	0.441
Parameter estimation	Estimate	*Std. Err.*	*t*	*cl* -95%	*cl* +95%
Intercept	0.073893	0.022305	3.312**	0.029045	0.118743
d_{con}	0.000226	0.000036	6.295**	0.000154	0.000298

We used the equation system proposed by Borcard et al. (1992) to fully partition the variation in β_w due to these three effects into nine components describing their independent and joint influences (see examples in Lobo et al. 2001, 2004, or Pereira and Pires da Fonseca 2003). The effect of pure connectivity appears as being almost the unique single effect in shaping faunistic differences (Figure 4). This analysis also reveals that the effect of environmental distances cannot be separated from that of the connectivity. The high amount of unexplained variance (more than 50%) puts into evidence that the here considered factors may not be the only ones affecting β_w patterns. Other effects, such as historical contingent events, or other environmental effects not included in our analysis, may underlie this pattern. Here, it is important to take into account that the spatial scale used (grain size) may not be the best to cover the effects of differences in landscape structure (usually studied at 1x1 and 10x10 km grain sizes) and environment (mostly studied at 10x10 and 50x50 km grain sizes) (see discussions at Whittaker et al. 2001, Willis and Whittaker 2002, and Blackburn and Gaston 2002). Moreover, the weight assignation to the different factors in the construction of the d_{env} matrix was arbitrary, and no assessment is available on its accuracy to describe the environmental differences important for mammal diversity.

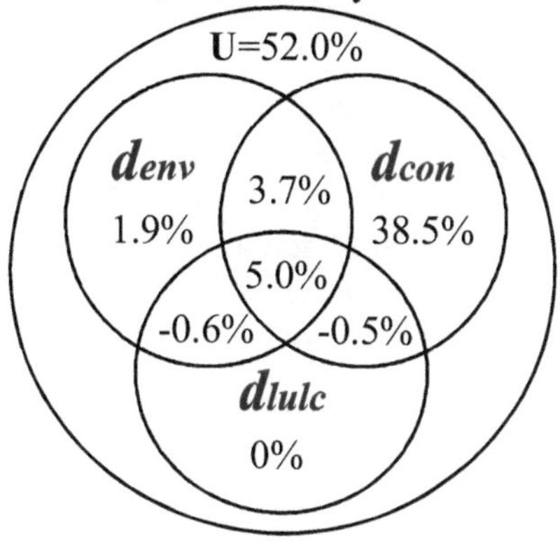

Fig. 4. Variation in β_w linked to the three variation factors studied (abbreviations following text). U is the unexplained variation. Explained variation values correspond to Ajusted R^2 values, corrected for negative values. The negative values of the joint variation of d_{lulc} and d_{env}, and d_{lulc} and d_{con}, may be due to synergic effects, but must be taken with caution due to the low explanatory value of this factor

In spite of these considerations, however being partial and preliminary, our analyses show a great effect of connectivity between pairs of areas in the replacement of faunas. Thus, we have used the original d_{con} data to develop a model that relates connectivity and β_w (see Table 2 and Figure 3).

Table 3. Significance and parameter estimation of the model relating the scores of β_w between UTM 100x100 grid squares and connectivity distances (β_w = Intercept + *estimate**d_{con}; see also Figure 3), developed from the data of the 50 pairwise comparisons of recent Iberian mammals. Estimate shows the parameter scores in the model, *Std. Err.* does for the standard error of these estimates, *t* for the *t* statistic, and *cl*-95% and *cl*+95% are the lower and upper confidence limits for the estimates. The rest of the abbreviations as in text and Table 1

	df	F	Mult. *R*	Mult. R^2	Adj. R^2
Model significance	48	39.621**	0.672	0.452	0.441
Parameter estimation	Estimate	*Std. Err.*	*t*	*cl* -95%	*cl* +95%
Intercept	0.073893	0.022305	3.312**	0.029045	0.118743
d_{con}	0.000226	0.000036	6.295**	0.000154	0.000298

10.4 Assessing connectivity, model performance, and β-diversity patterns in the Iberian mammals during the Neogene

In this section, we test if our model on recent mammals could be used to assess the relationship between connectivity and faunistic replacement between regions in the Iberian Neogene. Therefore, we try to characterize β-diversity patterns through this period from the data provided by well known sites, to identify possible episodes of isolation between regional faunas and/or massive migrations.

10.4.1 Could β-diversity patterns be extrapolated to the past using connectivity between areas?

Neogene (23 to 2 Ma) mammal record is extraordinarily rich in Spain where more than 200 fossil sites have been already unearthed. The data used in the present analysis come from an updated compilation of the regional taxonomic diversity recorded in Spain published by Nieto et al. (2003). Only macromammal species are considered, excluding Chiroptera, Rodentia, and Insectivora, despite the relevant information that microamammals may provide. Differences in the sampling methodology and effort as well as the fact that in most cases both records come from different localities, make difficult to join both data. Species is used as taxonomical unit for the present analysis since Spanish mammals have been intensively studied and their taxonomy is well established (see Nieto et al. 2003 for further discussion and references). Using a more stable taxonomic category, like genus, will reduce the diversity information providing only a minor increase in taxonomical stability. For our analyses, the regional diversity was estimated using basins as spatial units (see figure 5), and Mein's MN zones (Mein 1975) as temporal units. This zonation, based on the mammalian faunal composition, is the most commonly used temporal scale for the Neogene continental record of Europe.

We have obtained regional faunas by aggregating the species present in all the sites of the sedimentary basin during each biozone. To avoid extreme sampling biases, we have used for the analyses only those basins with data for 3 or more sites in the biozone. Then, we calculated β_w scores for all the possible pairwise comparisons of areas. All these scores, ranging from 0.4118 to 0.9333 (mean value = 0.7239 ± 0.1347), were much higher than the maximum β_w observed with recent faunas (see Table 3). The use of macromammals for the analysis, instead of using all mammals, should not account for such a pattern, as they are known to have broader distribution ranges (e.g. Blackburn and Gaston 1996), thus producing smaller expected β_w values. To test the influence of possible biases in the fossil record of the basins studied on this result, we have computed Spearman Rank correlation tests for the relationship of β_w and several variables used as proxies for the completeness of the record. This variables are the total number of species in the two basins compared (N_{sp}), their variance (Var_{sp}) and the number of sites per biozone (N_s). β_w is not significantly correlated to any one of the variables considered. This result suggests that the described pattern of high species replacement between basins can not be fully attributed to the incompleteness of the fossil record.

Fig. 5. Spatial location of the sedimentary basins used to group the Neogene depoits into different mammal faunistic regions. (Eb) Ebro; (Tj) Tajo; (Dr) Duero; (CT) Calatayud-Teruel; (VP) Vallés-Penedés; (Lv) Levantine basins; (Bt) Betic basins.

Table 4. Spearman rank-correlation test for the relationship among β_w and three variables used as proxies for fossil record completeness. N_{sp}= number of species in each pair of basins used for comparison; Var_{Nsp}= variance in the number of species of these two basins; N_s= number of sites in the biozone

	N	Spearman	t(N-2)	p-level
β_w / N_{sp}	19	-0.326	-1.421	0.173
β_w / N_s	19	0.214	0.901	0.380
β_w / Var_{Nsp}	19	0.276	1.185	0.252

Connectivity between sedimentary basins was assessed by calculating minimum cost pathways using the same technique and cost maps formerly described. Here underlies the assumption that geomorphology-mediated connectivity among areas has remained constant enough during the last 23 Myr in the Iberian Peninsula. It is well known that many geological changes have happened in the last 23 million years in the Iberian Peninsula, including four important orogenic events (Pickford & Morales 1994). The main effect of these geological and climatic changes has been a gradual transformation from the former endorheic basins to present exorheic ones. However, its main structural configuration, that is, the positions of large mountain chains and basins has remained constant, so the basins recognized in the Neogene can still be recognized today (except the Baetic mountain chain and Guadalquivir basin, appeared during the Neogene; see Rögl 1999). To what extent such similarities in the connectivity among areas are enough for the present analysis are indirectly tested by our results.

For each basin, one or more centroids were assessed using the 'center of mass' technique (that is, one or more subpolygons were identified for each basin, and the mean of the latitude and longitude of all their perimeter points was used as its centroid; see details in Bosque Sendra 1997). Then, connectivity pathways between each pair of basins were calculated between the two nearest centroids in the friction surface. The results of this analysis could aid to identify possible migration corridors (that is, areas of heavy transit of species) in the Iberian Peninsula (see Figure 6), because the lines that connect basins suggest the location of the pathway that minimizes the cost of migrations.

We have used our model ($\beta_W = 0.073896 + 0.000226 \cdot d_{con}$; see Table 2 and Figure 3) to predict the species replacement scores between these pairs of basins (see results in Table 4). As could be expected, given the great differences between the value observed at recent and Neogene faunas, the model was not able to predict correctly any β_w score, being predicted and observed scores not related (Spearman r = -0.2957; N = 19; $t_{(N-2)} = -1.2761$; $p = 0.2190$).

Thus, the preliminary assumption of the maintenance through time of the relationship between β_w-diversity and geomorphologic connectivity is not supported by our analyses. However, it is important to take into account that, in this case, lack of evidence could not be due to lack of pattern.

The basic assumption of this analysis is that geomorphology-mediated connectivity among areas has remained constant enough during the last 20 Myr in the Iberian Peninsula.

Table 5. Comparisons among mammal Neogene regional faunas. Biozones are the Mammal Neogene biotic periods, following Mein (1975; see text). Basins are the sedimentary basins of each comparison (codes as in Figure 5). β_w Obs. and β_w Prd. are, respectively, the observed species replacement scores, and the scores predicted by the model developed from recent data (see Table 2 and Figure 3). Resid. are the residuals from this model, *Std. Err.* the standard error for each β_w estimate, and *cl* -95% and *cl* +95% are its confidence limits

Biozone	Basins	d_{con}	β_w Obs.	β_w Prd.	Resid.	Std. Err.	Cl -95%	cl +95%
MN2	Tj-Dr	415.5	0.8519	0.1679	0.6840	0.0102	0.1473	0.1884
MN3	CT-VP	503.5	0.5385	0.1878	0.3507	0.0089	0.1699	0.2056
MN4	Tj-CT	426.8	0.9000	0.1704	0.7296	0.0100	0.1503	0.1906
	Tj-VP	757.0	0.6000	0.2451	0.3549	0.0108	0.2235	0.2667
	CT-VP	503.5	0.7619	0.1878	0.5741	0.0089	0.1699	0.2056
MN5	Tj-CT	426.8	0.7500	0.1704	0.5796	0.0100	0.1503	0.1906
MN6	Tj-CT	426.8	0.4118	0.1704	0.2414	0.0100	0.1503	0.1906
MN7/8	Dr-CT	692.4	0.6279	0.2305	0.3974	0.0095	0.2114	0.2496
	Dr-VP	1050.8	0.6800	0.3116	0.3684	0.0191	0.2731	0.3500
	CT-VP	503.5	0.6471	0.1878	0.4593	0.0089	0.1699	0.2056
MN9	Dr-CT	692.4	0.7447	0.2305	0.5142	0.0095	0.2114	0.2496
	Dr-VP	1050.8	0.6444	0.3116	0.3328	0.0191	0.2731	0.3500
	CT-VP	503.5	0.7108	0.1878	0.5230	0.0089	0.1699	0.2056
MN10	Tj-CT	426.8	0.7561	0.1704	0.5857	0.0100	0.1503	0.1906
	Tj-VP	757.0	0.7091	0.2451	0.4640	0.0108	0.2235	0.2667
	CT-VP	503.5	0.6552	0.1878	0.4674	0.0089	0.1699	0.2056
MN11	CT-Lv	313.9	0.5600	0.1449	0.4151	0.0126	0.1195	0.1703
MN13	CT-Lv	313.9	0.6585	0.1449	0.5136	0.0126	0.1195	0.1703
MN17	Lv-Bt	259.4	0.8889	0.1326	0.7563	0.0141	0.1042	0.1610

However, as previously stated, this assumption may not be true at small scales, a pattern that may include several local isolation events. Moreover, during the referred period there have been important geomorphologic changes, including four orogenic events (see Pickford and Morales 1994), that have not been taken into account in our analysis, due to the lack of GIS information on past geomorphology. Thus, actual d_{con} scores between basins could have been different during the past, being these geomorphologic differences part of the explanation for the observed lack of relationship.

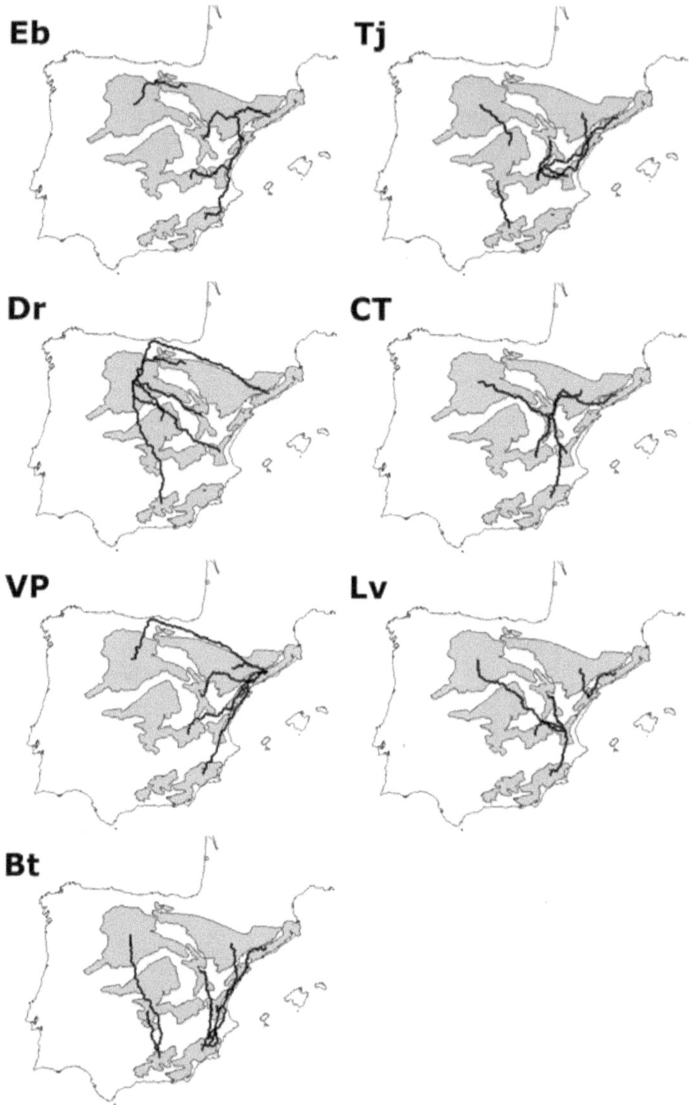

Fig. 6. Spatial location of the minimum cost pathways calculated for the movement among the centroids of the sedimentary basins (see text). Basin codes as in Figure 5

Interestingly, this is consistent with the lack of correlation between the scores of the connectivity distance between pairs of basins (d_{con}), and their corresponding β_w observed scores (Spearman r = -0.2957; N = 19; $t_{(N-2)}$ = -1.2761; p = 0.2190).As the correlation between d_{con} and the results of the model used to calculate β_w predicted scores is 1, the figures of the correlation are the same from those of the correlation observed and predicted β_w scores).

Currently we have no way to assess how this lack of relationship is due to differences in the connectivity among areas in the past, or to other differences in the studied faunas (see below). Thus, new data and/or good-quality extrapolations about the physical geography of the Iberian Peninsula are needed to unravel this Gordian knot. Latter analyses should try to overcome this drawback.

Together with the abovementioned geomorphologic changes, it is important to take also into account another two issues. On the one hand, the time averaging effect implicit in the use of biozones with large time extent may also play a fundamental role in such pattern. On the other, climatic variability through time and adaptive differences between different groups of mammals could be avoiding the extrapolation of recent patterns to the past (see further insights below).

The lack of relationship between β_w-diversity and connectivity, suggests that other factors not present in recent times were affecting Neogene mammal faunas. Since we have not included any proxy for environmental connectivity in our database a higher environmental diversity between sedimentary basins that is observed today is a possible explanation. Ecological barriers, not only topographical barriers, may produce isolation between sedimentary basins and promote divergence in their pool of species, thus leading to higher β_w-diversity values.

As a matter of fact, previous analyses have revealed increasing evidence of a variable heterogeneity of the faunas across the recorded sedimentary basins (Agustí 1999, Morales et al. 1999, Nieto et al. 2003). These differences can be recognized in many periods along the Neogene record of Spain but there are also several periods when faunal differences between basins seem to disappear. Heterogeneity and its temporal changes have been related to variation in environmental heterogeneity (Agustí 1999) but also it has been suggested that immigration events may have also affected the degree of heterogeneity.

Variations in β_w-diversity throughout the Neogene and significant climatic and migratory events are summarised in Figure 7. Although this pattern is obviously affected by the nature of our data, some tentative interpretations may be proposed. According to Agustí (1999), regional heterogeneity was established early in the Miocene, around 19 Ma. (MN3) when strong differences in the faunas can be appreciated between the Mediterranean basins and the interior basins.

Our results contrast with this hypothesis, since a low β_w-value is observed during MN3. As mentioned by us elsewhere (Nieto et al. 2003), these differences in the results may be caused by a poor record. MN3 is marked by an immigration event characterized by the arrival of the North American equid *Anchitherium* although the effects of this event are not reflected in our results.

On the contrary, the *Gomphotherium* landbridge (the connexion between Africa and Eurasia that caused a large faunal interchange, the *Gomphotherium* event), during the MN4, is associated with a high mean β_w-value and high variance.

Other immigration events are also associated with high β_w figures, like the immigration flow between North America and Eurasia during the MN9 (arrival of *Hipparion*) and between Africa and Eurasia during the Messinian Crisis (MN13).

The temperature and humidity curves proposed by Calvo et al. (1993) point out the relationship between climatic changes and changes in β-diversity. As can be seen in Figure 7, β-diversity follows the changes in the temperature curve during much of Early and Middle Miocene. High β-diversity scores coincide with the warm periods of MN4, MN5, MN9 and MN10, whilst low figures are found in the colder MN3 and MN6. However, this relationship is far from clear, with MN2 presenting high β-diversity despite its cold climate.

In contrast, the warm conditions on MN11 are linked to a low β-diversity, although this fact might be explained by the aridity of this period, which resulted in a generalization of savannah environment throughout all the Iberian Peninsula (Morales et al. 1999). However, despite its similar arid and warm conditions, MN5 presents high β-diversity figures, questioning the previous interpretation. It seems that different factors related to climate and migrations interact in a complex way to determine the observed patterns of β-diversity changes through time.

At the same time, the observed patterns may be affected by differences in the completeness of the fossil record. This also applies to the extreme β-diversity value of MN17, which can not be interpreted and should be supposed to be the result of their poor record, restricted to few localities.

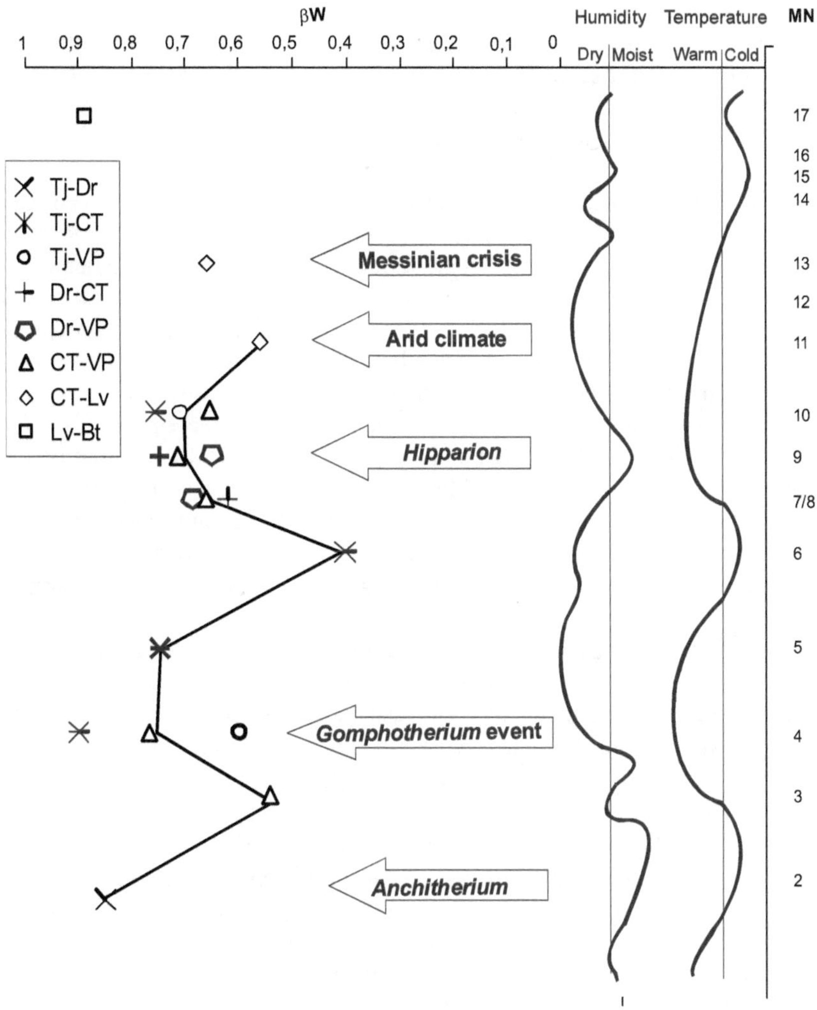

Fig. 7. Variations in β_w between basins throughout the Neogene, and significant climatic and migratory events. Continuous line connects mean values. Basin codes for the basins as in Figure 5. Palaeoclimatic humidity and temperature curves modified from Calvo et al. (1993)

10.5 Further insights

Our results from recent faunas show the importance of connectivity, and thus of species migration between areas, on the patterns of species replacement. Although the analysis here developed is partial and still preliminary, differences in environmental conditions seem to play a secondary role in the faunistic change between areas of the Iberian Peninsula in the present, when connectivity issues are taken into account. This should not be a surprise, as other authors have found inter-area migration to have an important effect on species replacement. As Duivenvoorden et al. (2002) state, a correct implementation of dispersal influence into the analyses of biodiversity patterns produces variations in the description of the factors affecting its geographic distribution. This aspect should be integrated into further macroecological work. Today, GIS tools provide a good method to describe patterns of connectivity (Verbeylen et al. 2003, Nikolakaki 2004, this work), or biological migrations and invasions (e.g. Macdonald and Rushton 2003, Dark 2004, Wiegand et al. 2004, Ming and Albrecht 2004), and new tools are developed to assess and analyze distance matrices (see Ferrier 2002), so it is time to have a new look to the patterns of β_w, using an spatially-explicit point of view.

Although our knowledge about mammal Neogene faunas is incomplete and spatial and temporally biased (see Nieto et al. 2003), several aspects arose from the developed analysis. No relationship can be appreciated between topographical connectivity and β-diversity among regions during the Neogene, despite they are highly correlated in the present. Higher β-diversity episodes seems to be related to major immigration events, which may affect different basins in a differential way, while lower β-diversity may be associated with environmental homogeneity, in opposition to recent pattern, where habitat and environment were less important. Although a test of our model using more complete checklists for past faunas is highly desirable, these kind of analyses are probably not affordable with present state of knowledge (past temporal resolution and geomorphologic data).

Therefore, more complete analyses on the relationship between faunistic and environmental differences in the present are also needed to confirm the here-found pattern of the dependence of β-diversity on connectivity. Among these analyses, the role of spatial distance *per se* should be evaluated, separating its effects from the purely attributable to connectivity-mediated distances. On the other hand, as mentioned before, the role of geomorphology may not be constant i) through time (varying due to climatic variation and geomorphologic changes; see Morales and Pickford 1994), and ii) in different groups of species (with different bioclimatic requirements). According to the first statement of this dichotomy, it would be advisable for further investigations to make an effort to develop better estimations on the connectivity distance between basins during the past. This includes the development of fine grain estimations of the geomorphologic configuration of the Iberian Peninsula at different time periods. Following the second statement, more precise studies of the role of connectivity on species dispersal should focus on the response of groups of species with similar bioclimatic requirements (e.g., dry-Mediterranean shrubland species, temperate forest species or Mediterranean forest species) to different climatic scenarios (glacial maximums, warm interludes, etc.). Today, we have a good knowledge on the environmental requirements of the Iberian Holocene mammal species, as well as good scenarios of the climatic and landscape variations through that period, as well as for future climatic changes. Before obtaining a knowledge good enough as to determine the role of connectivity in shaping Neogene faunas, the Quaternary seems to be a good benchmark to test its role through the past and today. The rich mammal fossil record of the European Quaternary may be a good target for such an study. Geomorphology may be considered constant for the last million years, being the single major factor affecting connectivity during the Pleistocene the eustatic changes in sea level, a variable that may be easily included in a GIS environment from topographic maps of land and seafloor elevation. Including these data will give temporal continuity to present analyses, maybe helping to understand the lack of agreement between recent and Neogene patterns. Moreover, the existence of several records for different time-periods during the last century (see an example at Araújo et al. 2004) may help us to investigate up to what point it is possible to extrapolate distributions from one time to another using connectivity, as well as to determine the importance of connectivity in modifying the purely environmentally-induced patterns in the distribution of single species and entire communities.

10.6 Acknowledgements

The authors wish to thank Dr. Ashraf Elewa for his kind invitation to participate in this book, to Paulo M. Pereira their helpful comments on regression techniques, and to Drs. Jorge Morales and P. David Polly for their reviews of this contribution. This work was supported by the Spanish

D.G.I. projects REN2001-1136/GLO, BOS2003-08938-C03-02 and BTE 2002-00410. JH was also supported by the Fundación BBVA project "Yámana - Diseño de una red de reservas para la protección de la biodiversidad en América del Sur Austral utilizando modelos predictivos de distribución con taxones hiperdiversos", the Portuguese FCT (Fundação para a Ciência e Tecnologia) Center CITA-UA and also by Project ATLANTICO, financed by European Community program INTERREG III B. JR is supported by a postdoctoral Grant from Fundación Atapuerca.

References

Adams J (1998) Europe during the last 150,000 years. In Adams JM, Faure H (eds) Review and Atlas of Palaeovegetation. Preliminary Land Ecosystem Maps of the World since the Last Glacial Maximum. Environmental Sciences Division (Oak Ridge National Laboratory). Available at http://www.esd.ornl.gov/projects/qen/nercEUROPE.html

Agustí J (1999) A critical re-evaluation of the Miocene mammal units in Western Europe: dispersal events and problems of correlation. In Agusti J, Rook L, Andrews P (eds) The evolution of Neogene terrestrial ecosystems in Europe. Cambridge University Press, Cambridge: 84-112

Araújo MB, Cabeza M, Thuiller W, Hannah L, Williams PH (2004) Would climate change drive species out of reserves? An assessment of existing reserve-selection methods. Global Change Biology 10: 1618-1626

Bhattarai KR, Vetaas OR (2003) Variation in plant species richness of different life forms along a subtropical elevation gradient in the Himalayas, east Nepal. Global Ecology and Biogeography 12: 327-340

Blackburn TM, Gaston KJ (1996) Spatial patterns in the geographic range sizes of bird species in the New World. Philosophical Transactions of the Royal Society of London Series B-Biological Sciences 351: 897-912

Blackburn TM, Gaston KJ (2002) Scale in macroecology. Global Ecology and Biogeography 11: 185-189

Borcard D, Legendre P, Drapeau P (1992) Partialling out the spatial component of ecological variation. Ecology 73: 1045-1055

Bosque Sendra J (1997) Sistemas de Información Geográfica. Segunda Edición. Editorial Rialp

Calvo JP, Daams R, Morales J, Lopez-Martinez N, Agustí J, Anadón P, Armenteros I, Cabrera L, Civis J, Corrochano A, Díaz-Molina M, Elizaga E, Hoyos M, Marín-Suárez E, Martínez J, Moissenet E, Muñoz A, Perez-García A, Pérez-González A, Portero JM, Robles F, Santiesteban C, Torres T, Van Der Meulen AJ, Vera JA, Mein P (1993) Up-To-Date Spanish Continental Neogene Synthesis and Paleoclimatic Interpretation. Revista de la Sociedad Geológica de España 6: 29-40

Clark Labs (2000) Global Change Data Archive Vol. 3. 1 km Global Elevation Model. CD-Rom, Clark University

Clark Labs (2001) Idrisi 32 Release 2. GIS software package. Clark Labs

Clustan Ltd (2002) Clustan Graphics Version 5.25. Clustan Ltd

Cornell HV, Lawton JH (1992) Species interactions, local and regional processes, and limits to the richness of ecological communities: a theoretical perspective. Journal of Animal Ecology 61: 1-12

Cristoffer C, Peres CA (2003) Elephants versus butterflies: the ecological role of large herbivores in the evolutionary history of two tropical words. Journal of Biogeography 30: 1357-1380

Dark SJ (2004) The biogeography of invasive alien plants in California: an application of GIS and spatial regression analysis. Diversity and Distributions 10: 1-9

Dobson A (1999) An introduction to Generalized Linear Models. Chapman & Hall/CRC

Duivenvoorden JF, Svenning J-C, Wright SJ (2002) Beta diversity in tropical forests. Science 295: 636-637

EEA (2000) NATLAN. Nature/land cover information package. European Environment Agency

Ferrier S (2002) Mapping spatial pattern in biodiversity for regional conservation planning: Where to from here? Systematic Biology 51: 331-363

Gower JC (1971) A general coefficient of similarity and some of its properties. Biometrics 27: 857-871

Henderson PA, Seaby RMH (2002) Species Diversity and Richness v3.0. Pisces Conservation Ltd

Hendry AP, Kinnison MT (2001) An introduction to microevolution: rate, pattern, process. Genetica 112-113: 1-8

Hengeveld R (1989) Dynamics of Biological Invasions. Chapman & Hall

Hortal J, Lobo JM (2005) An ED-based protocol for the optimal sampling of biodiversity. Biodiversity and Conservation

Hortal J, Garcia-Pereira P, García-Barros E (2004) Butterfly species richness in mainland Portugal: Predictive models of geographic distribution patterns. Ecography 27: 68-82

Hortal J, Rodríguez J, Nieto M, Lobo JM (in revision) Large-Scale Contemporary and Regional Determinants of Global Patterns in the Species Richness of Mammal Local Communities. Proceedings of the Royal Society of London B

IGN (1995) Geología. In Atlas Nacional de España. Centro Nacional de Información Geográfica, Instituto Geográfico Nacional

Koleff P, Gaston KJ, Lennon JJ (2003) Measuring beta diversity for presence-absence data. Journal of Animal Ecology 72: 367-382

Kozlowski J (1999) Adaptation: a life history perspective. Oikos 86: 185-194

Legendre P, Legendre L (1998) Numerical Ecology. Elsevier

Levine JM, D'Antonio CM (1999) Elton revisited: a review of evidence linking diversity and invasibility. Oikos 87: 15-26

Lichstein JW, Simons TR, Shriner SA, Franzreb KE (2002) Spatial autocorrelation and autoregressive models in ecology. Ecological Monographs 72: 445-463

Lobo JM, Halffter G (2000) Biogeographical and ecologic factors affecting the altitudinal variation of mountainous communities of coprophagous beetles (Coleoptera: Scarabaeoidea): A comparative study. Annals of the Entomological Society of America 93: 115-126

Lobo JM, Martín-Piera F (2002) Searching for a predictive model for species richness of Iberian dung beetle based on spatial and environmental variables. - Conservation Biology 16: 158-173

Lobo JM, Castro I, Moreno JC (2001) Spatial and environmental determinants of vascular plant species richness distribution in the Iberian Peninsula and Balearic Islands. Biological Journal of the Linnean Society 73: 233-253

Lobo JM, Jay-Robert P, Lumaret JP (2004) Modelling the species richness distribution for French Aphodiidae (Coleoptera, Scarabaeoidea). Ecography 27: 145-156

Macdonald DW, Rushton S (2003) Modelling space use and dispersal of mammals in real landscapes: a tool for conservation. Journal of Biogeography 30: 607-620

Magurran AE (1988) Ecological Diversity and Its Measurement. Princeton University Press

Martín-Piera F, Lobo JM (2003) Database records as a sampling effort surrogate to predict spatial distribution of insects in either poorly or unevenly surveyed areas. Acta Entomológica Ibérica e Macaronésica 1: 23-35

Mein P (1975) Proposition de biozonation du Néogène Méditerranéen a partir des mammifères. Trabajo de Neógeno y Cuaternario 4: 112

Ming P,Albrecht J (2004) Integrated Framework for the Simulation of Biological Invasions in a Heterogeneous Landscape. Transactions in GIS 8: 309-334

Morales J, Nieto M, Khöler M, Moyá-Solá S (1999) Large mammals from the Vallesian of Spain. In Agusti J, Rook L, Andrews P (eds) The evolution of Neogene terrestrial ecosystems in Europe. Cambridge University Press, Cambridge: 113-126

Niemela J, Spence JR (1991) Distribution and Abundance of an Exotic Ground-Beetle (Carabidae) - a Test of Community Impact. Oikos 62: 351-359

Nieto M, Hortal J, Martínez-Maza C, Morales J, Rodríguez J (2003) Variaciones en la riqueza de macromamíferos de España durante el neógeno. Graellsia 59: 299-317

Nikolakaki P (2004) A GIS site-selection process for habitat creation: estimating connectivity of habitat patches. Landscape and Urban Planning 68: 77-94

Olff H, Ritchie ME, Prins HHT (2002) Global environmental controls of diversity in large herbivores. Nature 415: 901-904

Pereira, P. M. and Pires da Fonseca, M. 2003. Nature vs. nurture: the making of the montado ecosystem. Conservation Ecology 7: 7 [Online].

Palomo LJ, Gisbert J (eds) (2002) Atlas de los mamíferos terrestres de España. Dirección General de Conservación de la Naturaleza-SECEM-SECEMU. Organismo Autónomo de Parques Nacionales, Madrid

Pickford M,Morales J (1994) Biogeography and palaeobiogeography of East Africa and the Iberian peninsula. Palaeogeography, Palaeoclimatology, Palaeoecology 112: 297–322

Rodríguez J (1999) Use of cenograms in mammalian palaeoecology. A critical review. Lethaia 32: 331-347

Rodríguez J, Nieto M (2003) Influencia paleoecológica en mamíferos cenozoicos: limitaciones metodológicas. Coloquios de Paleontología, Vol Ext 1: 459-474

Rögl F (1999) Mediterranean and Parathetys palaeogeography during the Oligocene and Miocene. In Agustí J, Rook L, Andrews P (eds) The evolution of Neogene Terrestrial Ecosystems in Europe. Hominoid Evolution and Climatic Change in Europe. Cambridge University Press, Cambridge, pp. 8-22

Sakai AK, Allendorf FW, Holt JS, Lodge DM, Molofsky J, With KA, Baughman S, Cabin RJ, Cohen JE, Ellstrand NC, McCauley DE, O'Neil P, Parker IM, Thompson, JN, Weller SG (2001) The population biology of invasive species. Annual Review of Ecology and Systematics 32: 305-332

Shigesada N, Kawasaki K (1997) Biological Invasions: Theory and Practice. Oxford University Press

Spector S (2002) Biogeographic crossroads as priority areas for conservation. Conservation Biology 16: 1480-1487

StatSoft I (2001) STATISTICA (data analysis software system). StatSoft, Inc

Verbeylen G, De Bruyn L, Adriaensen F, Matthysen E (2003) Does matrix resistance influence Red squirrel (Sciurus vulgaris L. 1758) distribution in an urban landscape? Landscape Ecology 18: 791-805

Vitousek PM, Walker LR, Whiteaker LD, Muellerdombois D, Matson PA (1987) Biological Invasion by Myrica-Faya Alters Ecosystem Development in Hawaii. Science 238: 802-804

Whittaker RH (1960) Vegetation of the Siskiyou Mountains, Oregon and California. Ecological Monographs 30: 279-338

Whittaker RJ, Willis KJ, Field R (2001) Scale and species richness: towards a general, hierarchical theory of species richness. Journal of Biogeography 28: 453-470

Wiegand T, Knauer F, Kaczensky P, Naves J (2004) Expansion of brown bears (Ursus arctos) into the eastern Alps: a spatially explicit population model. Biodiversity and Conservation 13: 79-114

Willis KJ, Whittaker RH (2002) Species diversity - Scale matters. Science 295: 1245-1248

Wilson MV, Shmida A (1984) Measuring beta diversity with presence-absence data. Journal of Ecology 72: 1055-1064

Index

FSC
www.fsc.org
MIX
Papier | Fördert gute Waldnutzung
FSC® C083411

Kolibri 360 GmbH
Ferdinand-Jühlke-Straße 7
99095 Erfurt, Deutschland
produktsicherheit@kolibri360.de